GUIDELINES FOR LABORATORY DESIGN: HEALTH AND SAFETY CONSIDERATIONS

Second Edition

LOUIS J. DIBERARDINIS
Massachussetts Institute of Technology

JANET S. BAUM
Architecture and Planning

MELVIN W. FIRST
Harvard School of Public Health

GARI T. GATWOOD
Consultant

EDWARD GRODEN
Consultant

ANAND K. SETH
Massachusetts General Hospital

A Wiley-Interscience Publication
JOHN WILEY & SONS, INC.
New York / Chichester / Brisbane / Toronto / Singapore

Copyright © 1993 by John Wiley & Sons, Inc.

Library of Congress Cataloging in Publication Data:
Guidelines for laboratory design : health and safety considerations /
   Louis J. DiBerardinis . . . [et al.]. —2nd ed.
      p.   cm.
   "A Wiley-Interscience publication."
   Includes index.
   ISBN 0-471-55463-4 (cloth)
      1. Laboratories—Design and construction.   2. Laboratories—Safety
   measures.   I. DiBerardinis, Louis J., 1947–
   TH4652.G85   1992
   727'.5—cd20   92-16785

Printed in the United States of America

10 9 8 7 6 5 4 3 2 1

CONTENTS

PART III: LABORATORIES SUPPORT SERVICES

PART IV: ADMINISTRATIVE PROCEDURES

PART V: HVAC SYSTEMS

PART VI: APPENDIXES

ABBREVIATIONS

AAALAC	American Association for Accreditation of Laboratory Animal Care
ACD	Allergic contact dermatitis
ACGIH	American Conference of Governmental Industrial Hygienists
ACS	American Chemical Society
ACZ	Adjustable current sensor inverter
AIA	American Institute of Architects
ALARA	As low as reasonably achievable
AMCA	Air Moving and Conditioning Association
ANSI	American National Standards Institute
ASME	American Society of Mechanical Engineers
ASSE	American Society of Safety Engineers
ASTM	American Society for Testing and Materials
AVI	Adjustable voltage inverter
BOCA	Building Officials and Code Administrators International
BSI	British Standards Institution
CADD	Computer Aided Design and Drafting
CAP	College of American Pathologists
CFC	Chlorofluorocarbon
CFR	Code of Federal Regulations
CMR	Code of Mass. Regulations
CO_2	Carbon dioxide
D.I.	Deionized (water)
DC	Direct current
DD	Double-Duct system (ventilation)

DEAE	(Diethylamino)ethanol
DEP	Massachusetts Department of Environmental Protection
DNA	Deoxyribonucleic acid
DOP	Dioctylphthate (filter testing)
DOT	Department of Transportation
DX	Direct expansion refrigeration cooling system
EIA	Electronic Industries Association
EPA	U.S. Environmental Protection Agency
FCCV	Fan-coil constant-volume reheat system (ventilation)
FCVAV	Fan-coil variable air volume reheat system (ventilation)
FM	Factory Mutual (casualty insurance group)
FRP	Fiberglass-reinforced polyester
GFIC/GFI	Ground fault interrupter
GSA	General Services Administration
HEPA	High-efficiency particulate air (filters)
HHS	U.S. Department of Health and Human Services
HVAC	Heating, ventilating, and air conditioning
I.D.	Inside diameter
IARC	International Agency for Research on Cancer
JCAH	Joint Council for Accreditation of Hospitals
LANL	Los Alamos National Laboratory
LD_{50}	Lethal dose that kills 50% of the exposed population, usually in animal toxicology studies
LED	Light-emitting diode
LEL	Lower explosive limit (gases)
MIT	Massachusetts Institute of Technology
N.C.	Normally closed (control valve)
N.O.	Normally opened (control valve)
NAS	National Academy of Sciences
NCCLS	National Committee for Clinical Laboratory Standards
NFPA	National Fire Protection Association
NIH	National Institutes of Health
NIOSH	National Institute for Occupational Safety and Health
NMR	Nuclear magnetic resonance (instrument)
NRC	U.S. Nuclear Regulatory Commission
NSC	National Safety Council
NSF	National Sanitation Foundation
NTP	National Toxicology Program
OMCVD	Organometalic chemical vapor deposition
OSHA	Occupational Safety and Health Administration
PI	Principal investigator
PVC	Polyvinylchloride
PWM	Pulse-width modulator
RH	Relative humidity

SAMA	Scientific Apparatus Manufacturers Association
SMACNA	Sheet Metal and Air-Conditioning Contractors National Association
TDS	Total dissolved solid (water)
TLV	Threshold limit value (air contamination)
TMR	Terminal reheat system (ventilation)
TRH	Constant-volume terminal reheat system (ventilation)
UL	Underwriters Laboratory
ULPA	Ultralow penetration aerosol (filters)
USP	U.S. Pharmacopeia
VAV	Variable air volume (HVAC system)
VRH	Variable-volume reheat system (ventilation)
VVTRH	Variable-volume terminal reheat system (ventilation)

UNITS

A	Ampere
AC/HR	Air changes per hour
cc	Cubic centimeter
cfm	Cubic feet per minute (air volume rate)
cm	Centimeter
°F	Degrees Fahrenheit
fpm	Feet per minute (air velocity)
ft	Feet
ft-c	Foot candles (illumination)
ft^2	Square feet
ft^3	Cubic feet
gal	Gallon
gpm	Gallons per minute
gsf	Gross square feet
h	Hour
hertz	Cycles per second (electrical)
in	Inch
in. w.g.	Inches water gauge (pressure)
lux	Illuminance
m	Meter
μCi	Microcurie (ionizing radiation)
$\mu g/m^3$	Micrograms per cubic meter
μm	Micrometer
mA	Milliampere
min	Minute

ml	Milliliter
mrem	Millirem
mW	Milliwatt
nsf	Net square feet
ohm-cm	Ohm-centimeter (water conductivity)
ppm	Parts per milliion (by volume)
psig	Pounds per square inch guage (pressure)
rad	Unit of ionizing radiation (dose)
V	Voltage
W/ft^2	Watts per square foot (illumination allowance)

PREFACE

The second edition of this book expands the different types of laboratories presented in the first edition and discusses the emerging design issues for renovated or newly constructed laboratories.

Two new types of laboratories are presented. One is a microelectronics laboratory. This is an extension of the chapter on clean rooms, which addresses clean room operations that use extremely hazardous materials. A chapter on printmaking has also been added to address those laboratories that may be involved in such activities as lithography, screen printing, and intaglio and relief processes. In addition, the chapter on the teaching laboratory has been expanded to reflect the trend toward microscale techniques and the resultant design considerations.

A new part has been added to the book discussing the support services that are needed for laboratory operations. These include: photographic darkrooms; support rooms for such activities as glassblowing, metal working and woodworking; and rooms to handle chemical, biological, and radiation waste.

The sections covering building and laboratory layouts and heating, ventilating, and air conditioning have been expanded to include a discussion of the latest trends in these areas. In particular, a new chapter entitled "Variable-Air-Volume Systems" has been prepared to discuss the advantages and disadvantages of this emerging technology.

The authors wish to gratefully acknowledge Mark Nielsen, who provided the illustrations, and Massachusetts General Hospital, which provided support for the illustrations. Special thanks to Pam Greenley and Laslo Veradi who reviewed the chapter on Microelectronics Laboratory, Don Haes who reviewed the chapter on Radiation Laboratory, Mark Lentz who reviewed the chapter on Background HVAC and Gordon Sharp who reviewed the chapter on Variable Air Volume Systems. The authors also wish to thank Margaret Mahoney, Joan Sullivan, Margaret Apruzzese, and Maureen Gatto, who assisted in the manuscript preparation.

INTRODUCTION

1. NEED FOR THIS BOOK

The construction of new laboratory buildings and the renovation of old ones require close communication between the laboratory users, project engineers, architects, construction engineers, and safety and health personnel. With a multitude of needs to be addressed, all too often safety and health conditions are overlooked or slighted and laboratories may be built with unanticipated safety and health hazards. It is clear that one of the principal objectives of laboratory design should be to provide a safe place in which scientists, engineers, and their staff can perform their work. To fulfill this objective, all safety and health considerations must be evaluated carefully and protective measures incorporated into the design wherever needed.

Over many years, chemists, physicists, biologists, research engineers, and their technicians and assistants have met with injury and death in their laboratories by fire, explosion, asphyxiation, poisoning, infection, and radiation. Injury and death have also resulted from more common industrial accidents, such as falls, burns, and encounters with broken glassware and falling objects. Emphasis on safety usually begins in well-organized high-school science classes and continues with increasing intensity and sophistication through colleges and graduate schools for the express purpose of educating scientists to observe safe laboratory procedures while learning and to carry this knowledge and experience into their careers. Often, however, the very laboratories in which they later practice their profession are obsolete or, when modern, fail to incorporate safe design principles. Unless such scientists have had the good fortune to observe well-designed laboratories, they may be ill equipped to assist architect-engineers with safety design when new or renovated laboratories are being prepared for their use. Few architect-engineers are specialists in laboratory safety, and they usually need and welcome the active participation of the scientists to whom the laboratories will be assigned for this type of assistance.

Because laboratory scientists tend to do their work alone or in very small clusters, the dire effects of serious breeches of good safety and health practices seldom result in numerous casualties and for this reason are poorly reported in the popular and professional news media. This may give the impression that the dangers of laboratory work have been exaggerated and perhaps lull scientists into a false sense of security. Accident statistics, however, confirm that laboratories can become dangerous workplaces. Careful thought for worker safety remains an essential part of the laboratory design process.

This book is designed to provide in a concise, easy-to-use format the information needed by architects and project engineers to design safe and efficient laboratories. It includes safety considerations that must be addressed to comply with governmental regulations as well as recognized good practice standards. Although the book emphasizes U.S. regulations, it is expected that application of the safety principles cited here will provide safe and efficient laboratories wherever they may be needed.

2. OBJECTIVES OF THIS BOOK

The purpose of this book is to provide reliable design information related to specific health and safety issues that should be considered when planning new or renovated laboratories. The objective will be approached within the framework of other important factors such as efficiency, economy, energy conservation, and design flexibility. Although precise specifications will be provided in some cases, the general intent is to review the relevant safety and health issues and then to recommend appropriate design action, including, where possible, a range of alternatives. In those cases where there are specific U.S. code requirements, the appropriate section of the code will be referenced. In many cases, consultation between project engineers, laboratory users, and industrial hygiene and safety experts will be required at one or more design stages. These instances will be noted in the text, and it is the hope of the authors that a relationship characterized by close cooperation and understanding will develop among these groups as a result of the use of this book.

The book seeks to address at the design stage the many issues that have a direct bearing on the occupational health and safety of those who work in science and engineering laboratories. It makes no attempt to address all the building structural service requirements that are normal architectural and engineering design considerations, nor does it intend to define good practice laboratory health and safety programs in operating laboratories. It recognizes that all these matters should have an important influence on design considerations and addresses them solely in that context.

It is always important for the project manager to communicate frequently with all laboratory users to keep current with their specific needs. Experience has amply confirmed that there is a steep learning curve whenever laboratory personnel enter into the design phase of their own laboratories, and changing requirements are the norm at the start. Because many safety considerations are specific to certain laboratories but absent in others, it is extremely important that the typical laboratory chosen for design purposes be identified unequivocally as the one the user needs.

When dealing with renovation projects, the original building layout must be evaluated very carefully to determine its compatibility with the needs of the intended occupants as well as with the good practice layouts recommended in this book. Therefore, it will be essential to review each laboratory design recommendation to investigate its compatibility with the building that has been selected for renovation. This must be done cooperatively with the user, the architect, and management because critical compromises are almost inevitable when an existing building is adapted to new uses.

3. HOW TO USE THIS BOOK

A. Subject Matter Organization

All of the technical matter in this book is divided among five parts and several appendixes in a manner designed to provide easy access to all the occupational safety and health information needed to complete a specific design assignment. This has been accomplished in two ways. *First,* Chapters 1 and 2 (Part I) contain general technical information that applies to all, or nearly all, laboratory buildings (Chapter 1) and laboratory modules (Chapter 2) regardless of the precise nature of the work that will be conducted in each. The purpose of placing all generally applicable information into two early chapters is to avoid repeating it when each of the distinctive types of laboratories is discussed in the individual laboratory-type chapters contained in Part II. The general technical information contained in Part I is also intended to present to the reader a unified body of design principles that will be instructive as well as easily accessible as a reference source.

The *second method* used to coordinate the information in the several technical chapters is the use of an invariant numerical classification system throughout the chapters in Parts I, II, III, and IV whereby indentical topics are always listed under the same numerical designation. For example, in every chapter in these four parts, all space requirements and spatial organization information will be found in sections numbered 2 after the chapter number. Therefore, sections numbered 1.2 are in Chapter 1 and are concerned with technical aspects of building layout, whereas sections numbered 2.2 are in Chapter 2 and refer to the general technical aspects of laboratory module layout. Similarly, these numbers have been assigned to the chapters that cover each unique type of laboratory. For easier understanding by the reader, the numerical classification system that is used in each chapter throughout Parts I through IV of the book is summarized in Section 3B, "Book Organization."

Part II contains information on detailed specifications, good practice procedures, and cautionary advice pertaining to 20 specific types of commonly constructed laboratories for academic and industrial research and for educational purposes. Some laboratories are intended for general purpose usage (for example, undergraduate chemistry teaching), whereas others are intended for very specific and well-defined research activities (for example, work with biological hazards). Where appropriate, graduated levels of usage are recognized in the complexity of the facilities that are described; for example, general chemistry teaching laboratories for high-school, college, and graduate-school instruction.

The safety and health design recommendations for each laboratory are based on the operations that are to be performed, as well as on the materials and equipment that will be used. It is recognized that laboratory usage patterns tend to change over time, and therefore it is prudent to try to provide for unique functions with as much design flexibility as possible. In some cases, a predictable changing pattern of usage may call for what we refer to as a *general purpose laboratory*. We have therefore treated the general purpose laboratory as one of the special laboratory design categories.

No attempt has been made to treat every conceivable type of laboratory in a separate chapter because some are highly specialized and have too restricted a range of usage to make it worthwhile (for example, total containment biological safety laboratories), whereas others are offshoots of one or more of the laboratories that are described in detail, and the transference of information will be obvious. Where gaps of coverage remain, it is hoped that the general principles enunciated in Part I, plus the specific information contained in Part IV, will provide adequate guidance for those confronted with a need to design and construct unique and innovative types of laboratories.

Part III contains three chapters that are concerned with support facilities commonly needed by laboratories of all types. They include photographic suites, research model shops, and special waste-handling facilities designed for the collection, temporary storage, consolidation, and shipping of chemical, biological, and radioactive wastes. All aspects of hazardous wastes are rigidly regulated by the U.S. Environmental Protection Agency and the U.S. Nuclear Regulatory Commission. In the field of biological hazards, the U.S. National Institutes of Health and Centers for Disease Control issue guidelines that have the practical force of regulations.

Part IV contains administrative matters pertaining to bidding procedures, final acceptance inspections, and energy conservation considerations. Although energy conservation strategies are discussed and incorporated throughout the book, they are treated as a major topic in Part IV. They emphasize (a) heating, cooling, and ventilation systems that minimize the discharge of uncontaminated air (b) recommendations for the installation of exhaust air devices, (for example, fume hoods) that discharge the least air volumes consistent with safety and (c) the use of fully modulated HVAC systems that supply tempered air consistent with exhaust requirements, but no more (see Chapter 33). It is anticipated that all ordinary energy conservation measures associated with the laboratory structure will be familiar to architects and engineers and that they will be incorporated into the design of the building. Therefore, these important energy conservation methods will not be discussed in this book. Instead, only those conservation techniques closely associated with the functioning of occupational health and safety matters will be covered.

Part V is devoted to a number of general and specific topics associated with heating, ventilating, and air-conditioning laboratories. Background information is presented on comfort perception and on important system components such as fans and filters. The major emphasis of Part V, however, is on laboratory hoods of all kinds and on variable air volume systems designed to conserve energy by automatically reducing emissions of air from the laboratory when exhaust air services are not needed at their maximum design capacity.

Part V also contains commonly consulted specifications, consensus standards, good practices, and institutional bid documents and procedures found to be useful in the design of well-functioning HVAC systems.

Part VI contains (a) appendixes related to universally used laboratory safety items (emergency shower and eyewash stations, check valves to prevent backflow of wastes into potable water supplies, and warning signs) and (b) a matrix table intended to be used as a handy checklist for health and safety design items and to inform the reader where pertinent sections are located in the book.

B. Book Organization

This book is organized with sufficient flexibility to guide the user in the design of a complete new, multistory laboratory building, as well as in the renovation of a single laboratory module. It is arranged in a format that allows the user to start with the building description and then to proceed in a logical sequence to the development of each individual laboratory module. The safety and health considerations that need to be addressed in every laboratory design assignment are explained and illustrated in five broad categories.

(1) Guiding Concepts. This section defines each type of laboratory by (a) the nature of the tasks normally performed there, (b) the special materials

and equipment used, and (c) the nature of the requirements that contribute to making this laboratory unique. In some instances, hazardous or specialized materials and equipment that should not be used in a particular laboratory are listed to aid in making certain that the architect, project engineer, and laboratory user all have the same laboratory type under consideration. When the laboratory type first selected has serious contraindications for some of the projected activities, an alternative type that does not have such exclusions should be identified and those design guidelines followed.

(2) Laboratory Arrangement. This section discusses and illustrates the area requirement and spatial organization of each type of laboratory with special regard to egress, equipment and furniture locations, and ventilation requirements. Typical good practice layouts are illustrated and a major effort is directed toward calling attention to those that are clearly undesirable. The location of exhaust hoods, biological safety cabinets, clean benches, and items of similar function are given special attention.

(3) Heating, Ventilating, and Air Conditioning (HVAC). This section describes the desirable elements of a laboratory HVAC system that is designed for comfort and safety. Wherever unique requirements have been recognized because of the critical nature of the work or equipment, they are given special consideration and definition in a special requirements section. Usually, minimum performance criteria are specified in bid documents, but it should be recognized that somewhat better performance ought to be provided by the design to allow for inevitable system deterioration while in use because health and safety equipment installed in laboratories and laboratory buildings must perform its design function with a high level of reliability throughout an assigned service life that may be the same as the life of the building. Some loss of efficiency and effectiveness over long time periods is usual for machinery and structures. Therefore, a factor to account for normal deterioration must be included in procurement specifications so that, at the end of the expected service period, performance will be adequate to accomplish the assigned health and safety functions with an adequate margin of safety. Procurement documents often specify minimum performance criteria equal to end-of-life health and safety requirements, thereby assuring less-than-desired performance over a major fraction of the service life of the items. Thus, minimal acceptance criteria for procurement must take into account (1) lesser performance experienced after installation than contained in manufacturers' performance tables developed under ideal test conditions plus (2) an additional factor for normal deterioration during long-term usage to assure acceptable health and safety protection initially and over the life of the facility. Special attention is given to providing adequate makeup air for exhaust-ventilated facilities and to the pressure relationships between laboratories, offices, and

corridors. When construction requirements for laboratory systems differ substantially from those that apply to ordinary HVAC installations, the differences are made explicit and appropriate codes and standards are cited in the text.

(4) Loss Prevention, Industrial Hygiene, and Personal Safety. This section presents checklists of items that must be evaluated for their inclusion during the design stages. They encompass a wide variety of safety devices and safety design options intended to protect workers and property. The important subjects of handling dangerous substances and disposal of laboratory waste, including animal waste and animal carcasses, is included. In many cases these items can be attended to later, but usually only at greater expense.

(5) Special Requirements. This section deals with the unique aspects of each identified laboratory type. Not all of the noted special requirements may be needed exclusively for safety reasons, but their presence in a laboratory may affect overall safety considerations and be important for that reason. This section evaluates their potential impact and presents appropriate safety measures when required.

C. Codes and Standards

Governmental and code requirements that pertain to specific safety items are stated and the sources referenced. In the United States, the major codes and regulations that must be met are those of the latest editions of the Occupational Safety and Health Administration (OSHA), the National Fire Protection Association (NFPA), and the Building Officials and Code Administrators, International, Inc. (BOCA) or equivalent national building code. In addition, there are many local codes, ordinances, and state laws that must be observed. In the absence of specific regulations or code requirements, numerous safety-related topics have been treated with special detail because we are of the opinion that our recommendations will have an important impact on improved safety in areas not now adequately addressed elsewhere. In these instances, considerable pains have been taken to justify the recommendations. Whenever possible, alternative recommendations are made to permit flexibility of design and construction, especially for renovation projects where physical constraints are encountered frequently. Even when no specific recommendations have been made, a checklist of items to be considered is often presented. When additional interpretation of recommendations or further explication of design cautions is considered desirable, it is highly recommended that the project engineer and architect work closely with industrial hygiene and safety professionals in an endeavor to design and build the safest laboratories possible.

D. Information Sources

Applicable federal and state regulations, codes and standards, textbooks, and published articles on the safe design of laboratories have been referenced throughout the book to provide the user with more detailed information. Close communication with industrial hygiene and safety professionals throughout the planning phases has been recommended. In the absence of qualified staff personnel, competent professional guidance may be obtained in several ways.

1. Consultants. The American Industrial Hygiene Association* and the American Society of Safety Engineers† maintain current lists of consultants, which can be obtained on request.

2. Governmental Agencies
- (a) Many state Departments of Occupational Health (or Industrial Hygiene) receive federal assistance for the express purpose of providing professional help for occupational health and safety needs, and they can be called or visited for advice.
- (b) Regional offices of the National Institute for Occupational Safety and Health (in the U.S. Department of Health and Human Services), as well as the Occupational Safety and Health Administration (in the U.S. Department of Labor), can also be requested to provide answers to specific health and safety issues and interpretation of federal regulations.
- (c) Local fire departments usually review large renovations and new building plans with respect to fire regulations.
- (d) Regional offices of the Environmental Protection Agency and State Departments of Environmental Protection are prepared to interpret regulations regarding permissible emissions to air and water and disposal of hazardous solid wastes.

Note. Because it is expected that this book will be used by people with diverse technical backgrounds (for example, architects, engineers, laboratory scientists, and nontechnical administrators), many terms and concepts have been defined and explained in a degree of detail that may seem excessive to one or another of these professional groups. Should this occur, we hope that the reader will be patient and understanding of the knowledge limits of others on the design team.

---

* American Industrial Hygiene Association, 2700 Prosperity Avenue, Merrifield, VA.
† American Society of Safety Engineers, 1800 E. Oakton Street, Des Plaines, IL 60018-2187; (312)692-4121.

PART I

COMMON ELEMENTS OF LABORATORY DESIGN

The two chapters that make up this part of the book address the several elements of the design process, starting with essential decisions regarding building size and function, progressing through structural and modular design choices, and on to individual laboratory requirements for space, utilities, clustering, and the auxiliary facilities that allow laboratories to function safely and productively.

For the most part, the subjects covered in these two chapters apply to all laboratories and to all buildings primarily devoted to housing laboratories. Therefore, they cover the general principles of modern laboratory building and laboratory module design with a special focus on the health and safety of those who will occupy the finished structures and work there. The diverse nature of present-day science and engineering activities calls for unique features and equipment in most special-function laboratories. These special needs generally call for additions to the basic requirements covered in the first two chapters, rather than substitutions, and these special requirements are covered in considerable detail on an individual laboratory-by-laboratory basis in Part II.

The information contained in this Part is directed, for the most part, to new construction because of the difficulty in attempting to anticipate all the varieties of structural constraints that can be associated with the renovation of existing buildings and laboratories. Nevertheless, it is anticipated that most of the material can be applied as well to renovations and reconstructions. Certainly, the same modern principles of laboratory health and safety protection serve new and renovated laboratory facilities alike and can be applied to both.

CHAPTER 1

BUILDING CONSIDERATIONS

1.1 GUIDING CONCEPTS

This chapter deals principally with alternative building layouts for the design and construction of new laboratory buildings. The advantages and disadvantages of a variety of alternative building designs are presented, along with the preferred choices. Laboratory requirements based on one or another of the preferred building layouts are discussed. During the useful life of a building, laboratories may be renovated several times. Therefore, as much flexibility as possible has been provided so that the health and safety concepts given here may be applied to renovation of existing buildings as well as to original construction. Facilities undergoing simple renovation need not be substantially revised just to meet the requirements contained in this chapter if no safety hazards are present, but consideration should be given to following as closely as possible the precepts contained in this chapter when substantial modifications are to be made. Because renovations of laboratories may be carried out within building layouts that are less than ideal for the purpose, careful consideration and application of health and safety requirements will be required. Nevertheless, most safety and health requirements can be applied to many different laboratory and building layouts, and it should always be possible to meet essential safety requirements.

Guidelines for Laboratory Design: Health and Safety Considerations, Second Edition
By Louis DiBerardinis, Janet S. Baum, Gari T. Gatwood, Anand K. Seth, Melvin W. First, and Edward F. Groden.
ISBN 0-471-55463-4   Copyright © 1993 by John Wiley & Sons, Inc.

1.2 BUILDING LAYOUT

1.2.1 The Building Program

The architect and project engineer, with the assistance of the laboratory users, develop the building program from analysis of data collected on (1) the number and types of personnel who will occupy the building, (2) the research, teaching, and industrial functions to be housed, and (3) the interrelationships of functions and personnel.

1.2.1.1 Program Requirements

DESIGN GOALS. A building program is a written document that describes and quantifies the design goals for a building. The goal of a good program is to define a building that will have ample space, function safely, and realistically meet an owner's needs and budget. A building program is prepared by a project team of users and management from within the institution or corporation, often with the assistance of a consultant programmer, who has experience in the type of building proposed. The program describes for whom and where the building will be constructed, and what the building functions will be for the owners and users. Architects and engineers use a building program to learn for whom they are designing the facility, what spaces and facilities are required, and where functions should be located in relation to each other.

There are three primary types of building programs, based on the owner's project team objectives. (1) A conceptual program, used to test feasibility of a building or renovation project, also can be used for fund raising and convincing potential funding sources of the seriousness and utility of the project. A conceptual program quantifies net usable area or gross area for each department or generic space type. Generic space categories are laboratory, laboratory support, and specialized areas that include office and administration, personnel support, and building support. (2) An outline program, used to list specific rooms types, quantities, and areas of each, can be used as a tool for recruiting additional research scientists. (3) A detailed program, the most common program document, is used to control cost and to build consensus within the proposed group of research occupants. A detailed functional program describes architectural, mechanical, electrical, plumbing, and fire protection performance criteria for building functions that must be accommodated. A detailed program identifies areas of special concern for safety, such as high hazard areas that use flammable, toxic, and pathogenic materials or processes; the program also includes the waste removal implications of these sensitive materials. The detailed functional program does not need to be written with any preconceived formal design philosophy in mind, except as may be required to incorporate health and safety guidelines. A detailed functional program is intended for owners and users to enable them to evaluate the planning and architectural design that the design team ultimately develops. Table 1-1 lists the tasks for completing the three types of programs described below.

POPULATION NUMBERS AND PROGRAMS. The programmer or project team consults with department heads, principal investigators, administrators, and laboratory managers for information on population numbers and functions. When the persons who will be responsible for managing the laboratories and occupants are not yet known, the project team consults the administrators of the organization who will hire these primary staff members to establish what functions will be carried out in the building. When no better information is available to the project team, allocations of the major divisions of space can be estimated on the basis of the occupancy patterns of well-functioning buildings of similar purpose. It may be expected that information from such indirect sources will be (a) nonspecific to the actual project and (b) a less precise estimate of needs than information obtained directly from future occupants and administrators.

To estimate a laboratory building population when specific numbers are unobtainable, an alternative population model can be constructed based on an understanding of the most commonly observed laboratory working groups as shown in Table 1-2. This example is based on a typical research laboratory at a medium-sized academic institution. Staffing for different laboratory types and other sizes of research organizations may differ from the example given here. The illustrative tables do not show statistics for very small teams

TABLE 1-1. Program Tasks and Sequence of Program Documents

Sequence and Task	Conceptual	Outline	Detailed
1. Existing facility analysis or comparison to similar facility	Yes	Yes	Yes
2. Interview management and users, if identified	Yes	Yes	Yes
3. Establish space standards	Yes	Yes	Yes
4. Develop room type list		Yes	Yes
5. Diagram room types			Yes
6. Determine number and area of each room type	Yes[a]	Yes	Yes
7. List room performance specifications and data			Yes
8. Project building net area	Yes	Yes	Yes
9. Project building gross area	Yes	Yes	Yes
10. Describe basic mechanical, electrical, plumbing systems		Yes	Yes
11. Estimate cost of construction	Yes	Yes	Yes

[a] Note: Room type list for conceptual program has only generic room types: laboratory, laboratory support, administration, personnel support, building support.

TABLE 1-2. Alternative Research Team Population Estimate for a Conceptual Program[a]

Team Members	Small Team (full time)	Medium Team (full time)	Large Team (full time)
Principal Investigator	1	1	1
Research Assistant	1	2	2–3
Post Doctoral Student	0	1	1–2
Technician	1	1	1–2
Graduate Student	1	1	2–3
Secretary or Admin.	$\frac{1}{4}$	$\frac{1}{2}$	1
TOTAL TEAM POPULATION	$4\frac{1}{4}$	$6\frac{1}{2}$	8–12

[a] Applies to most experimental science activities and disciplines.

(i.e., less than 4) or for the very large teams of over 12 to 25 that are encountered in some laboratory settings. Area allotments for very large and small groups can be extrapolated from the values shown in Tables 1-2 through 1-5. The next step is to establish research area standards. This is best done by calculating the existing laboratory area population density and analysis of occupancy patterns, assessing the adequacy of existing conditions, and then setting realistic but safe area goals for the new facility. Here, also, an alternative model, based on commonly observed laboratory settings, must be resorted to when there are no preexisting facilities to work from. It is the experience of the authors that when a new facility is well planned and well constructed, demand on it can go far beyond conservative estimates that were established during the programming phase. Within a few years, occupancy of a new laboratory can reach 120% to 150% of the original population envisioned in the building program.

The example of area standards per researcher shown in Table 1-3 is based on a number of typical biomedical and molecular biology laboratory facilities. As noted later in Table 1-5, different activities and disciplines have different area requirements. Animal facility area requirements may be the most difficult to estimate because animal facility area demand depends upon three major factors: (1) specific types of research programs (including breeding), (2) the number of different species of research animals, and (3) the number of animals in each species to be accommodated. The net area planned for animal facilities per researcher may vary from 0 to 150 nsf and sometimes even more. Very careful consideration must be given to animal facility demand in order to develop a facility of appropriate size within a new or renovated laboratory building.

The conceptual program requires similar estimates for three non-laboratory space categories: administration, personnel support, and building support. If the new laboratory building is one of a number of similar buildings on a well-established campus or industrial complex, institutional administra-

TABLE 1-3. Sample Principal Investigator Area Standards for Biomedical and Molecular Biology Laboratories [Net sq. ft. Area (nsf) Assigned per Researcher]

Area Category	Small Team	Medium Team	Large Team
PI Office	30	20	15
Secretary Office	30	20	12 to 15
Staff Offices[a]	30	30	30
Modular Laboratory	130	130	120 to 130
Dedicated Lab Support	40	40	30
Shared Lab Support	40	40	30
Common Lab Support	20	20	20
Animal Facility	varies	varies	varies
Subtotal Lab Area	320	300	257 to 270
Administration	varies	varies	varies
Personnel Support	varies	varies	varies
Building Support	5	5	5
Total Net Area	325	305	262 to 275

[a] Note: Staff office area may be added to modular laboratory area per researcher, when writing desk work stations are provided within the laboratory. Thirty net square feet (nsf) ea. represents a minimal desk carrel arrangement of four persons per standard office area of 120 net square feet.

tion may be located in an entirely separate building. When administration is located within a building that is principally devoted to laboratories, the rooms that will be needed can be listed and the area estimated. The estimated area can be divided by the total full-time research population to derive the area per researcher figure. Personnel support requirements can be estimated in a similar fashion. When certain needed facilities exist nearby (e.g., a cafeteria), they may not need to be duplicated in a new or remodeled laboratory building. Building support, however, is in a different category, and every laboratory building requires adequate areas for materials handling, maintenance, and storage. When a loading dock and temporary storage room(s) for daily deliveries and shipments are not conveniently close, alternative facilities must be provided for these activities. Dedicated storage rooms for maintenance equipment and supplies are as important as storage space for scientific apparatus and materials. The paragraphs on room types and functions (see below) should be consulted for further discussion of building support requirements. A minimum area of 5 nsf per researcher for building support is assumed in Table 1-3.

Area standards are used to determine the amount of research and building net area for each research team, as shown in Table 1-4. Two primary factors which distinguish area standards among the various experimental science activities and disciplines are the recommended area per researcher for (1) a modular laboratory unit and (2) laboratory support categories, namely, dedi-

TABLE 1-4. Sample Team Area Standards for Biomedical and Molecular Biology Laboratories [Net Area (nsf) Assigned per Research Team]

Area Type	Small Team	Medium Team	Large Team
PI Office	120	120	120 to 150
Secretary Office	120 (shared)	120 (shared)	120
Staff Offices	120	180	240 to 300
Modular Laboratory	520	780	960 to 1300
Dedicated Support	160	240	240 to 300
Shared Lab Support	160	240	240 to 300
Lab Support Services	80	120	160 to 200
Animal Facility	varies	varies	varies
Total Team Research Net Lab Area	1280	1800	2080 to 2670
Administration	varies	varies	varies
Personnel Support	varies	varies	varies
Building Support	20	30	40 to 50
Total Net Area per Research Team	1220	1830	2120 to 2720

cated, shared, and support services. Definitions of these and other major space categories in a laboratory building follow:

1. *Laboratory* is a category of net assignable area, in which diverse mechanical services and special supply and exhaust ventilation devices are available. Laboratories are often modular, that is, designed on a standardized room size or a precise multiple of that.

2. *Laboratory support area* is a category of net assignable area that contains the same services and ventilation as the laboratory area but may or may not conform to the same modular laboratory configuration. Dedicated lab support area is assigned to a principal investigator and may adjoin the modular laboratory unit or may be elsewhere. Shared lab support is assigned to and used by more than one principal investigator or a department. Lab support services may be assigned to a department, but they function as a specialized resource by researchers throughout the building.

3. *Administration area* is a category of net assignable area that contains only standard commercial electrical, telecommunication, and office ventilation services. Ventilation air from these areas may be recirculated.

4. *Personnel support area* is a category of net usable area that is similar in function to administration areas, but may contain added mechanical and HVAC services to provide for special needs.

5. *Building support area* is a category of net usable area or gross area that may contain special mechanical and HVAC to provide for special needs.

Typical research area standards are shown in Table 1-5 for a number of commonly encountered laboratory types. There are fewer functional differences in allocation of office space accountable to the various disciplines than there are differences influenced by an institution's culture. In some academic and industrial settings, the size and qualities of an office precisely indicate the individual researcher's status to the square inch!

Another way to look at area allocations per person is based on work category. For example, if a principal investigator (P.I.) is assigned a total of 2000 nsf for his or her program needs, the laboratory population estimate will be eight based on the following area allocations: P.I., 650 nsf; two research assistants, 250 nsf each; one postdoctoral student, 200 nsf; two technicians, 200 nsf each; one graduate student, 130 nsf; and one secretary, 120 nsf. When using this method, the results must be cross-checked against functional performance criteria in a more detailed room list program.

When the function of each team or group of teams is known, the conceptual building program will include a listing of the projected laboratory population by laboratory group or special function and these population data will be used later for the design of each laboratory and each floor of the building. When developing a conceptual program, the concluding step is to establish the total population and, from this, the net program area and gross building area. The total net building area is calculated by estimating the number of research teams of each category and multiplying that figure by the required area for each team.

ROOM TYPES AND FUNCTIONS. An outline type of building program provides a list of all proposed room types with information that relates the nature of the research, equipment, and activities that will take place within them. This program has more specific information than does a conceptual program. However, as in the conceptual program, there are five general area and

TABLE 1-5. Sample Research Area Standards for a Variety of Experimental Science Laboratories [Net Area (nsf) per Full Time Researcher]

	Laboratory Area Categories			
Primary Activity	Office min.-ave.	Laboratory min.-ave.	Lab Support min.-ave.	Total nsf[a] min.-ave.
Molecular Biology	57-90	120-130	80	257-300
Tissue Culture	57-90	95-130	95	247-315
Analytical Chemistry	57-90	110-150	20-35	187-275
Biochemistry	57-90	130-175	60-80	247-345
Organic Chemistry	57-90	150-190	40-50	247-330
Physical Chemistry	57-90	170-200	30-40	257-330
Physiology	57-90	150-170	20-40	227-300

[a] Note: The totals do not include area allocations for animal facilities, administration, personnel, or building support.

function categories, not including structural and mechanical spaces: (1) laboratories, (2) laboratory support facilities, (3) administration, (4) personnel support facilities, and (5) building support. The American Educational and Government Institution Survey (AEGIS, 1987) publishes a list of approved room type names for government buildings and federally supported colleges and universities.

There are many special laboratory types: general chemistry, physics, controlled environment, animal, teaching, and so on. Many types are discussed in considerable detail in Part II of this book. The various office spaces that are directly involved in research activities, such as those assigned to principal investigators, staff, and secretaries and research team conference rooms, are included in laboratory room types. Other offices are included under administrative facilities. Laboratory room types that may be used for teaching must be clearly designated as such, because many states have separate, special codes governing the construction of teaching facilities.

Laboratory support facilities include the following: equipment and storage rooms that may be located near the laboratories they serve; balance and special instrument rooms; data processing facilities; glassware washing rooms; sterilization facilities; preparation rooms for media and solutions; sample processing and distribution rooms; machine shops; electronics shops; darkrooms; fluorescent microscopy rooms; and electron microscopy suites.

Administration facilities that do not directly support research program activities include private offices, group offices and secretarial pools, business offices, personnel records offices, and data processing offices that are assigned to administration of the building or to general administration of the institution. Other administrative facilities include libraries, conference rooms, seminar rooms, and mail rooms.

Personnel support facilities include reception areas, toilets, change rooms, locker and shower rooms, health and first-aid offices, lounges, meeting rooms, dining facilities, kitchens, and recreation areas that are indoors.

Building support facilities include stock rooms, shipping and receiving areas, chemical or flammable liquid storerooms, and storerooms for radioactive, chemical, or biological hazardous wastes. Some types of special materials handling and storage rooms are discussed in Part III.

SHARED AND PROPRIETARY FACILITIES. Questions concerning the use of centralized versus proprietary facilities assigned to an individual investigator or research team must be answered before it is possible to estimate the number of each type of support room and laboratories that will appear on the room-type list. The outline building program addresses the major issue of which facilities will be repeated on each laboratory floor, those that will be shared by occupants of a department, those that will be common to all occupants of the building, and those that will be provided outside the building.

For example, controlled environment rooms may be provided on each floor, but only one radiation laboratory may be provided for all members of a department. A shipping and receiving dock is an example of a single facility for an entire building. An example of a support facility that is best located exterior to a laboratory building in a separate structure is a flammable chemical storage facility. Although it is often more economical to build centralized laboratory support facilities rather than to duplicate them for each department or laboratory group, costs for operating and administering centralized services must be taken into account.

In the programming phase, the planning team resolves all issues of centralized versus proprietary facilities with principal occupants and the owner's representatives. After that, the number of each type of room can be estimated. This task should not be deferred until the design phase, because changes can have a serious impact on building area after the building construction budget is set. However, some minor adjustments to the specific number and distribution of a few support rooms may be permitted during the design phase without serious consequences.

Planners and architects use a number of terms to characterize area data. Terms adopted by the American National Standards Institute (ANSI Z65.1-1990) and the American Society for Testing and Materials ("Standard Classification for Building Floor Area Measurements," which has not yet been released) are frequently referenced. Definitions of a few useful area data terms follow.

NET ASSIGNABLE AREA. Net assignable area is the floor area, excluding interior partitions, columns, and building projections, that lies within the walls of a room or the total area of all rooms and spaces on the room type list assigned to or available for assignment to a specific occupant, group, or function, such as to a principal investigator or to a department. Net area may refer to the total assigned floor area within all rooms and spaces on the room-type list under all categories except personnel and building support.

NET USABLE AREA. Net usable area is floor area that is assigned or available for assignment to a specific occupant or function and includes interior walls, columns, and building projections and excludes public circulation areas such as corridors, stairs, elevators, and vertical shafts. Net usable area may also define the total floor area within the building's exterior wall enclosure that includes floor area taken up by the structure and partitions but excludes circulation areas and vertical shafts.

DEPARTMENTAL GROSS AREA. Departmental gross area is the floor area within the exterior wall enclosure assigned to a specific group that includes secondary circulation within the department's boundaries, interior walls, columns, and building projections.

GROSS AREA. Gross area is the total building floor area. It includes the area occupied by the structure, exterior walls, partitions, and vertical shafts, plus all usable area. Interstitial space is a floor constructed between any two occupied floors of a building that is dedicated to mechanical, electrical, and plumbing distribution systems. Interstitial space is not included in building gross area.

TOTAL AREA CALCULATIONS. All building programs should contain guidelines for the architects and engineers regarding the optimal amount of net assignable area by function and location within the projected total floor area. Laboratory building configurations show variations of 40% to 80% net assignable laboratory area on a typical floor in these examples. They are shown later in Figures 1-4 through 1-10. Total net area must be converted to gross area in order to estimate the amount of actual building area that will be constructed to accommodate all the programmed functions. The conversion factor used to make this calculation is called the *net-to-gross ratio*. Net-to-gross ratios vary from 45% for intensive chemistry laboratory buildings with a high proportion of laboratory area, to 65% for the most efficient laboratory buildings. Buildings containing a low proportion of laboratories to nonlaboratory areas may achieve even higher efficiencies. Buildings which don't have mechanical penthouses or expansive mechanical equipment rooms, when utilities such as steam, hot water, and process chilled water are supplied from an external source, experience higher net-to-gross ratios than lab buildings which are mechanically free-standing. Net-to-gross factors should be very carefully considered.

FLEXIBILITY OF ROOM CATEGORY DEFINITIONS. The amount of laboratory area available in a building can be increased at a later time by converting nonlaboratory areas, such as offices, stockrooms, and personnel support areas, into laboratories. However, to do this safely, efficiently, and at least cost, advance planning is required to provide a latent reserve capacity to significantly increase the delivery of building systems used for heating, ventilating, and air conditioning, electrical services, and piped utilities. Demand and capacity standards for ventilation, cooling, electricity, water, waste drainage, gas, and so on, are far greater for laboratories than for nonlaboratory areas. Normal engineering diversity factors for electrical and cooling capacity do not apply for laboratory use. The constantly operating electrical load is very high for laboratory equipment, some of which is very rarely turned off. This, in turn, puts a greater and more sustained demand on building cooling equipment. Therefore, the building program should (a) identify all nonlaboratory rooms and spaces that are likely to be converted to laboratories when the need arises and provide them with the reserve capacity or (b) specify the proportion of nonlaboratory area that should be engineered for future conversion to laboratories.

RELATIONSHIPS BETWEEN SPACES AND FUNCTIONS. The next task in developing an outline program is to inform the architects and engineers of the important relationships between the parts of the building that were identified above. The outline building program does not need to specify what is on every floor; that information is developed and organized in the planning process. However, after the building site has been selected, conceptual floor plans may be provided in a detailed program. There are five sets of questions that have proven helpful for establishing important relationships between spaces and functions:

1. What is the organizational structure of the institution or corporation for whom the building is being designed? Should room assignments and groupings reflect this hierarchy, or is some other pattern preferred?

2. Do materials, processes, and waste products contained or produced in one area affect the function or pose a hazard for any other area or function? If the answer is affirmative, what arrangements can be made to reduce or eliminate conflicts? Are there appropriate rooms assigned to specialized waste handling and storage? Where should they be located with regard to the areas of waste generation and pathways used for waste removal?

3. How close should laboratory support facilities be to the laboratories they serve? Are there critical relationships that affect health, safety, or efficiency?

4. How close should offices be to laboratories? Should offices be within laboratories, contiguous with laboratories, across the hall, or in a separate wing? What are the health, safety, and efficiency implications of each locations?

These last questions generally brings up one of the more contentious issues in laboratory building layout. Some researchers insist that their offices be located in or adjacent to their labs, whereas other researchers prefer to have their offices outside the laboratory zone. Offices in laboratory zones are readily used by researchers for laboratory functions, whether these offices are equipped to safely support laboratory functions or not. Especially in academic settings, there is pressure on researchers to acquire new equipment or projects even when appropriate space or funds for renovations are not available. Proximity of the office to the lab usually dominates the other criteria, such as adequate power, piped utilities, and appropriate ventilation. In the worst cases, offices fit-out by researchers have additional power supplied by electric extension cords and piped utilities by rubber tubing for gas and water, which are strung across corridor ceilings from nearby labs. Offices have their doors propped open to improve ventilation by exhausting fumes into corridors and capturing more cool air for equipment that would otherwise overheat. Extremely serious health and safety hazards are generated by these conditions.

When offices are located in a laboratory area, whether across the hall from, adjacent to, or within the laboratory, ventilation requirements for these offices should meet laboratory standards, in which general exhaust and 100% outside supply air are provided. In addition, electric panels should be sized to accommodate future increased power demands when these offices are converted to laboratory use. There is a high initial construction cost to provide piped utilities and laboratory waste water drains to offices, but if future flexibility and safety are priorities, it is a reasonable investment. These criteria should be included in the building program list of performance requirements.

5. Do certain laboratories, or the mechanical services to them, need to be isolated from other building functions or services for reasons of health and safety, or as a necessary part of their procedures and equipment operation?

Answers to these questions provide additional information to assist the architects and engineers in the preparation of a description of critical adjacencies that may be documented in text, charts, and diagrammatic floor plans. Figure 1-1 illustrates a matrix format, similar to that of a road mileage map,

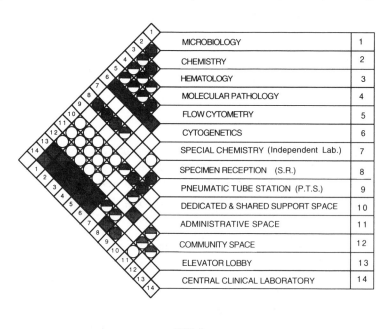

MICROBIOLOGY	1
CHEMISTRY	2
HEMATOLOGY	3
MOLECULAR PATHOLOGY	4
FLOW CYTOMETRY	5
CYTOGENETICS	6
SPECIAL CHEMISTRY (Independent Lab.)	7
SPECIMEN RECEPTION (S.R.)	8
PNEUMATIC TUBE STATION (P.T.S.)	9
DEDICATED & SHARED SUPPORT SPACE	10
ADMINISTRATIVE SPACE	11
COMMUNITY SPACE	12
ELEVATOR LOBBY	13
CENTRAL CLINICAL LABORATORY	14

KEY

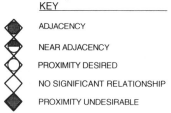

ADJACENCY

NEAR ADJACENCY

PROXIMITY DESIRED

NO SIGNIFICANT RELATIONSHIP

PROXIMITY UNDESIRABLE

FIGURE 1-1. Adjacency matrix (Satellite Clinical Laboratory).

and Figure 1-2 shows a bubble diagram format that conveys adjacency information in a graphic fashion, representing, for example, a clinical laboratory.

The ultimate type of program that can be developed for a building is called a *detailed functional program*. It contains all the information in an outline program but differs from the outline program in two significant ways. The detailed functional program adds complete information, to the extent that it is known at the time, on the performance specifications for the following systems in each individual room or generic room type:

Lists of requirements for mechanical, electrical, and plumbing systems

Fire protection and safety equipment

List of potential hazards

List of probable chemicals usage

List of probable major equipment

Storage requirements for chemicals and equipment

Number and type of fume hoods, special exhaust requirements

Number of work stations and types of benches

Figure 1-3 shows a typical room data collection form for a research building program. The detailed program may include schematic plans of each room or room type, and drawings of the major wall elevations may also accompany each room data collection sheet. In the case of a program for a building that will be occupied by personnel who will be relocated from another laboratory building, not only can an equipment list be developed, but photographs and installation specifications can be obtained of the equipment to be relocated plus manufacturer's cut sheets and specifications for new equipment.

The final task that may be considered for any program type is to estimate the probable cost of construction, perhaps the most critical information in the document. Should an administration set the construction budget without regard for the documented requirements and for the size and complexity of the project, the success of the building function, its safety and appropriateness will be put at risk. Unless the group doing the programming (professional architects and engineers or an in-house team) has available to them up-to-date, extensive cost data on laboratory buildings, professional cost estimators should be hired to do this task. Professional cost estimators, with experience in laboratory building construction, can review the program description, whether conceptual, outline, or detailed, and develop a construction cost estimate. The estimator should be asked to provide design assumptions and the percent range of accuracy. A range of plus or minus 20% is not unusual during programming. Obviously accuracy improves in the design phase with development of more advanced plans and specifications.

For the design team, the building program contains performance specifications and states the owner's building function and goals. For the owner, the

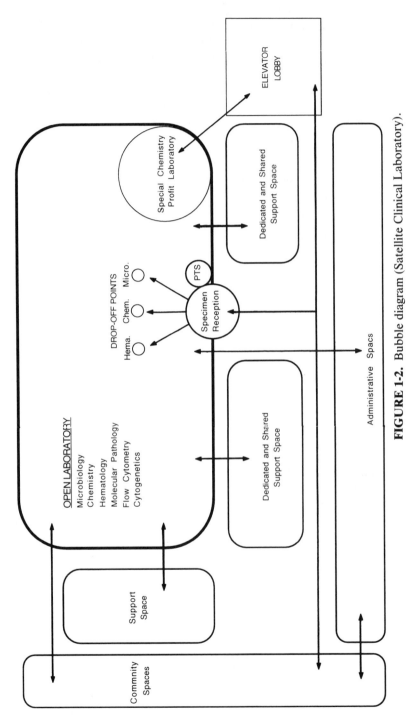

FIGURE 1-2. Bubble diagram (Satellite Clinical Laboratory).

24

SAMPLE ROOM DATA COLLECTION FORM FOR A RESEARCH BUILDING PROGRAM   P 1

DATA
NO. DATE _____ PROJECT NAME _____ ROOM ID. _____

ROOM TYPE _____

A IDENTIFICATION

1 BUILDING _____
2 ROOM ID NUMBER _____
3 DIVISION _____
4 DEPARTMENT _____
5 ASSIGNED TO _____

B ROOM DESCRIPTION

6 ACTIVITIES _____

7 GENERIC TYPE   laboratory _____   lab office _____   admin. _____   personnel _____
                       lab support _____   building support _____   personnel support _____
8 ROOM TYPE _____
9 NET AREA _____
10 HAZARD RATING   fire _____   biohazard _____
                   explosion _____   radiation _____
                   toxicity _____   reactivity _____

C FUNCTIONAL RELATIONSHIPS

11 LAB ADJACENCIES _____
12 FLOOR ADJACENCIES _____
13 BUILDING ADJACENCIES _____

D SAFETY EQUIPMENT

14 FIRE EXTINGUISHE   type _____   size _____
15 FIRE BLANKET
16 EMERG EYEWASH   tempered _____   gpm _____
17 DELUGE SHOWER   tempered _____   gpm _____
18 SPILL KIT   type _____
19 SAFETY GLASSES   type _____
20 OTHER _____

E LAB FURNISHINGS

TYPE	L.F.	OUTLETS	VOLTS	WATER	GAS	AIR	VACUUM
21 HIGH BENCH							
22 LOW BENCH							
23 SINK							
24 SHELVING							
25 OTHER STG							
26 DESK							

FIGURE 1-3. Sample room data collection.

SAMPLE ROOM DATA COLLECTION FORM FOR A RESEARCH BUILDING PROGRAM   P 2

DATA
NO. DATE   PROJECT NAME _____ ROOM ID. _____

ROOM TYPE _____

F MECHANICAL STANDARDS

27	TEMPERATURE	summer	_____	winter	_____
28	HUMIDITY	summer	_____	winter	_____
29	AIR CHANGES	total	_____	fresh	_____
30	PRESSURIZATION	positive to	_____		
		negative to	_____		
31	FILTRATION	supply	standard _____	special	_____
		exhaust	standard _____	special	_____
32	EXHAUST	fume hoods	cfm _____	type	_____
		biosafety cabinets	cfm _____	type	_____
		local exhaust	cfm _____	type	_____
		glove box	pressure _____	type	_____
		recirc. allowed	yes _____	no	_____
33	VIBRATION	isolation	standard _____	special	_____
34	NOISE CONTROL	decibel limit	_____		

G ELECTRICAL POWER

35	NORMAL POWER	circuits _____	volts _____	outlets _____	breaker _____
36	EMERGENCY POWE	circuits _____	volts _____	outlets _____	
37	CONDITIONED	circuits _____	volts _____	outlets _____	
38	UPS	circuits _____	volts _____	outlets _____	
39	SPECIAL	circuits _____	volts _____	outlets _____	
40	MASTER ELEC KILL SWITCH	_____	_____		

H DATA & COMMUNICATIONS

41	TELEPHONE	no. units	_____	type	_____
42	DATA	no. outlets	_____	wiring _____	face plate _____
43	INTERCOM/PUBLIC ADDRESS				
44	SECURITY ALARMS	type	_____	wiring _____	display _____
45	ALARM/SENSORS	high/low temp.	_____		
		humidity	_____		
		power on/off	_____		
		special	_____		

I LIGHTING

46	FOOTCANDLES	_____	
47	SPECIAL LAMPS	_____	
48	SWITCHING	half-load	_____
		reostat	_____
		darkroom	_____
49	TASK LIGHTING	type	_____

FIGURE 1-3. (*Continued*)

SAMPLE ROOM DATA COLLECTION FORM FOR A RESEARCH BUILDING PROGRAM   P 3

DATA
NO. DATE   PROJECT NAME   ROOM ID. _____

ROOM TYPE _____

J FIRE PROTECTION

50 WET SPRINKLER   _____
51 DRY PREACTION   _____
52 OTHER   _____
53 SMOKE DETECTORS   _____
54 SPECIAL ALARMS   _____

K PLUMBING

55 PROCESS WATER   hot/cold   _____
56 POTABLE WATER   hot/cold   _____
57 CENTRAL PURE   m. ohms rating   _____
58 LOCAL POLISHER   m. ohms rating   _____
59 CHILLED WATER   gallons/min   _____   temperature _____
60 LAB SINKS   size   _____   number _____
61 CUP SINKS   number   _____
62 FLOOR DRAINS   _____
63 EMERG. SYSTEMS   eyewash fountains _____   deluge shower _____
64 STEAM   low/high pressure _____
65 COMPRESSED AIR   _____
66 CENTRAL VACUUM   _____
67 CENTRAL NITROGEN   _____
68 CENTRAL CO2   _____
69 CYLINDER GASES   type   _____   pressure _____   hazard _____

L MAJOR EQUIPMENT LIST

	EQUIP. TYPE	NO.	MFG. & MODEL	SIZE & WT	BTU/hr.	AMPS	VOLT	PHASE	WATER	DRAIN	AIR VAC GAS
70											
71											
72											
73											
74											
75											
76											
77											
78											
79											
80											
81											
82											
83											
84											
etc.											

FIGURE 1-3. (*Continued*)

building program is a document against which the architectural design can be evaluated for adherence to the stated goals and specifications. To further assist the architects and engineers, the laboratory procedures and safety manuals used by the building owners should accompany the building program to alert the designers to particular health and safety concerns of the owners. When such documents have not yet been prepared, borrowed manuals for similarly engaged facilities or appropriate sections of this manual may be substituted.

1.2.2 Planning

A building program gives the architects and engineers information they need to plan the building. Planning, the first phase of the design process, includes all the issues to be described in Sections 1.2.2.1 through 1.2.2.7. The end product of the planning process is a set of schematic design drawings and specifications for materials. Architects customarily show the layout of each floor level and indicate egress pathways; they also present their initial concepts of what the building will look like, including height, shape, volume, and materials to be used. The location on the site, including vehicular, service, and pedestrian entry points, will also be shown in schematic design drawings. Engineers will prepare their estimated load calculations and will locate primary equipment. They will also develop concepts for all mechanical, electrical, and plumbing systems, including fire protection, laboratory water and waste treatment, safety control systems, energy conservation, and emergency power. They will identify locations for primary equipment and mechanical rooms, as well as for utilities distribution, air intake and exhaust outlets, utility entry points, and sewage connection. A structural system may also be proposed in schematic design.

Schematic design drawings will show the size and location of every room listed in the building program by type and by floor level. This is a good time to develop a detailed layout of the generic modular laboratory (which is the basic building block of most laboratory buildings) if it wasn't proposed in the program document. Preliminary room and area assignments for the proposed occupants of the building may be shown on the schematic design drawings when the final occupants are known. Owners, occupants, architects, and engineers should communicate intensively and frequently during the planning process to exchange information, to discuss issues brought forward by the generation of design ideas, to resolve problems that were not foreseen during development of the building program, and to make decisions. To establish effective lines of communication within the owner's group, an individual or a small committee comprised of future laboratory users—along with other occupants (if it is a mixed-use building), facility management (including safety representation), and administrators—should be designated to act as the owner's representative. The designated person(s) should have the time and temperament to learn the opinions and concerns of the people who will occupy the new facility. When future occupants are not identified, the owner representative should contact a wide range of other people in the organization (or on campus) who have an interest in the new facility. The owner representative will receive oral and written communications from the design team on a day-to-day basis and distribute this information to elicit opinions upon which decisions can be made. Another function of the owner representative is to help build consensus, to learn what occupants really need, and to transmit this information to the design team and to the management group, or building committee, who will make the final decisions.

The schematic design phase is the time to explore options in the layout, building mechanical systems, and architectural design, because the design is still in transition. The phase that follows is called *design development*. At this time, greater detail and a clearer definition of the final building is incorporated into the agreed upon concept. Cost estimating is usually done at the conclusion of schematic design and design development activities. Adjustments in building area, materials, or methods of construction are generated during this time to keep the project within budget. In later stages of the design, especially in production of construction documents, small refinements are possible, but changes in basic concepts and selected materials generate difficulties and substantial costs. Late changes are expensive to execute and time-consuming; they put the project schedule at risk.

1.2.2.1 Building Spatial Organization. While the optimal building enclosure configuration, site location, and expression are being designed, architect and engineers will generate concepts for the internal organization of the building spaces. Internal organization of a laboratory building is comprised of six major patterns of spatial definition:

1. Circulation of people and materials (Section 1.2.2.2).
2. Generic laboratory modules (Section 1.2.2.3).
3. Distribution of mechanical equipment and services (Section 1.2.2.4).
4. Structural system (Section 1.2.2.5).
5. Site regulations (Section 1.2.2.6).
6. Building enclosure configuration (Section 1.2.2.7).

These patterns of spatial definition often conflict when they are not considered together during the planning phase and are not thoroughly coordinated by the architectural and engineering team during later design phases. Although structural and mechanical engineers can design many solutions to fit a building enclosure, a comprehensive concept is needed to fulfill building program requirements and an optimized solution for function, safety, and flexibility. All systems should complement one another to provide a safe and healthful work environment at reasonable cost. Drawing techniques, such as those available in computer-aided design and drafting (CADD) and manual overlay methods, allow superimposition of the design layouts prepared by mechanical and electrical engineers and those prepared by the architect to detect physical conflicts. The conceptual compatibility of all of the systems that will define the structure depends on the knowledge and creativity of the design team.

1.2.2.2 *Circulation of People and Materials.* A major determinant of spatial definition is the circulation of people and materials within the building and around it. Health and safety issues of circulation are primarily concerned with (a) emergency egress of building occupants and (b) access to the building and its internal parts by emergency personnel such as firemen and police. The major reference sources used in this book for guidance are the National Fire Protection Association (NFPA, 1992), Occupational Safety and Health Administration (OSHA, 1992), Americans with Disabilities Act of 1991, and Building Officials and Code Administrators International, Inc. (BOCA, latest edition). These codes and standards define and specify all the components for building egress and access that pertain to safety. Specific sections of note are the following:

1. *NFPA Bulletin 101, 1991, Life Safety Code,* Chapter 5, "Means of Egress"; Section 13-2, "Business Occupancies, Means of Egress Requirements"; and Section 14-2, "Industrial Occupancies, Means of Egress Requirements."
2. OSHA Health and Safety Standards, *29 CFR 1910.35,* Chapter XVII, 1991 ed., Subpart E, "Means of Egress."
3. Americans with Disabilities Act of 1990, *28 CFR, Title III, Part 36,* "Nondiscrimination on the Basis of Disability by Public Accommodations and in Commercial Facilities, Final Rule," Subpart D, "New Construction and Alterations," Paragraph 4.3 ("Accessible Route") and Paragraph 4.3.1.0 ("Egress").
4. *BOCA Code,* 1990, Section 600.

There are several other nationally recognized building codes including the *Uniform Building Code,* by International Conference of Building Officials, and *Southern Building Code Congress International,* plus state or city codes that also address building egress and access. They may differ somewhat from the specific citations referred to in this manual, but take precedence when they are more restrictive.

Special attention should be given to ease of access and emergency evacuation for disabled persons. The Americans with Disability Act of 1990, *28 CFR, Part 36,* describes and discusses specific issues of accessibility and safety for disabled persons in commercial facilities. The act contains extensive guidelines on design standards that are based, in part, on ANSI *Standard 117.1-1986* "Providing Accessibility and Usability for Physically Handicapped People." Certain state and local jurisdictions may require more stringent standards for accessibility; these codes must be checked to make sure that a new construction, addition, or renovation complies if the lab building is in the category of "public accommodation." Private commercial laboratories, in certain aspects, may have fewer accessibility requirements, if no disabled persons are employed there. Buildings that are accessible facilitate employment of disabled persons and their integration into the workforce. No architectural barriers should be designed and constructed at main entrances to buildings, other doorways, public toilet rooms, elevators, drinking fountains, or public telephones. Surrogates for visible signs, highly legible signs, and acceptable door and elevator hardware should be provided for those who are visually impaired. Alarms, warnings, and controls detectable by both disabled and unimpaired people are required.

1.2.2.3 Laboratory Module. A single laboratory module is defined as a basic unit of space of a size commonly referred to as a *two-person laboratory*. Formulation of the internal organization of the laboratory building begins with a decision on the dimensions of the laboratory module. This task redirects the planning focus from the large scale of the total facility down to the small scale of a single laboratory module.

1.2.2.3.1 Laboratory Width. Criteria for the individual laboratory work area were studied in detail by the Nuffield Foundation, United Kingdom, over three decades ago (Nuffield, 1961). Their time/motion efficiency studies led to specifications for optimal dimensions of a standard laboratory aisle that were derived from ergonomic factors related to reaching across and above work surfaces. The laboratory aisle is a space, usually flanked by an array of work surfaces, equipment, benches, and utilities, where laboratory personnel have access to the work area and, essentially, spend their workday. The aisle between benches, work surfaces, or equipment should be a minimum of 60 in. (150 cm) so that one person can pass behind another working at a bench. The maximum clearance, in a research laboratory, should be no more than 72 in. (180 cm) because aisles wider than this tend to get clogged with free-standing equipment and other obstructions set up by researchers to make maximum utilization of the space close at hand. The average width of an array of benches and equipment on both sides of an aisle varies widely. Typically, a bench, with utility chase behind, may be 30 in. from the face of the bench to the wall. Free-standing equipment typically ranges from 2 ft to 3 ft wide; when there is a built-out utility chase behind, total width can be over 42 in. Table 1-6 assumes an average 36 in. width.

To assure that there is sufficient space for a single-aisle laboratory at each module, the dimension for the thickness of a standard interior partition should be added to the clear width of each module. This provides flexibility to divide a large laboratory into individual single-module laboratories, or an entire laboratory floor into individual laboratories of varied sizes, without

TABLE 1-6. Laboratory Width by Number of Modules in a Building Unit

No. of Modules	1	2	3	4	5	6
Number of parallel rows of						
Aisles	1	2	3	4	5	6
Benches or equipment	2	4	6	8	10	12
Utility strips	2	4	6	8	10	12
Width of parallel rows						
Aisles–60 in. wide	5'-0"	10'-0"	15'-0"	20'-0"	25'-0"	30'-0"
Equipment–30 in. wide	5'-0"	10'-0"	15'-0"	20'-0"	25'-0"	30'-0"
Utilities–6 in. wide	1'-0"	2'-0"	3'-0"	4'-0"	5'-0"	6'-0"
Total constructed width, center to center						
Walls 4 $\frac{5}{8}$ in. GWB[a]	11'-4 $\frac{5}{8}$"	22'-9 $\frac{1}{4}$"	34'-1 $\frac{7}{8}$"	45'-6 $\frac{1}{2}$"	56'-11 $\frac{1}{8}$"	68'-3 $\frac{3}{4}$"
Walls 5 $\frac{5}{8}$ in. CMU[b]	11'-5 $\frac{5}{8}$"	22'-11 $\frac{1}{4}$"	34'-4 $\frac{7}{8}$"	45'-10 $\frac{1}{2}$"	57'-4 $\frac{1}{8}$"	68'-9 $\frac{3}{4}$"

[a] GWB is a partition type with 3 $\frac{5}{8}$" metal studs with one layer of $\frac{1}{2}$" gypsum wall board on each side.
[b] CMU is a partition type constructed with 6" nominal width concrete masonry units (concrete blocks).

reducing the recommended aisle width. Special-purpose laboratories, such as controlled-environment laboratories and pilot plants, have space requirements that may not conform to a standard laboratory module. Refer to Section 2.2 (Chapter 2) for more information on the internal organization of a laboratory module.

Existing structures converted to laboratory use may not have structural module dimensions consistent with division into the recommended module dimensions. To overcome this difficulty, adjustments may have to be made on the depth of benches or in the distribution of utilities to the laboratory floor, but under no circumstances should the clear aisle width be reduced below 60 in. or 150 cm. Table 1-6 presents laboratory widths for a variety of modular arrangements.

1.2.2.3.2 Laboratory Length. Length of a laboratory module is governed by several variables: the overall width of the building enclosure, the structural span, and the area allotment for a standard module. Laboratory module length is generally 20 to 30 ft for efficient operation of the laboratory. Laboratory length in excess of 35 ft may generate egress problems. The problem, however, can be overcome in deeper laboratory modules. Refer to Figure 2-4 for an example of an alcove module arrangement which draws the primary exit toward the middle of the full length of the module, a layout that reduces the distance to the exit.

1.2.2.3.3 Laboratory Unit. The National Fire Protection Association defines a laboratory unit as ''an enclosed space used for experiments or tests.'' It may include an assembly of a number of laboratory modules, corridors, contiguous accessory spaces, and offices into a larger space category (departmental gross area; see definition in Section 1.2.1.1). Entire buildings may be defined as a laboratory unit. The National Fire Protection Association's *Standard on Fire Protection for Laboratories Using Chemicals* (NFPA 45, 1991), offers criteria for planning laboratory units. Two important tables in this bulletin are ''Maximum Quantities of Flammable and Combustible Liquids in Laboratory Units Outside of Approved Flammable Liquid Storage Rooms'' (reproduced here as Table 1-7) and ''Construction and Fire Protection Requirements for Laboratory Units'' (reproduced here as Table 1-8). Table 1-7 establishes fire hazard classifications of laboratory units on the basis of the number of gallons of flammable liquids stored within the laboratory unit per 100 nsf area. The concept of Table 1-7 is that the amount of flammable liquids allowed within a laboratory unit is directly proportional to its size, up to a defined maximum quantity.

Table 1-8 lists minimum fire separation requirements for laboratory units in each hazard classification. From Table 1-8, the maximum or optimal size of a laboratory unit can be estimated. In actual practice, hazard classification of laboratory units is often made after the building is occupied, rather than in the planning or programming phase. Consideration always must be given in the

TABLE 1-7. Maximum Quantities of Flammable and Combustible Liquids in Laboratory Units Outside of Approved Flammable Liquid Storage Rooms[a]

Laboratory Unit Class	Flammable or Combustible Liquid Class	*Excluding* Quantities in Storage Cabinets and Safety Cans[b]			*Including* Quantities in Storage Cabinets and Safety Cans[b]		
		Maximum Quantity[c] per 100 ft² of Laboratory Unit (gal)	Maximum Quantity[d] per Laboratory Unit (gal)		Maximum Quantity[c] per 100 ft² of Laboratory Unit (gal)	Maximum Quantity[d] per Laboratory Unit (gal)	
			Unsprinklered	Sprinklered[e]		Unsprinklered	Sprinklered[e]
A[f] (High hazard)	I	10	300	600	20	600	1200
	I, II, and IIIA[g]	20	400	800	40	800	1600
B[h] (Intermediate hazard)	I	5	150	300	10	300	600
	I, II, and IIIA[g]	10	200	400	20	400	800
C[h] (Low hazard)	I	2	75	150	4	150	300
	I, II, and IIIA[g]	4	100	200	8	200	400

[a] See description of flammable liquid storage room in Section 4-4 of NFPA 30, *Flammable and Combustible Liquids Code*. See description of storage cabinet in Section 4-3 of NFPA 30.

[b] For SI units: 1 gal = 3.785 L; 100 ft² = 9.3 m².

[c] For maximum container sizes, see Table 7-2, NFPA 45.

[d] Regardless of the maximum allowable quantity, the maximum amount in a laboratory unit shall never exceed an amount calculated by using the maximum quantity per 100 ft² of laboratory unit. The area of offices, lavatories, and other contiguous areas of a laboratory unit are to be included when making this calculation.

[e] Where water may create a serious fire or personnel hazard, a nonwater extinguishing system may be used instead of sprinklers.

[f] Class A laboratory units shall not be used as instructional laboratory units.

[g] The maximum quantities of Class I liquids shall not exceed the quantities specified for Class I liquids alone.

[h] Maximum quantities of flammable and combustible liquids in Class B and Class C instructional laboratory units shall be 50% of those listed in the table.

Source: National Fire Protection Association, Quincy, MA, 1991.

TABLE 1-8. Construction and Fire Protection Requirements for Laboratory Units[a]

Laboratory Unit Fire Hazard Class	Area of Laboratory Unit (Square Feet[d])	Nonsprinklered Laboratory Units				Spinklered Laboratory Units[b]	
		Construction Type I and II[c]		Construction Types III, IV, and V[c]		Any Construction Type[c]	
		Fire Separation[e] from Nonlaboratory Area	Fire Separation[e] from Lab Units of Equal or Lower Hazard Classification	Fire Separation[e] from Nonlaboratory Areas	Fire Separation[e] from Lab Units of Equal or Lower Hazard Classification	Fire Separation[e] from Nonlaboratory Areas	Fire Separation[e] from Lab Units of Equal or Lower Hazard Classification
A	Under 1000	1 h	1 h	2 h	1 h	1 h	NC/LC[c,f]
	1001–2000	1 h	1 h	N/A[f]	N/A	1 h	NC/LC
	2001–5000	2 h	1 h	N/A	N/A	1 h	NC/LC
	5001–10,000	N/A[f]	N/A	N/A	N/A	1 h	NC/LC
	10,001 or more	N/A	N/A	N/A	N/A	N/A	N/A
B	Under 20,000	1 h	NC/LC[c,f]	1 h	1 h	NC/LC[g,i]	NC/LC
	20,000 or more	N/A	N/A	N/A	N/A	N/A	N/A
C	Under 10,000	1 h	NC/LC[g,i]	1 h	NC/LC[g,i]	NC/LC[g,i]	NC/LC[g,h]
	10,000 or more	1 h	NC/LC	1 h	1 h	NC/LC[g,i]	NC/LC

[a] Where a laboratory work area or unit contains an explosion hazard, appropriate protection shall be provided for adjoining laboratory units and nonlaboratory areas, as specified in Chapter 5, NFPA 45.

[b] In laboratory units where water may create a serious fire or personnel hazard, a nonwater extinguishing system may be substituted for sprinklers.

[c] See Appendix B-4, NFPA 45.

[d] For SI units: 1 ft^2 = 0.929 m^2.

[e] Fire separation = from one level of rated construction to another.

[f] N/A, no allowed; NC/LC, noncombustible/limited combustible construction. (See Appendix B-4, NFPA 45).

[g] May be ½-h fire-rated combustible construction.

[h] Existing combustible construction is acceptable.

[i] Laboratory units in educational occupancies shall be separated from nonlaboratory areas by 1-h construction.

Source: National Fire Protection Association, Quincy, MA, 1991.

planning process to establish a reasonable subdivision of space and use of fire protective construction to limit spread of fire, fumes, and other hazards that may threaten life or cause extensive property damage.

Other laboratory hazards such as highly toxic or radioactive chemicals, gases under high pressure, and highly pathogenic materials also bear hazard classifications. They are established by the Environmental Protection Agency for chemical hazards, by the Department of Transportation (DOT) for high-pressure gases, and by the National Institutes of Health in cooperation with the Centers for Disease Control for pathogenic materials, among others. Ideally these categories are identified during the programming or planning phase to assure that proper layout, construction materials, ventilation, and safety equipment are provided in the budget as well as in the design.

1.2.2.4 Distribution of Mechanical Equipment and Service. Mechanical engineers, in consultation with the architect and project engineer, design the distribution of ventilation air, mechanical equipment, and piped utilities. Plans for recommended laboratory layouts and distribution of services are presented in Figures 1-4 to 1-10. These layouts are categorized by the location at each module of the vertical shafts that contain risers of electric conduit, piped utilities, and ventilation ducts, as follows:

Figure 1-4. Evaluation of utility shaft adjacent to service corridor.

Figure 1-5. Evaluation of utility shaft at rear laboratory wall.

Figure 1-6. Evaluation of utility shaft at exterior wall.

Figure 1-7. Evaluation of utility shaft at interior wall.

Figure 1-8. Evaluation of utility shaft between modules.

Figure 1-9. Evaluation of central utility shaft.

Figure 1-10.Evaluation of central utility shaft in laboratory service core.

Laboratory buildings require a great amount of energy to supply conditioned air to a comfortable temperature, and exhaust ventilation air. Vertical distribution of ducts is an efficient design approach for exhaust systems. However, continuity of vertical shafts from floor to floor and through the entire building to exhaust fans and equipment on the roof is necessary for this organization. Therefore, functions that require a large undivided floor area (such as an auditorium) cannot be located above laboratory floors or sandwiched between laboratory floors without the loss of net usable floor area and possibly serious interference with intended functions. Vertical shafts, of adequate dimensions, located at each module allow an additional laboratory hood or other special exhaust system to be installed there in the future. The same flexibility is available for piped utilities to each module through vertical shafts. The floor area occupied by vertical shafts ranges between 1% and 10% of the net usable area on a typical laboratory floor.

Building supply air is usually provided from a combination of vertical and horizontal systems. Supply air is filtered and conditioned through an air-handling unit that may be located anywhere between the basement and a penthouse on the roof. Primary air supply risers are distributed in vertical shafts. From there, air is distributed to each space, by horizontal duct runs. Horizontal supply and exhaust air duct runs require vertical space that is unobstructed by structural frame members, between floors, so that comfortable ceiling heights can be maintained. In general, the more systems that require horizontal distribution, the more clear ceiling height is required. Use of interstitial floors, constructed solely for utility and air distribution, is an extreme example of a horizontal distribution system. Following is a discussion of the building layouts shown in Figures 1-4 through 1-10.

There are advantages and disadvantages associated with placement of vertical mechanical shafts near the service corridor (Figure 1-4). An important advantage is that the laboratory hood will be located away from the door and adjacent to the vertical shaft, favoring a short, energy-efficient connection to the exhaust riser. Desks, which generally represent the area of lowest potential hazard in the laboratory, can then be placed at the public access corridor wall, near the primary exit. These relative locations of fume hoods and desks comply with the concept of hazard zoning illustrated in Figure 1–11 and discussed further in section 2.2.2. This arrangement permits maintainance and servicing of pipes and ducts in the vertical shafts to be done from a service corridor outside the laboratory modules, reducing the frequency of maintenance personnel entering laboratories to perform routine maintenance. The service corridor itself provides a second egress from each laboratory module and a separate fire-zone for emergency egress if it is constructed according to code with adequate, unobstructed width and appropriate ventilation, along with fire-rated wall, ceiling, floor, and door assemblies. The laboratory layouts shown in Figure 1-4 are optimal because of the greater life safety characteristics afforded by the separate emergency exit into a separate fire-zone. In other layouts, a second egress may be permitted through adjacent laboratories, of equal or less hazard, but this arrangement is not ideal, because operations and hazard levels in laboratories change. A major disadvantage of a layout of a service corridor with laboratories along both sides is that no direct natural light can be brought into the laboratories. Laboratories are strictly interior spaces, even if there is borrowed light from public access corridors on exterior walls into laboratories with windows. The quality of the work environment is diminished without natural light and views to the outside.

Figure 1-5 shows the vertical shaft zone arrayed at the rear of back-to-back laboratories. These layouts have many of the advantages and all the disadvantages of layouts in Figure 1-4. However, if the area between back-to-back laboratories (Figure 1-5) is a large utility shaft, it is not likely to meet all the requirements of a legal egress route. Secondary exits, as in the following layouts, would lead to other laboratory spaces.

When mechanical shafts are placed adjacent to a public access corridor

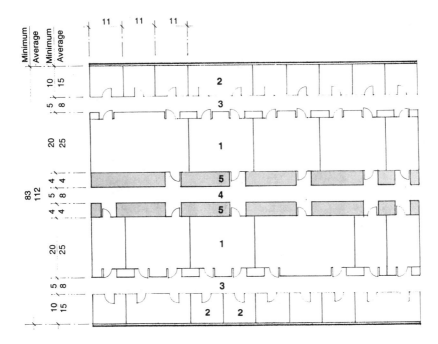

PLAN A

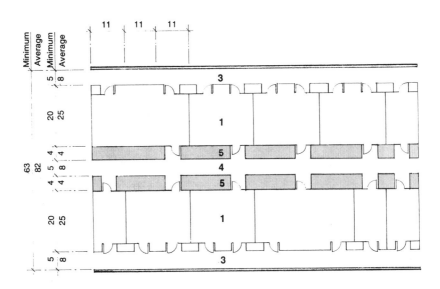

PLAN B

KEY
1 Laboratory
2 Office
3 Staff Corridor
4 Service Corridor
5 Mechanical Shafts

FIGURE 1-4A. Utility shafts at service corridor.

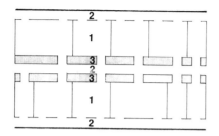

ADVANTAGES

short horizontal exhaust duct from fume hoods
fume hoods are in a protected position at end of aisles
major traffic within the lab is near public corridor egress
writing stations are near public corridor egress
second egress is into separate fire zone
pipes and ducts are accessible from service corridor

DISADVANTAGES

no natural light directly into laboratories
outswinging doors obstruct public corridor
low net to gross area efficiency

KEY

1 Laboratory
2 Corridor
3 Mechanical Shafts

FIGURE 1-4B. Evaluation of utility shafts at service core.

(Figure 1-7), maintenance and repairs can be done from that corridor rather than from within the laboratory. The second exit from the laboratory, through another room, can be located away from the exit to the corridor to establish two separate paths of egress. The width of the shaft itself forms an alcove against which the laboratory door can swing outward and yet not project too far into the corridor. An expansive window area on the exterior wall is possible because it can be kept free of most mechanical services.

The primary disadvantage of locating mechanical shafts at an interior public access corridor wall is that the fume hood, which should be away from the primary egress, will be away from the riser. Therefore, a run of horizontal duct must be used to connect the chemical fume hood, or other special exhaust application, to the riser in the shaft. The second exit, although it can be well positioned in the laboratory, may open into another laboratory or office rather than into a corridor in a separate fire zone, as was possible in the first example, Figure 1-4. Another slight disadvantage of mechanical shafts at the public corridor wall is that laboratory entries can only occupy the spaces between shafts.

When vertical mechanical shafts are at an exterior wall, chemical fume hoods can be located adjacent to shafts at the rear of the laboratory as shown in Figure 1-6. This arrangement provides a threefold advantage: First, there is a very short horizontal run for the duct connection to the exhaust riser in the shaft; second, the fume hood is away from the primary exit; and third, desks can be arranged at the corridor wall away from potential hazards.

However, without an adjacent service corridor, maintenance must be done to risers and ducts inside mechanical shafts located within laboratories from the laboratories, and this is a distinct disadvantage. Welding, as well as other repair activities that pose an increased risk in a laboratory environment, should be scheduled when the laboratory is not occupied. A fire watch should be maintained during welding activities in a laboratory building.

Another disadvantage is that the second exit from such laboratories is likely to be near to and open into the same public access corridor as the

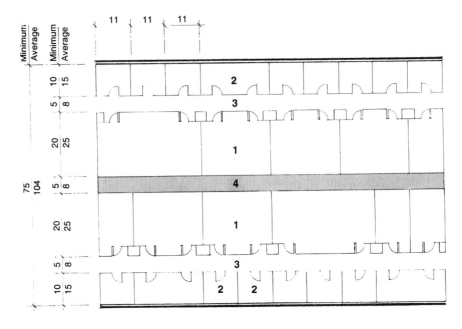

PLAN   C

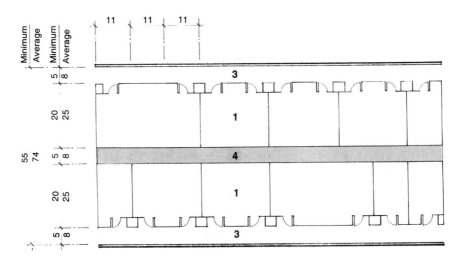

PLAN   D

KEY

1   Laboratory
2   Office
3   Corridor
4   Mechanical Shafts

FIGURE 1-5A. Utility shafts at rear wall of laboratory.

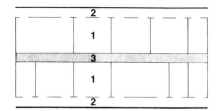

ADVANTAGES

short horizontal exhaust duct from fume hoods
fume hoods are in a protected position at end of aisles
major traffic within the lab is near public corridor egress
writing stations are near public corridor egress
pipes and ducts are accessible from service corridor

DISADVANTAGES

second egress is not in a separate fire zone
no natural light directly into laboratories
outswinging doors obstruct public corridor

KEY
1 Laboratory
2 Corridor
3 Mechanical Shafts

FIGURE 1-5B. Evaluation of utility shafts at rear wall of laboratory.

primary exit. When shafts occupy most of the area of the exterior wall, window area is restricted. Large glass areas to bring natural light into laboratories are highly desirable, although operable windows are undesirable, because they interfere with ventilation and energy conservation plans and with efficient operation of laboratory hoods.

Locating vertical mechanical shafts between modules (Figure 1-8) is primarily used in nonlaboratory buildings converted to laboratory use or for laboratory renovations where there is inadequate shaft space. When the structural grid of the building does not conform to the recommended laboratory module dimensions, shafts between modules can be sized to make up the difference. When locations for penetrations from floor to floor are limited due to the existing structural system, shafts between modules may be the only practical option. A very short horizontal duct connection to the exhaust riser in the shaft will be needed, but the fume hood can still be placed away from the primary exit. Window area on the exterior wall is not restricted.

When shafts are placed between modules (Figure 1-8), the laboratories are not as flexible for reconfiguring a series of modules into larger labs as are other layouts. Other disadvantages of shafts placed between laboratory modules include (a) a need to service pipes and ducts within the laboratory and (b) the restriction that single-module laboratories can have utilities available along one wall only.

A central mechanical shaft (Figure 1-9) has the advantage of producing a relatively low ratio of mechanical utility shaft area to net laboratory area, improving overall building efficiency. Chemical fume hoods can be placed adjacent to shafts at the rear of the laboratories. All pipes and ducts for a cluster of up to four laboratories accessible from any one of the four central utility shafts permit greater variation in module length than for any other layout. Because a large central shaft may need space for four or more dedicated exhaust risers at each floor, in addition to other piped utilities, there is a limit to the number of floors that a single shaft can accommodate unless the shaft area increases as it rises to the top of the building, through many laboratory floors collecting additional risers at each level.

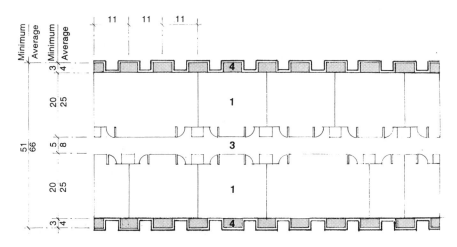

PLAN E

KEY

1   Laboratory
2   Office
3   Corridor
4   Mechanical Shafts

FIGURE 1-6A. Utility shafts at exterior wall.

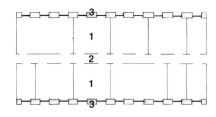

KEY

1   Laboratory
2   Corridor
3   Mechanical Shafts

ADVANTAGES

short horizontal exhaust duct from fume hoods
fume hoods are in a protected position at end of aisles
major traffic within the lab is near public corridor egress
writing stations are near public corridor egress
natural light directly into laboratories

DISADVANTAGES

second egress is not in a separate fire zone
second egress near primary egress
outswinging doors obstruct public corridor
pipes and ducts are accessible from laboratory or building exterior

FIGURE 1-6B. Evaluation of utility shafts at exterior wall.

A variation on the central mechanical shaft layout is location of shafts within a core area on the laboratory floor (Figure 1-10) that distinguishes the zone for special laboratory support and the zone for generic modular laboratories. An advantage to laboratory building layouts with mechanical shafts located in a center core is improved building flexibility. These shafts can be made large enough to walk into in order to install new risers and make repairs and are accessible from corridors. They are suitable for high-rise buildings. Layout of the core zone can economically allow for specialized support room layouts and mechanical services, because rooms in the core are closest to the utility shafts. The mechanical shaft located in the support core area is effective when fume hoods are manifolded, as well as installed with dedicated risers to fans on the roof. A disadvantage in layouts in Figure 1-10 is that considerable horizontal duct runs are required from fume hood locations within labs to the shafts across public access corridors.

1.2.2.5   Structural System. The structural engineer is guided in the design of the structural system of a laboratory building by the organization and dimensions of the laboratory modules on a typical floor. The module lines show where dead loads from walls, benches and equipment will occur. Vertical shafts for distribution of mechanical services show where major penetrations in the floor slabs will occur. From data collected on equipment that has been identified through the building program process, the spaces that pose structural or construction difficulties can be determined. For instance, special design and construction will be required in areas that contain equipment that is unusually heavy, vibration-producing, or vibration-sensitive. Areas that must be isolated from building vibrations have to be structured differently than a typical laboratory floor, as will areas through which it will be necessary to move extremely heavy equipment. Structural damage, or even failure, may occur in areas not designed for heavy loads, even though the excessive load is of short duration. Heavy loads are commonly encountered in both (a) radiation source equipment that is shielded with lead bricks and (b) large filtration tanks for some types of water purification systems.

It is desirable to have as much space as possible above finished laboratory building ceilings (or equivalent free space when there is no finished ceiling) to use for the large number of ducts, pipes, and electrical services that are characteristic of modern laboratory facilities. Deep solid beams reduce unobstructed clearance, and truss-type structural members (such as bar joists) do not improve conditions very much. During design there is always competition between structural requirements and the needs of the mechanical trades for the ceiling space above the occupied work zone. A reasonable balance between the requirements of efficiency and economy for all construction components must be struck. In this decision, however, it must be kept in mind that in laboratory buildings the cost of mechanical and electrical systems is typically 40% to 60% of the entire construction cost. Ample floor to floor heights simplify mechanical systems installation, reduce cost, and allow greater opportunities for horizontal distribution.

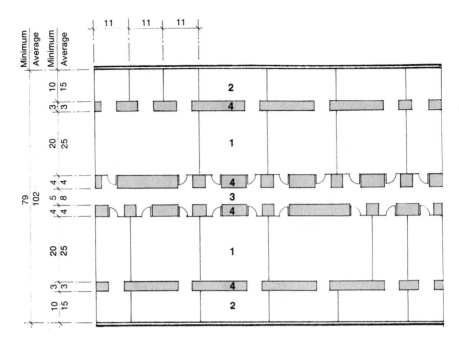

PLAN   F

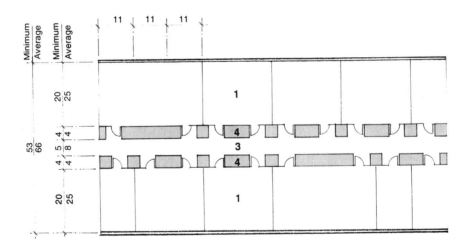

PLAN   G

<u>KEY</u>

1   Laboratory
2   Office
3   Corridor
4   Mechanical Shafts

FIGURE 1-7A. Utility shafts at interior wall.

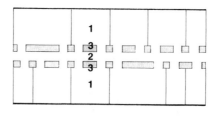

KEY

1 Laboratory
2 Corridor
3 Mechanical Shafts

ADVANTAGES

fume hoods are in a protected position at end of aisles
writing stations are near public corridor egress
pipes and ducts are accessible from public corridor
natural light directly into and views from laboratories
outswinging doors shielded in alcove

DISADVANTAGES

longer horizontal exhaust duct from fume hoods
second egress is not in a separate fire zone
peninsula benches may shift major traffic within lab toward fume hood zone

FIGURE 1-7B. Evaluation of utility shafts of interior walls.

However, some laboratory types such as microelectronics, optical research, surgical research, and electron microscopy have very stringent requirements for structural stability and vibration isolation. Greater-than-usual structural mass and stiffness, combined with building foundation stability, can contribute to reduction in structure-borne vibrations that are generated within and outside laboratory buildings. On the other hand, reduction of effects from vibration-generating factors such as air-borne noise produced by people and machines, high-velocity air currents or turbulence within a room and within ducts, turbulent flows of fluids in building piping, and vibrations caused by people walking and carts moving along building corridors adjacent to sensitive areas can be handled by vibration suppression measures. Good location on the site and thoughtful building layout, particularly of mechanical rooms, elevators, and utility distribution pathways, help to reduce vibration. Attention to construction details of machine, duct, and pipe mountings, as well as use of local vibration isolating devices at sensitive equipment locations, will further reduce effects of vibration. Acoustical experts and engineers who are specialists in structural dynamics can analyze potential trouble spots in the site and building and can establish design criteria for vibration control, and they can also develop strategies to reduce internally generated noise and vibration. In some laboratories and laboratory support areas, such as machine shops or glasswashing rooms, where excessive noise levels gener-

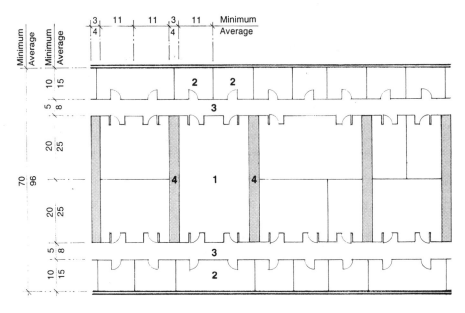

PLAN   H

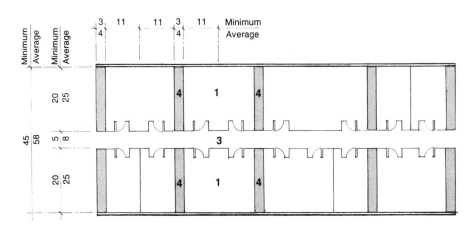

PLAN   I

KEY
1   Laboratory
2   Office
3   Corridor
4   Mechanical Shafts

FIGURE 1-8A. Utility shafts between modules.

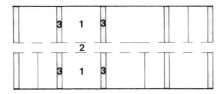

KEY
1 Laboratory
2 Corridor
3 Mechanical Shafts

ADVANTAGES

short horizontal exhaust duct from fume hoods
fume hoods are in a protected position at end of aisles
major traffic within the lab is near public corridor egress
writing stations are near public corridor egress
natural light directly into and views from laboratories
applicable for conversion of existing building to laboratory use

DISADVANTAGES

second egress is not in a separate fire zone
second egress near primary egress
outswinging doors obstruct public corridor
pipes and ducts are accessible from laboratory
when alignment of peninsula benches is parallel to public corridor distance to egress increases
low net to gross area efficiency

FIGURE 1-8B. Evaluation of utility shafts between modules.

ate a concern for health and safety, acoustical engineers may be consulted during the design phase to devise building construction methods that will isolate or reduce noise at the source.

1.2.2.6 Site Regulations. The site for a new laboratory facility lies within the boundaries of municipal, regional, and national jurisdictions. Each of these governing units enacts regulations governing land use and construction methods within its boundaries that will affect construction and location of the proposed facility. Local zoning ordinances, for example, often contain criteria for the following planning concerns: fire district regulations, building use classification, building height restrictions, allowable floor area ratio to site area (site coverage), clearance and easements within and around site boundaries, number of parking spaces required on the site, and guidelines on the use of utilities, such as sewer and water. Some state regulatory and municipal agencies require permits to exhaust contaminated air from laboratory exhaust ventilation systems. Other jurisdictions issue permits for, inspect, and regulate sewer discharge from laboratory buildings. On the other hand, when a building is equipped with automatic sprinklers, most local zoning and building codes are less restrictive on allowable building height and total floor

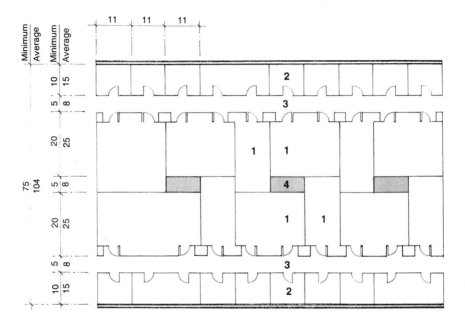

PLAN   J

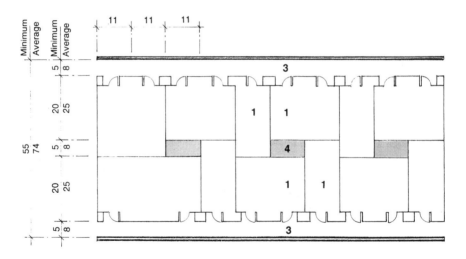

PLAN   K

KEY

1   Laboratory
2   Office
3   Corridor
4   Mechanical Shafts

FIGURE 1-9A. Utility shafts at central locations.

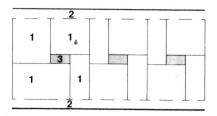

KEY

1 Laboratory
2 Corridor
3 Mechanical Shafts

ADVANTAGES

second egress is not in a separate fire zone
short horizontal exhaust duct from fume hoods
fume hoods are in a protected position at end of aisles
major traffic within the lab is near public corridor egress
writing stations are near public corridor egress
applicable for conversion of existing building to laboratory use

DISADVANTAGES

outswinging doors obstruct public corridor
pipes and ducts are accessible from laboratory
no natural light directly into laboratories
pipes and drains have long horizontal runs to some benches

FIGURE 1-9B. Evaluation of utility shafts at central locations.

area. This book recommends that laboratory buildings be so equipped throughout.

Approvals by local governing boards and regulatory agencies are generally required before construction can begin. Therefore, a thorough search of all applicable codes, regulations, and ordinances early in the planning phase is prudent. Preliminary plan review with agencies that issue critical permits are advised, because these discussions allow the owner and designers to anticipate problems and make adjustments to conform with officials' interpretation of the codes.

1.2.2.7 Building Enclosure. The final factor that influences laboratory building layout is the building enclosure that will provide the total building area required, comply with the building program, and meet zoning requirements. The volumetric and geometric characteristics of the structure include perimeter shape, number of floors, building height, site coverage, and orientation to sun and prevailing winds. For renovations, building enclosure, total building area, and the structural system are known and relatively fixed. Therefore, the assignment of functions to various existing spaces and the proposed modifications to those spaces are conducted in a manner that will fulfill the building program goals in the most effective way. In renovation and

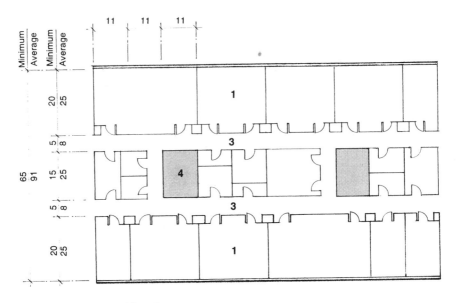

PLAN   L

KEY
1   Laboratory
2   Office
3   Corridor
4   Mechanical Shafts

FIGURE 1-10A. Utility shafts at central core M.

building conversion to laboratory use there is generally less design flexibility than in new construction.

1.3   GUIDING PRINCIPLES FOR BUILDING HEATING, VENTILATING, AND AIR-CONDITIONING SYSTEMS

Laboratory building ventilation is needed to provide an environment that is safe and comfortable. It is accomplished by providing controlled amounts of supply and exhaust air plus provisions for temperature, humidity, and velocity control. Good laboratory local exhaust ventilation captures toxic contaminants at the source and transports them out of the building by means of ductwork and a fan in a manner that will not contaminate other areas of the building by recirculation from discharge points to clean air inlets, or by creating sufficient negative pressure inside the building to subject inactive hoods to down drafts. The building supply ventilation system may provide all of the makeup air needed for local exhaust air systems in addition to taking care of all comfort requirements, or there may be a separate supply system

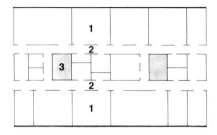

KEY

1 Laboratory
2 Corridor
3 Mechanical Shafts

ADVANTAGES

second egress is not in a separate fire zone
fume hoods are in a protected position at end of aisles
major traffic within the lab is near public corridor egress
writing stations are near public corridor egress
pipes and ducts are accessible from public corridor
applicable for conversion of existing building to laboratory use
natural light directly into laboratories

DISADVANTAGES

long horizontal exhaust duct from fume hoods
outswinging doors obstruct public corridor
pipes and drains may have long horizontal runs to some benches

FIGURE 1-10B. Evaluation of utility shafts at central core M.

for each function. Chapter 27 contains background information on HVAC system design.

1.3.1   Laboratory and Building Pressure Relationships

Laboratory safety requires a careful balance between exhaust and supply air volumes, as well as concern for the quantity of each. Even within the same laboratory type, requirements can vary depending on the hazard rating of the materials being used, the quantity of hazardous materials that will be handled, and the nature of the laboratory operations. Communication with laboratory users at an early stage in planning will help to identify and locate sites of known and potential hazards, making it possible to provide adequate facilities to meet additional needs.

Ventilation systems for laboratories can be divided into three main categories, based on function:

1. *Comfort ventilation* is provided to the laboratory by a combination of supply and return airflows through ceiling and wall grilles and diffusers. The main purpose is to provide a work environment within a specified tempera-

ture, air exchange, and humidity range. Part or all of the comfort ventilation may be provided by special systems installed for health and safety purposes.

2. *Supply air systems* provide the require makeup air removed by the health and safety local exhaust ventilation systems. Comfort ventilation air may be supplied by one system, and makeup air for health and safety exhaust systems may be supplied by another, or the two supply systems may be combined.

3. *Health and safety exhaust ventilation systems* remove contaminants from the work environment through specially designed hoods and duct openings.

Components of a ventilation system include fans, ducts, air cleaners, inlet and outlet grilles, sensors, and controllers. Automatic fire dampers are usually required when air ducts pass through fire barriers and are advisable in any case when work with large quantities of flammables is contemplated.

1.3.2   Dedicated and Branched Air Systems

Building supply air may be provided through a central system that serves all areas—for example, offices, storage areas, and public areas as well as all laboratories. Alternatively, there may be separate systems dedicated exclusively to laboratory use. Although combined systems are acceptable for supply air, hoods and all other health and safety local exhaust facilities should be vented through individual dedicated exhaust ducts and fans. Comfort air exhaust systems may be, and commonly are, combined. The potential for recirculating contaminated air, particularly after a spill or accidental release, may make recirculating comfort exhaust air from laboratory spaces less attractive.

Although individual dedicated exhaust ducts and fans are recommended for each major local exhaust ventilation facility, manifold systems that combine exhaust ducts from several hoods may be used to reduce the number of small stacks on the roof or to reduce shaft space for mechanical services. Manifold systems would not usually be considered for a renovation project unless a sound duct system of adequate capacity was already in place. The advantages and disadvantages of using manifolded exhaust duct systems versus single hood exhaust systems depend on numerous factors.

The advantages of a manifold system are:

1. Lower initial cost because the number of ducts, fans, and stacks will be fewer.
2. Better atmospheric dispersion of the exhaust plume due to the enhanced momentum effect of a larger air mass exiting from the stack.
3. Less maintenance required for fewer installations.
4. Reduced pipe shaft space needs within the building.

The disadvantages of a manifold system are:

1. Different exhaust streams may not be compatible. Generally this is not a problem when the quantity of contaminants is low and the dilution volumes in the system are high.
2. A need to add a booster fan for each individual branch that requires addition of an air-cleaning device (e.g., hoods used for radioactive isotopes) when the manifold system as a whole does not.
3. Branch ducts require careful air balancing and a control mechanism to maintain design air flow distribution (see Chapters 32 and 33).
4. Inability to shut off individual systems without the addition of sophisticated control systems that automatically rebalance supply and exhaust volumes for each laboratory connected to a single multifacility exhaust system.

Before selecting a manifold system, the above advantages and disadvantages should be carefully considered for the specific applications. If a manifold system is to be used, the following requirements should be included in the design.

1. Easy access to a straight duct run in each branch to allow for air flow measurements.
2. A monitor at each exhaust point to indicate correct flow.
3. Easy access to all control valves for inspection and repair.
4. Adequate exhaust capacity for each space served.
5. Maintenance of desired directional airflow requirements for each space at all times.
6. Negative pressure throughout manifold plenums should be reasonably uniform to maintain design air flow from branch ducts.
7. A standby or cross-connected exhaust fan that can be put into service rapidly in the event of a fan failure. Alternative methods to maintain manifold suction may satisfy this requirement.
8. Emergency power to the exhaust fans.
9. Training for maintenance personnel and laboratory users in the proper use and care of manifold systems. This criterion is often overlooked and is very critical.

1.3.3 Constant Volume and Variable Air Systems

It is advisable, in all cases, to maintain constant pressure relationships between laboratory rooms (greatest negative pressure), anterooms, corridors, and offices (least negative pressure) to avoid intrusion of laboratory air into other areas of the building. This becomes particularly important in the event of an accidental release of a volatile chemical in a laboratory. For laboratories in which hazardous chemicals and biological agents are used, pressure gradients that decrease incrementally from areas of high hazard to areas open to public access are an essential part of the building's health and safety protective system. Even in laboratories where hazardous materials are not employed, animal holding rooms, animal laboratories, autopsy rooms, media preparation rooms, and similar facilities are likely to generate unpleasant odors. Graded air-pressure relationships are usually relied upon to prevent release of foul-smelling air from these rooms. The reverse pressure relationship is required for germ-free and dust-free facilities, such as operating room and white room (clean room) laboratories.

As a rule, in rooms and spaces requiring only comfort ventilation, the pressure relationships relative to the laboratory spaces are intended to be maintained constantly; this calls for invariant airflows. Health and safety ventilation systems may also be designed for continuous, invariant service. Such an arrangement is advantageous when the number of exhaust-ventilated health and safety facilities is small and when the health and safety air-supply systems have been combined with the comfort system. However, when health and safety exhaust ventilation represents the major proportion of the total air circulation requirements for the entire building, energy conservation measures call for an ability to shut down these services when not needed. To be effective, there must be two separate supply and exhaust air systems: (1) a comfort ventilation system that provides constant and invariant design temperature, humidity, air exchange, and room pressure conditions and (2) one or more tempered makeup modulating air-supply ventilation systems that are individually coupled to specific exhaust air facilities so both may be turned on and off simultaneously to avoid disturbing the room pressure relationships established by the comfort ventilation system. For a more thorough discussion on variable air-volume systems design, control, and components, refer to Chapter 33. It is sometimes advantageous to provide a constant volume system to some parts of a laboratory building and variable systems to others. The nature of the installed facilities and the intensity of laboratory usage will determine the advisability of hybrid ventilation systems.

1.3.4   Supply Ventilation for Building HVAC Systems

1.3.4.1 Supply Air Volume. All air exhausted from laboratories must be replaced with supply or infiltration air. An equivalent volume of replacement or makeup air is essential to provide the necessary number of air changes needed to facilitate safe working conditions and to maintain design pressure relationships between rooms and other spaces for health and safety protection. When laboratory health and safety exhaust ventilation requirements are not dominant, total building air-conditioning needs for maintaining heating, cooling, and ventilation loads may dictate the supply air volume to each room. Total supply air volume cannot be calculated until the amount of air to be exhausted has been determined.

A variable air-volume supply system will meet the needs of the space by reducing or increasing air quantities caused by changing space conditions and orientation. Interior spaces and those facing north will require reduced air flow while those facing south will require increased airflow to meet the requirements for air conditioning.

1.3.4.2 Supply Air Velocity, Temperature, and Discharge Location. The location and construction of room air outlets and the temperature of the air supplied are critical. High-velocity air outlets create excessive turbulence that can disrupt exhaust system performance at a hood face. In addition, comfort considerations make it necessary to reduce drafts. Therefore, the supply air grilles should be designed and located so that the air velocity at the occupant's level does not exceed 50 feet per minute (fpm). There is no single preferred method for the delivery of makeup air; each building or laboratory must be analyzed separately. When supply air is used only for replacing health and safety exhaust ventilation air, as distinguished from temperature control, it is desirable that the air be discharged near the exhaust facility when the space is air-conditioned. However, when the room is not air-conditioned, there is no advantage associated with close proximity of makeup air to discharge points.

1.3.4.3 Air Intakes. Outside air intakes must be located so as to avoid bringing contaminated air into the building air-supply systems. Examination of likely contaminant sources, such as air exhaust points and stacks, should be conducted before outside air intakes are selected. A minimum distance of 30 ft from air-discharge openings to air intakes is recommended to reduce fume reentry problems, but it is good practice to design for the maximum feasible separation. Outside air intakes located at ground level are subject to contamination from automobile and truck fumes, whereas air intakes at roof level are subject to contamination from laboratory exhaust stacks or high stacks serving offsite facilities in the vicinity. When buildings contain more than 10 stories, it is advisable to locate the air intakes at the midpoint of the building. Difficult sites may require wind tunnel tests to investigate the fume reentry problem under simulated conditions. Computer modeling programs are also available to assist in stack-discharge design and air-intake location.

1.3.5 Air Discharges

For roof-mounted laboratory hood exhaust installations, the stack on the positive side of the fan should extend at least 10 ft above the roof parapet and other prominent roof structures in order to discharge the exhaust fumes above the layer of air that clings to the roof surface and prevents contami-

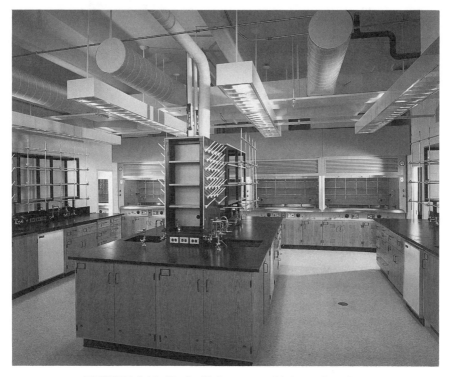

FIGURE 1-11. Typical laboratory with hazard zoning.

nants from displacing upward. This arrangement will help to avoid reentry through nearby air-intake points. To further assist the exhaust air to escape the roof boundary layer, the exhaust velocity should be at least 2500 fpm and there should be no weather cap or other obstructions to prevent the exhaust discharge from rising straight upward (see Figure 1-12).

1.3.5.1 Exhaust Fans. All exhaust fans should be installed on the building roof to maintain the exhaust ducts inside the building under negative pressure as a health protection measure. This arrangement makes it certain that should duct leakage occur, it will be inward. In cases where exhaust air ductwork has to be at positive pressure relative to the building interior, special care must be exercised by frequent pressure testing to ensure that the ductwork is airtight. Many types of exhaust fans are manufactured, but only a few meet all the requirements of a good exhaust ventilation system. Double-belted centrifugal utility-type exhaust fans are generally preferred because they are very reliable, have desirable pressure–volume characteristics, are widely available, and are easily adaptable to roof mounting and attachment of a stack of suitable cross section and height for proper discharge of exhaust fumes. For

FIGURE 1-12. Fume hood stacks at Wellesley College

critical exhaust air systems a direct-connected fan and motor, illustrated in Figure 2-9, avoids failure from slackening or loss of fan drive belts. Such an installation can be further simplified by selecting a weatherproof motor and eliminating the motor and drive housing. The material of construction for the fan, including protective coatings, should be selected to withstand corrosive and erosive conditions characteristic of the exhaust gases that will be handled. Considerations of life expectancy and maintenance availability will influence the final selection. It is important to specify fans manufactured and rated in accordance with standards established by the Air Moving and Conditioning Association (AMCA).*

A variable air-volume exhaust fan is a necessary adjunct to a VAV supply system and will conceptually be the catalyst that produces the signal to the supply fan that more or less air is required to maintain space pressure relations or when a chemical fume hood is turned on or off, producing a signal of a change of status. The fan control is maintained by using a static pressure controller to maintain a constant static pressure in the duct system(s). Variable air-volume systems and controls are discussed in Chapter 33.

Backwardly inclined fans with self-limiting horsepower characteristics have been widely used for general building ventilation purposes. Fan housings are usually constructed of steel and bonderized. When used for exhaust air contaminated with low concentrations of corrosive elements, the interior

* 30 West University Drive, Arlington Heights, IL 60004.

of the fan and connecting ducts are often coated with baked primers and finishes especially formulated to meet corrosion resistance standards. For severe corrosive service, especially when high humidity is also present, rigid PVC (polyvinyl chloride) or FRP (fiberglass-reinforced polyester) construction is essential. FRP is preferred because of its superior resistance to breakage and vibration cracking. It is necessary to add fire-retardant chemicals to the polyester resin. When the system is located on the roof and discharges straight upward without a rain cap, a drain connection should be placed at the bottom of the fan housing. (See Section 2.3.5 for additional design information.)

1.3.6 Supply Air Cleaning

All building supply air, including all portions of recirculated comfort air, should be cleaned according to the requirements of the space. The correct degree of filtration is important because excessive filtration results in a greater pressure drop through the system, thereby increasing operating costs, whereas insufficient filtration results in contamination of critical work areas or excessive maintenance costs from rapid soiling.

Many filter media are available, each providing a specified degree of air cleanliness. The choice depends on the need. Filters are clarified by the American Society of Heating, Refrigeration, and Air-Conditioning Engineers in Chapter 25 of HVAC Systems and Equipment handbook (ASHRAE, 1992) as throwaway or renewable. Throwaway filters are used once and discarded. They are rated as low efficiency (35% dust removal), medium efficiency (85% dust removal), or high efficiency (95% dust removal). The performance characteristics of a number of throwaway dry media filters used for air cleaning are shown in Chapter 30. Renewable filters are seldom used for building air cleaning. Electrostatic precipitators are also used for cleaning building supply air. They are designed for 85% or 95% atmospheric dust collection efficiency. Electrostatic precipitators are always made as cleanable units, the interval between cleaning being more or less than three months, depending on the dirtiness of the outside air. Cleaning involves washing the dust-collecting plates with detergent and water. Units may be purchased for manual or mechanical cleaning. Electrostatic precipitators generate small amounts of ozone during normal operation, and their use is counterindicated where this gaseous compound would be considered an interference with the work to be undertaken in the new laboratories. Should this be the case, it would also be necessary to remove the same compound from the air introduced into the building, because ozone regularly occurs in outdoor air in most parts of the United States. Ozone, sulfur dioxide, and most hydrocarbons that are normally present in urban air can be removed from ventilation air by passing it through gas-adsorption activated carbon beds after particle filtration or after treatment by electrostatic precipitators.

Cleaning of used comfort ventilation air prior to discharge to the atmosphere is seldom, if ever, attempted, nor is it necessary. This is not always true for health and safety system exhaust air. Filtration, liquid scrubbing, and gas adsorption may be needed to prevent the emission of toxic, infectious, and malodorous gases and aerosols to the atmosphere. Exhaust air cleaning is discussed in Section 2.3.5.1.

1.3.7 Supply and Exhaust Ducts

All ductwork should be fabricated and installed in accordance with Sheet Metal and Air-Conditioning Contractors' National Association Inc.'s standards (SMACNA, 1985). Ducts should be straight, be smooth inside, and have neatly finished joints. All ducts must be securely anchored to the building structure. The usual material for supply and exhaust ducts in comfort ventilation systems is galvanized steel. More corrosion-resistant materials, such as stainless steel, fiberglass-reinforced polyester, and PVC, are frequently used for health and safety system exhaust ducts. See Table 1-9 for information on chemical resistance of materials used for exhaust ducts and plenums. It should be noted that the NFPA code requires a building to have sprinklers where PVC piping is used for exhaust ducting.

It is essential that all ducts be constructed and installed in a leak-tight manner if the system is to function as the designer intended it to. This is especially important for exhaust ventilation ducts, which usually operate under far higher negative pressures than do comfort ventilation systems; hence the leakage through even small gaps in longitudinal seams and joints leads to a significant drop in system efficiency. The seams and joints of stainless steel ducts are usually welded airtight. Plastic pipes are constructed without longitudinal seams and the joints are sealed with plastic cements of appropriate composition. It is easier to construct airtight systems with round ducts than rectangular ducts. Rectangular ducts should be avoided at any cost in the health and safety exhaust air systems because they cannot be made airtight by any practical method. For noncorrosive material-containing systems, seams and joints may be sealed with long-lasting ventilation duct tape. Whatever method is selected, it is essential that ducts be made leak-tight if they are to give satisfactory service.

The current poor state of ventilation system engineering and installation is exemplified by the frequent inclusion of redundant "trimming dampers" in all duct runs. They are thought to be necessary because unanticipated changes often occur after the engineering design and the production of the final specification documents have been completed. This frequently occurs when construction costs exceed the budget and changes must be made to reconcile the two. People who bear the financial responsibility for building projects tend to minimize the effects of cost reduction changes on design integrity and seldom consult the design engineer to find out if the original design will be affected adversely; therefore, many design engineers believe it is thought to be prudent to design duct systems with balancing dampers in order to assure

TABLE 1-9. Table of Chemical Resistance of Materials Used for Exhaust Ducts

Type of Piping	Acids[a]		Alkalies[a]		Organic Solvents[a]	Flammability[b]
	Weak	Strong	Weak	Strong		
Aluminum[c]	N	N	N	N	N	G
Asphalt-coated steel[d]	Y	Y	Y	Y	N	G
Epoxy-coated steel	Y	Y	Y	Y	Y	G
Galvanized steel[e]	N	N	N	N	Y	G
Epoxy glass Fiber Reinforced[f]	Y	Y	Y	N	Y	SL
Polyester glass fiber reinforced[g]	Y	Y	Y	N	Y	SL
Polyethylene fluorocarbon[h]	Y	Y	Y	Y	Y	SE
Polyvinyl chloride[i]	Y	Y	Y	Y	N	SE
Polypropylene[j]	Y	N	Y	N	N	SE
316 Stainless steel[k]	Y	Y	Y	Y	Y	G
304 Stainless steel[k]	Y	N	Y	N	Y	G

[a] N—attacked severely; Y—no attack or insignificant.
[b] Flammability: G—good fire-resistant; SE—self-extinguishing; SL—slow burning.
[c] Aluminum is not generally used because it is subject to attack by acids and alkalies.
[d] Asphalt-coated steel is resistant to acids, subject to solvent and oil attack.
[e] Galvanized steel is subject to acid and alkali attack under wet conditions.
[f] Epoxy glass fiber reinforced is resistant to weak acids and weak alkalies and is slow burning.
[g] Polyester glass fiber reinforced can be used for all acids and weak alkalies but is attacked severely by strong alkalies and is slow burning.
[h] Polyethylene fluorocarbon is excellent material for all chemicals.
[i] Polyvinyl chloride is excellent material for most chemicals and is self-extinguishing but is attacked by some organic solvents.
[j] Polypropylene is resistant to most chemicals and is self-extinguishing but is subject to attack by strong acids, alkalies, gases, anhydrides, and ketones.
[k] Stainless steel 316 and 304 are subject to acid and chloride attack varying with the chromium and nickel content.

that the design airflow rates will be achieved. This should be avoided where possible. Although it is customary to balance ventilation systems after the building is completed, it is often the last construction step. It would be much better to test and balance the system as early as possible or, in any case, before the duct systems are concealed above suspended ceilings and behind cabinets, hoods, and similar large pieces of laboratory furniture.

Ducts are excellent conductors of sound. Special care must be expended to secure them so as to avoid vibration. In addition, they should be isolated from fans and other noise-generating equipment with the use of vibration-reducing connections and the installation of acoustical traps in the ducts between noise and vibration sources and the point of discharge to occupied areas.

1.3.8 HVAC Control Systems

Controllers are essential to ensure that all HVAC systems in a laboratory building will operate in a safe and economical manner. Control systems are needed for temperature, humidity, air exchange, and pressure regulation.

Control systems can be electric, electronic, or pneumatic or can operate on system high-pressure air. Variable air-volume systems require unique control systems such as static pressure controllers, inlet vane dampers on centrifugal fans, feather blades on vane axial fans, variable speed drives, and solid-state rectifiers. Variable air-volume systems and controls are discussed in Chapter 33.

1.3.8.1 Temperature Control. The temperature in most laboratory buildings does not require close regulation—that is, no better than ±3°F. Worker efficiency and productivity may be affected adversely when ambient temperature control permits temperatures to exceed 85°F in summer or to fall below 60°F in winter.

The comfort range is determined by combining dry bulb temperature, relative humidity (RH), and air velocity to derive a value called *effective temperature* (ASHRAE, 1977). An effective temperature of 77°F is considered to be a very desirable condition. This is achieved in winter by 68°F with 35% RH (±5%), and in summer by 78°F with 50% RH (±5%). More information on comfort indexes and comfort standards is given in Chapter 28.

1.3.8.2 Humidity Control Although close humidity control is not required in most buildings, some degree of humidity control should be included to provide comfort and avoid extremes. Chapter 28 should be consulted for additional details on humidification systems.

1.3.9 Air Exchange Rates

Recommended air exchange rates for public areas and for commercial and industrial workplaces are contained in the ANSI/ASHRAE ventilation standard 62-1989 (ANSI/ASHRAE, 1989). In most organized communities, the applicable building code will prescribe minimum ventilation rates for indoor air quality for a variety of building usages. Where they exist, they will take precedence as minimum requirements. They are the lowest air exhaust rates that must be maintained in each occupied room, even when the health and safety exhaust ventilation systems are turned off.

In crowded areas where smoking is permitted, as in waiting rooms, eating places, and group offices, the minimum air exchange rates required by the building codes will be inadequate to please a high percentage of nonsmokers; in addition, outside air rates of 30–60 cfm (cubic feet per minute) per active smoker will be needed to keep tobacco smoke concentrations close to background levels (ANSI/ASHRAE, 1989).

1.3.10   Fan Rooms (Equipment Rooms, Mechanical Rooms)

Because of the nature of the work that is required to maintain the equipment within the fan room, the design of the room is important from the standpoint of ensuring the owner of delivering a complete system that performs as designed and that will continue to perform during the life of the building. All too often the space provided for a fan room is too small to accommodate the equipment that is required to provide the services to the building. Therefore, the mechanical contractor viewing the mechanical drawings installs the equipment that is required to provide the services to the building. The equipment is usually installed double-tiered or in a fashion that makes it easiest for the sheetmetal contractor because his ducts are in locations which dictate the fan location. The piping contractor not having the necessary room installs the piping to make it easiest for him. The control contractor, being the last in and of course having the smallest pipe, can install his equipment and piping with ease. Often these contractors are not concerned with the need to maintain the equipment. Is it accessible for changing a motor, bearings, filters, and all other maintenance which may be required over its useful life? It is a nightmare when you open the door and find the following situation: There is no space; you cannot stand up in many locations; you cannot identify the equipment because you cannot see it; and what's more it is almost impossible to get to it. The obstacle course is ready for you and everyone on the design team is betting that you can handle it and are up to the task.

The layout of the fan room should be considered as important as the rest of the building, with input from those responsible for maintaining the building. Easy access to all equipment is required.

1.3.11   Glass-Washing Rooms

Glass-washing rooms are a part of a laboratory facility for a department or for several different departments all using the same facility in a laboratory building to perform the necessary washing and sanitizing of glassware used in research.

The glass-washing facility requires large amounts of hot water which is usually generated in the facility or remotely located close to the facility because of the intermittent use. Most laboratories require that the glassware be washed, sterilized, and ready for use the following morning, or the washing and sterilization is done early in the morning in order that the glassware is ready for use in their experiments that day.

The normal procedure is to wash the glassware in hot water and provide either distilled or pure water for the final rinse in order to provide exceptionally clean glassware, removing all traces of residue from previous experiments. This requires exceptionally large amounts of thermostatically controlled hot water, and the method of heating is usually steam. It also requires an adequate source of pure water which is used as the final rinse.

Adequate-sized drain piping should be provided to remove the volume of water that is discharged to drain usually by a butterfly valve which is normally a part of the glassware washer.

Because of the high heat load in the glass-washing area, the need for conditioned air is required year round so that those involved in this practice can perform their work in relative comfort. Autoclave rooms are also in this category, and the comfort of the operators should be considered.

Cage-washing areas also require large amounts of hot water but have another problem which requires attention at the outset—namely, providing a floor surface that is impervious to the strong acids that are used in the sterilization of the cages.

1.3.12   Commissioning

Commissioning is a term being used for final acceptance of mechanical/ electrical systems in the building. The process is fully described in Chapter 25.

1.4   LOSS PREVENTION, INDUSTRIAL HYGIENE, AND PERSONAL SAFETY

1.4.1   Emergency Electrical Considerations

The primary electric feed to laboratory buildings should be as reliable as possible. For example, separate and distinct feeds connected to a common bus and then to two separate transformers with network protectors should be installed, and each transformer should be large enough to carry the building load so that loss of any one line will not interrupt building power. When such practice is not possible, some other fail-safe electrical connection designs should be used. Even with this type of reliable service, it is often necessary to provide emergency electrical power because any of the primary electrical service components (transformers, main feeders, etc.) may fail and then emergency power will be required. Each building should have its own emergency power source that is adequate for all egress lighting and other life-safety requirements, as defined in *National Electrical Code* (NFPA 70, 1990) and *Safety to Life from Fire in Buildings and Structures* (NFPA 101, 1991). Several critical systems in laboratories may have to be connected to emergency electrical power for continuity of operation as well as for safety concerns. Items that should be connected to emergency electrical power are the following:

1. Fire alarm systems.
2. Emergency communication systems.
3. One elevator for buildings over 70 ft in height.
4. Egress pathway lighting.
5. Emergency lighting in rooms.
6. Egress signs that require lighting.
7. Fire pump, when it is electrically driven and not backed up by another driver.
8. Exhaust fans connected to critical health and safety exhaust ventilation systems.
9. Makeup air systems serving critical exhaust systems.
10. Heating systems and controls to prevent building from freezing during temperature extremes.
11. Emergency smoke evacuation systems.
12. All other systems whose continuing function is necessary for safe operation of the building or facilities during an emergency period.

In general, diesel-driven generators are preferred because they are readily available, easily maintainable, and easy to start (in less than the 10 s mandated by NFPA 70). Gas turbines are available in smaller sizes and may be satisfactory. However, turbine starting is sometimes difficult for large size generators.

The emergency generator should be connected to the selected load with a series of transfer switches. The transfer switches should automatically turn over to emergency power when normal power fails. Annunciation through a local or remote panel should be provided to let operators of the building know which transfer switch has changed modes. The generator control board should have an ammeter installed so operators can see the load on the generator and manually select other loads when necessary.

If emergency electrical power distribution is run throughout the facility, it must be run on a distribution system that is separate and distinct from the primary electrical distribution system. This is to prevent concurrent cable failure of both primary and emergency power in case of a fire or other emergency condition.

1.4.2 Construction Methods and Materials

According to BOCA (BOCA, 1991), laboratory buildings engaged in education, research, clinical medicine, and other forms of experimentation are included in the category of Class A building usage. Therefore, all pertinent sections of the BOCA code should be followed in the design and construction of all laboratory facilities, with special emphasis on fire safety for unusual as well as all ordinary hazardous conditions. Specifically, the provisions of Article IX (Fire-Resistive Construction Requirements) that govern the design and use of materials and methods of construction necessary to provide fire resistance and flame resistance must be followed. Flame resistance is defined in the BOCA code as "the property of materials, or combinations of component materials, which restricts the spread of flame as determined by the flame-resistance tests specified in the Basic Code." Some of the specific subjects covered in Article IX of the BOCA code are enclosure walls, fire walls and fire wall openings, vertical shafts and hoist ways, beams and girders, columns, trusses, fire doors, fire windows and shutters, wired glass, fire-resistance requirements for plaster, interior finish and trim, and roof structures. The purpose of the requirements of the BOCA code is to provide a building that will allow its occupants to move freely from the building in case of fire and unusual smoke conditions.

The general philosophy of all interior building design with respect to the combustible properties of construction materials should embrace the idea of eliminating those materials responsible for rapid flame spread and heavy smoke generation. Materials used in research buildings provide more than ordinary cause for concern, because the sources of fire initiation in laboratories are many times more numerous than for most other building uses.

1.4.3 Safety Control Systems for Laboratory Experiments

Provisions should be made for automatic or remote shutdown of well-defined portions of a building's services that provide energy to experimental operations having the capability to threaten parts of the building or personnel within the building, or to produce undesired effects should the experiment get out of hand while attended or unattended. This type of control should be used for the most sensitive types of operations where uncontrolled failure could result in a major loss of building or equipment, the release of highly toxic substances into the environment, or personal injury. A study committee composed of designers, users, and representatives of the health and safety professions should be set up to determine areas of need that will benefit by application of laboratory control systems. See Section 1.4.5.2.

1.4.4   Fire Detection, Alarm, and Suppression Systems

Costs of installing fire detection and suppression systems are very high after the construction of any type of laboratory building. Therefore, consideration should be given to this need during the design stage of new and renovated laboratories. Sprinklers are considered the best fire control.

1.4.4.1   *Fire and Smoke Detection.*   Laboratories should be equipped with a heat-sensitive detection system as a minimum. A standard sprinkler system will qualify even with its inherently slower detection ability. The use of ionization detection systems in laboratories is not recommended because they detect products of combustion (other than heat and chemicals) that the detector may be sensitive to as well as open flames. Ionization detection systems also fail when devices are placed near high-velocity air movement. In addition, ionization detectors are prone to sense fires that are not a threat and may fail to see threatening fires sooner than more stable thermal detectors. There are several light-emitting diode (LED)-operated visual smoke detectors available which can detect smoke at low levels when very early warning of smoke is desired. For reliability reasons, incandescent lamp detectors should be avoided. Also available are forced-air-sampling particulate-counting detectors which will detect early smoke. Rate-of-temperature-rise fire detectors are probably the best and most reliable of the available devices; however, when compared with twisted-pair thermal detectors, they do not cover as much area per unit of cost. However, twisted-pair thermal detectors cost more to repair and reset after fire or damage and are prone to damage through improper use. (We have seen investigators hang materials from the wires as one would from a clothes line, causing them to short and sound an alarm.) Fire detectors must meet the UL standards and be installed and spaced in accordance with *Automatic Fire Detectors* (NFPA 72E, 1990).

1.4.4.2   *Fire Suppression*

1.4.4.2.1   *Fixed Automatic Systems.*   All laboratory buildings should be designed with a complete water sprinkler system in accordance with *Installation of Sprinkler Systems* (NFPA 13, 1989). Wherever unusual hazards exist, special design of the system will be necessary. When water is contraindicated for fire suppression because of the presence of large amounts of water-reactive materials such as elemental sodium, other complete fire suppression systems must be used in those areas. They include (a) Halon 1301 for electronic, high-voltage, or computer laboratories and (b) total flooding with dry chemical for highly flammable chemical storage areas. A competent fire protection or safety engineer should participate in these decisions. All automatic fire suppression systems should be connected to the building central alarm system.

The vertical standpipes used for the water sprinkler system should also serve fire-hose cabinets on each floor of the laboratory building. Hoses should have a $1\frac{1}{2}$-in. diameter, and vertical risers should be so spaced that the maximum length of hose to reach a fire will not exceed 50 ft. Longer hose runs may lead to loss of fire control because a hose length exceeding 50 ft is difficult to turn on and use by persons lacking hands-on fire training.

1.4.4.2.2 Hand-Portable Extinguishers. Hand-portable fire extinguishers should be located in halls and main exitways, as well as within individual laboratory units. A clean fire extinguishing agent, such as carbon dioxide (CO_2), is usually preferred for use in laboratories. Within the laboratory, 5-lb units are the minimum recommended size.

Fire extinguishers installed in halls and exits should be sized and installed in accordance with *Portable Fire Extinguishers* (NFPA 10, 1990). The larger units will generally be multipurpose dry chemical extinguishers that will be more suitable for fighting a large laboratory fire. Small fires in the laboratory can be handled adequately with a CO_2 unit without having to cope with the mess and required cleanup of dry chemicals.

1.4.5 Alarm Systems

1.4.5.1 Fire Alarms. A Class A supervised fire alarm or signaling system should be installed throughout the laboratory building in accordance with *Installation, Maintenance and Use of Protective Signaling Systems* (NFPA 72, 1990). It should have all manual pull stations, sprinkler alarms, and heat-sensing detectors connected to it. All other detection systems should alarm locally, and connection to the class A system should be considered only after reviewing the false alarm potential.

1.4.5.2 Laboratory Experiment Alarms. Provisions should be made for a three-tier alarm system in all laboratories where experiments or operations need to be monitored for failure. The system should be designed to provide a communications link between the laboratory and a central station in the building that is monitored at all times, or, at the very least, when there are unattended operations in laboratory units. In general, a three-tier alarm system consists of the following parts:

1. A local alarm for room occupants that is audible and visual.
2. An audible alarm outside the laboratory door to pinpoint the location of the problem.
3. Remote annunciation to a constantly attended location.

The use of remote annunciation is critical in a large facility because it may be the only means of alerting service personnel to the problem. Remote annunciation is most critical during weekends and normally unoccupied periods. It is strongly recommended that alarms to all electromechanical equipment connected to laboratory safety systems be annunciated to a central location.

The use of microcomputer technology may be advantageous when more than one kind of alarm condition must be monitored, for example, fire alarms and HVAC alarms.

1.4.5.3 Other Service Alarms. Alarms may be needed to indicate failure of exhaust fans and makeup air systems, as well as for fire, loss of pressure, loss of temperature, presence of toxic gases, and other conditions that often require monitoring. In addition, whenever building services that are not normally monitored could cause loss through flood, fire, explosion, or release of hazardous materials in the event of their failure, a separate monitoring system with three-tier alarms should be installed. Alternatively, equivalent facilities could be added to the laboratory experiment monitoring and alarm system.

1.4.6 Hazardous Waste

A designated area must be provided to collect, consolidate, and store hazardous chemical, biological, and radioactive wastes in preparation for disposal. General waste collections, such as those from janitors' operations, should consist solely of paper, glass, and other nonhazardous refuse; waste chemicals should never be included. General waste should be collected and stored in an area of the building not associated with the chemical, biological, and radioactive waste storage areas. The waste storage areas should be within the main laboratory building or in an external facility. (See also Chapter 23 for more information about hazardous laboratory waste management.)

1.4.7 Chemical Storage

In addition to providing for the storage of a few days' supply of chemicals in each laboratory unit, there should be a central chemical storage room for bulk supplies. This room should be sized to hold enough materials to assure continued operations without interruption. The purchasing department can assist in determining what this quantity should be. Compressed gas cylinders should not be stored in the central storage room unless there is a subroom with high rates of ventilation. Good floor drainage should also be provided for compressed gas storage areas.

Shelves, cabinets, drawers, hoods, and special storage equipment should be installed to provide separation of incompatible chemicals. Shelves should have a $\frac{3}{4}$- to $\frac{1}{2}$-in. lip on the edges to prevent containers from falling off due to vibrations. Earthquake restraint bars should be installed at edges of shelves from which containers could fall. Four types of storage should be provided:

1. Shelves for nonhazardous chemicals.
2. Cabinets for controlled substances and very toxic materials.
3. Cabinets with fire ratings for storage of flammable liquids (see "Steel Storage Cabinets for Flammable Liquids," *Factory Mutual Approval Guide 1992* (FM, 1992) and OSHA 1910.106 (OSHA, 1992) for steel and wood cabinet requirements).
4. Storage hoods or other exhaust-ventilated facilities for unsealed toxic chemicals. Fume hoods should never be used as a major storage facility. When the number of items that require ventilated storage exceeds a few, a separate exhaust ventilated facility for storing unsealed toxic chemicals should be provided. For one possible solution see the CHEM SAFE models 2 and 3 manufactured by REI in Greenland, New Hampshire (Figure 1-13).

If operations in the chemical stockroom will involve the transfer of flammable liquids from container to container, the requirements of the National Fire Protection Association's Flammable and Combustible Liquids Code (NFPA 30, 1990) must be met. These include provisions for bonding and grounding of containers and explosion-proof electrical equipment.

The chemical stockroom should be protected with a standard water sprinkler system except where large quantities of flammable liquids or water-reactive chemicals dominate the stockroom. In those cases, an appropriate system, such as total flooding dry chemical, should be used. A 6A 60BC fire extinguisher, a fire blanket, and an emergency eyewash facility should be located within the stockroom, and a deluge safety shower should be nearby in a hall or corridor. For additional reading on chemical storage, see *Safe*

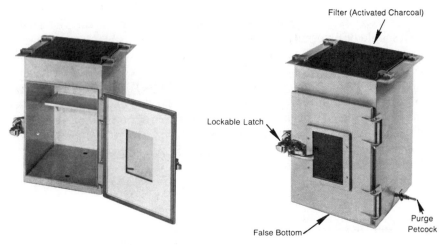

FIGURE 1-13. Passive chemical storage boxes.

Storage of Laboratory Chemicals (Pipitone, 1984) and *Hazardous and Toxic Materials* (Fawcett, 1984).

1.4.8   Compressed Gas Storage and Piping

When the laboratory management elects to pipe gases from a central compressed gas dispensing facility instead of placing cylinders in each laboratory unit, the location of the central facility and an outline of the design features must be included as an integral part of the building design.

A central gas cylinder farm should be located in a room with an outside wall for explosion venting in the ratio of one square foot of venting surface for each 40 to 60 cubic feet of room volume [see *Venting of Deflagrations* (NFPA 68, 1988)] or be housed in a room attached to the outside of the building. The room should be adequate in size to store in segregated locations (1) full cylinders and (2) empty cylinders awaiting removal for refilling, as well as the manifolds necessary for piping the gases. Ventilation in this room should be adequate to vent heat from the sun load on the roof and walls and to remove gases leaking from a failed regulator or valve. Air changes should be a minimum of 10 per hour.

Rigid and secure supports for gas tanks should be provided, and they should be designed to provide storage flexibility. Compressed gas cylinders for a high-pressure laboratory would be likely to be located within that laboratory, or close to it, to avoid any loss of high discharge pressure in the piping system that occurs for the general laboratory gases when piped from a central location.

The gas piping system should be of stainless steel with low-pressure reducers at the gas cylinder farm and orifice restrictions wherever pipe diameter exceeds $\frac{1}{4}$ in. to limit accidental flow into any area. The piping system should be external to the building when this is feasible. Internal piping and exterior piping, alike, should be exposed to view wherever possible. Excess flow check valves may also be installed to control runaway flow conditions of toxic or flammable gases (see Appendix III). Double-walled gas piping should be considered for highly toxic and flammable gases such as arsine and hydrogen, used in microelectronics. [See *Cleanrooms* (NFPA 318, 1991) and *Standard for the Storage, Use and Handling of Compressed and Liquified Gases in Portable Cylinders* (NFPA 55, 1993). See also Chapter 19, "Microelectronics Laboratory."]

1.4.9   Fuel Gas

A fuel gas shutoff must be provided for the entire building.

1.4.10 Hazardous Materials, Equipment, and Procedure Signs

Personnel within a laboratory building or about to enter a laboratory building need information regarding the operations, materials, risks, or special situations within. This information is most important to emergency response personnel, such as fire fighters and police or ambulance personnel, so they can carry out their functions safely and efficiently, usually in a time of stress. Many communities and cities have ordinances requiring certain signs such as those for flammable storage, which must be complied with. An acceptable system of signs is described in *Identification of the Fire Hazards of Materials* (NFPA 704, 1990). It was developed around nonlaboratory users of chemicals, but there may be an ordinance that requires compliance with NFPA 704 for laboratories. When a specific code is not mandated, we recommend adoption of a sign system that can better protect emergency response personnel in laboratory situations, such as the system shown in Appendix IV.

1.5 MISCELLANEOUS SERVICES

1.5.1 Lighting Power Limits

A lighting power limit is the maximum power that may be available to provide the lighting needs of a building. Separate lighting power limits should be calculated for the interior and exterior of the building. To establish lighting power limits, the following procedure should be used. The intent is to impose energy conservation discipline to minimize overlighting or overuse of electrical power. These levels are based on Massachusetts Building Code (Mass, 1990). Other states may have different requirements.

Energy Efficient Design of New Buildings (ASHRAE Standard 90.1-1989), *Energy Conservation in Existing Buildings—Commercial* (ASHRAE Standard 100.3-1985), *Energy Conservation in Existing Facilities—Industrial* (ASHRAE Standard 100.4-1989), and *Energy Conservation in Existing Building—Institutional* (ASHRAE Standard 100.5-1991) should also be reviewed.

For interiors—lighting limits:

1. Determine the use categories for the various parts of the building from Table 1-10.
2. Multiply the maximum power limit for each category by the gross floor area included in that category.
3. Add the total number of watts for each area to arrive at the total lighting power limit for the building.
4. In open-concept office spaces in excess of 2000 ft^2 without defined

TABLE 1-10. Lighting Limit (Connected Load) for Listed Occupancies

Type of Use	Maximum W/ft^2
Interior	
Category A: Classrooms, office areas, mechanical areas, museums, conference rooms, drafting rooms, clerical areas, laboratories, kitchens, examining rooms, book stacks, boiler rooms, combined kitchen and dining facilities, libraries, valence and display case lighting	2.00
Category B: Auditoriums, waiting areas, restrooms, dining areas, working corridors, book storage areas, active inventory storage, stairways, locker rooms, filing areas of offices, shipping and receiving areas	1.00
Category C: Corridors, lobbies, elevators, inactive storage areas, and foyers	0.50
Category D: Indoor parking	0.25
Exterior	
Category E: Building perimeter: wall-wash, facade, canopy	5.00 (per linear foot)
Category F: Outdoor parking	0.10
Lighting switching: In all areas exterior to the building lighting fixtures should be capable of being switched automatically for nonoperation when natural light is available.	

egress or circulation pattern, 25% of the area can be designated as category B in Table 1-10.

5. In rooms with ceiling height in excess of 20 ft, a power allowance (in watts per square foot) of an additional 2% per foot of height can be used, up to a maximum of twice the limit in Table 1-10.

For building exteriors calculate lighting limits for the following areas:

6. Facade lighting: multiply the limit given in Table 1-10 by the number of linear feet in the building perimeter.
7. Parking: multiply the value in category F in Table 1-10 by the area to be illuminated.

Allowable exceptions:

8. Task lighting need not be included in the lighting power limit.
9. Process lighting in the laboratories need not be accounted for in this lighting calculation.

1.5.2 Lighting Level Guide

Suggested minimum lighting levels are presented in Tables 1-11 and 1-12. These values can be compared with those calculated from Table 1-10. When large differences occur, the level of choice will depend on the nature of the room occupancy. The lighting level should be adequate for the occupants to perform their designated task. At the same time, an overlit area can be energy wasteful. Properly lit corridors and stairways also give a feeling of security to users. Table 1-13 can be used for special categories.

1.5.3 Plumbing

1.5.3.1 Sinks. Sinks should be constructed of materials such as stainless steel or epoxy resins which are resistant to chemical and other spillage. The

TABLE 1-11. Application of Lighting Categories

Operation of Area	Illuminance Category[a]
Administration areas and offices	E
Cartography, detailed drafting, designing	F
Accounting, bookkeeping, tabulating, business machines	E
Executive offices, general offices, reading and transcribing pencil handwriting, active filing, mail sorting	E
General offices with easier visual tasks, intermittent filing, and mechanical production	D
Conference rooms, inactive file areas, general reading of books and periodicals	E
Laboratory areas	E
Microanalytical, critical, or delicate operations; close work; etc.	F
General analytical, routine analytical, physical testing	F
Engine laboratories, equipment test areas, fume hoods	E
Pilot plant and process areas (indoor)	E
Process equipment	C
Main operating aisles	B
Sampling points	C
Feed and product handling	D
Secondary aisles	B
Control panel	E
Console or desk	E
General building areas	B
Stockrooms, reception areas	B
Corridors, hallways, stairways	B
Washrooms, toilets	B

[a] Suggested values for non-VDT tasks. Other applications should be analyzed using Tables 1-12 and 1-13.

TABLE 1-12. Illuminance Categories and Illuminance Values for Generic Types of Activities in Interiors

Type of Activity	Illuminance Category	Ranges of Illuminances		Reference Work-Plane
		Lux	Footcandles	
Public spaces with dark surroundings	A	20–30–50	2–3–5	General lighting throughout spaces
Simple orientation for short temporary visits	B	50–75–100	5–7.5–10	
Working spaces where visual tasks are only occasionally performed	C	100–150–200	10–15–20	
Performance of visual tasks of high contrast or large size	D	200–300–500	20–30–50	
Performance of visual tasks of medium contrast or small size	E	500–750–1000	50–75–100	Illuminance on task
Performance of visual tasks of low contrast or very small size	F	1000–1500–2000	100–150–200	
Performance of visual tasks of low contrast and very small size over a prolonged period	G	2000–3000–5000	200–300–500	Illuminance on task, obtained by a combination of general and local (supplementary lighting)
Performance of very prolonged and exacting visual task	H	5000–7500–10000	500–750–1000	
Performance of very special visual tasks of extremely low contrast and small size	I	10000–15000–20000	1000–1500–2000	

Source: IES Lighting Handbook, Applications Volume (1987).

TABLE 1-13. Weighting Factors to be Considered in Selecting Specific Illuminance Within Ranges of Values for Each Category

a. For Illuminance Categories A through C

Room and Occupant Characteristics	Weighting Factor		
	−1	0	+1
Occupants' ages	Under 40	40–55	Over 55
Room surface reflectances[a]	Greater than 70%	30–70%	Less than 30%

b. For Illuminance Categories D through I

Task and Worker Characteristics	Weighting Factor		
	−1	0	+1
Workers' ages	Under 40	40–55	Over 55
Speed and/or accuracy[b]	Not important	Important	Critical
Reflectance of task background[c]	Greater than 70%	30–70%	Less than 30%

[a] Average weighted surface reflectances, including wall, floor and ceiling reflectances, if they encompass a large portion of the task area or visual surround. For instance, in an elevator lobby, where the ceiling height is 7.6 meters (25 feet), neither the task nor the visual surround encompass the ceiling, so only the floor and wall reflectances would be considered.

[b] In determining whether speed and/or accuracy is not important, important or critical, the following questions need to be answered: What are the time limitations? How important is it to perform the task rapidly? Will errors produce an unsafe condition or product? Will errors reduce productivity and be costly? For example, in reading for leisure there are no time limitations and it is not important to read rapidly. Errors will not be costly and will not be related to safety. Thus, speed and/or accuracy is not important. If however, prescription notes are to be read by a pharmacist, accuracy is critical because errors could produce an unsafe condition and time is important for customer relations.

[c] The task background is that portion of the task upon which the meaningful visual display is exhibited. For example, on this page the meaningful visual display includes each letter which combines with other letters to form words and phrases. The display medium, or task background, is the paper, which has a reflectance of approximately 85 percent.

Source: IES Lighting Handbook, Applications Volume (1987).

drain should have a removable, cleanable strainer to prevent solid materials from getting into the drainage system.

1.5.3.2 Liquid Wastes In some laboratories, special waste-handling systems will be necessary, including dilution tanks under sinks or a central building dilution tank, with a monitoring system to measure the pH of all waste and automatically add acid or caustic to control the pH before it enters the municipal sewer system. An accurate chart log of the pH before and after treatment will assure compliance with discharge regulations. In other locations, it may be possible to install a central collecting station whereby all such waste is accumulated and then reclaimed. Sometimes it may be necessary as a matter of policy for occupants in the laboratory to put waste into a reusable container for later consolidation and disposal.

1.5.3.3 Water Pressure Sufficient water pressure should be available for all building needs. Separate piping loops are necessary for the sprinkler system and for potable water; the latter category includes drinking fountains, emergency eyewash fountains, deluge showers, and lavatory sinks and water closet water. Antiscalding temperature regulating devices should be installed in service hot water supply lines. For eyewash and deluge shower specifications, see Appendixes I and II. For standpipes in locations where municipal water supply does not provide sufficient pressure, separate water pressure booster systems are necessary. In locations where municipal water supply is not present or where the quantity or quality is not adequate, separate water storage systems will be necessary. For example, a laboratory building being built in desert regions may require a complete self-contained water system.

1.5.3.4 Drinking Water Protection Laboratory buildings need protection of their drinking water systems to prevent contamination from laboratories. This requires separation of the laboratory water system within the building from the water systems used for drinking, kitchens, toilet rooms, emergency showers, and eyewashes. For example, the need to conserve and protect entire municipal drinking water supplies from contamination due to backsiphonage or backpressure has been recognized by the Massachusetts Department of Environmental Protection (DEP) regulation 310 CMR 22.22 (CMR, 1990), which describes the need and the approved method for protecting state, city, town, and local drinking water systems from any possible degradation caused by cross-contamination. A double check valve, reduced-pressure backflow preventer with a relief valve and open drain is the only acceptable method approved by regulation 310 CMR 22.22. It offers the best available backflow protection and can be used on toxic and nontoxic systems.

Testing of backflow prevention equipment is required semiannually in Massachusetts. An additional reduced-pressure backflow preventer installed in a bypass arrangement is required to enable these tests to be done without loss of water service to the building.

The drinking water system inside the building should be protected in the same way as the municipal supply, by using reduced-pressure, backflow preventers and a bypass line to avoid loss of service during semiannual testing. The domestic hot water supply system requires similar treatment to provide the same kind of protection to building occupants.

1.5.4 Support Services

When designing laboratory buildings it is important to consider the health and safety issues related to laboratory support service personnel, for example, maintenance, housekeeping, and security. Adequate space must be provided for housing these people and their equipment. These areas should be provided with the same health and safety features as the other laboratory areas, including adequate ventilation, fire protection, lighting, and emergency egress. In addition, there may be some special considerations. For example, nonslip floor surfaces should be provided in glass-washing rooms, janitor closets, and similar areas where floors are frequently wet. Adequate general ventilation and work space must be provided in mechanical and fan rooms. Provision must also be made for routine maintenance of laboratory ventilation systems, and adequate access must be provided.

1.5.5 Electrical Harmonic Currents

Engineers have long been aware of the potential problems in building electrical systems by harmonic currents, but these problems were less noticeable prior to extensive use of computers and other electronic devices.

A systems report published by Atkinson, Koven, Feinberg Engineers (Atkinson, 1991) provided a good description of the problem. Electrical systems found in most buildings are designed for traditional linear loads. Linear loads (such as motors, electric heaters, incandescent lighting, and fluorescent lighting with standard wire-wound ballasts) consume current on a continuous (linear) sinusoidal basis. When this type of load is balanced across a typical three-phase, four-wire power source, the return currents of each phase cancel each other out in the neutral conductor and there is no risk of transformer overload or wires overheating.

In these times, however, due to the proliferation of solid-state devices (e.g., data processing units, personal computers, variable speed motor drives and electronic ballasts), nonlinear electrical loads result in the creation of harmonic currents in the distribution system. The explanation is as follows:

Solid-state devices draw current in pulses. The frequency with which the pulses occur and their waveshapes are referred to in terms of harmonics of the fundamental frequency (60 hertz). Generally, the pulses appear in the third, fifth, and seventh harmonics (180, 300, and 420 hertz). The third harmonic current is the predominant contributor to the overall system current waveform distortion. The fifth and seventh harmonics have a lesser impact. These third harmonic currents do not cancel in the neutral. The neutral conductor can be subjected to extremely high currents (even in excess of the phase lag currents), causing hazards such as transformer overload and overheating of neutral wires and bus bars. This situation can place an excessive amount of stress on the electrical power systems, causing equipment failure and/or reduction in the system's life expectancy.

The solutions may include:

1. Install special electronic filters or transformers.
2. Oversize the common neutral in a three-phase, four-wire circuit.
3. Add a separate neutral conductor from each branch circuit to the electric panel.

In upgrade of an existing electrical system, a qualified electrical engineer using a harmonic analyzer should be retained to evaluate the extent of the problem and recommend any solutions.

1.5.6 Steam Quality

Steam is often used in laboratory buildings for heating, humidification, sterilization, and glass- and cage-washing activities.

The steam quality and its content could become a concern and should be evaluated. Steam quality is defined thermodynamically. A 100% steam quality is saturated steam. Steam of lesser quality would contain moisture droplets.

Steam additives are mostly boiler treatment compounds and are added in the steam system to prolong the life of the boiler tubes, piping, and other auxiliary systems. These chemicals raise pH. They may possess some toxic properties. The most common compounds used in boiler treatment are amines, namely, morpholine, cyclohexylamine, and (diethylamino)ethanol (DEAE). These amines minimize the effect of dissolved gases such as carbon dioxide and sulfur dioxide on metals in boilers, feed water heater, and piping.

Poor steam quality sometimes leaves residue condensate on items being sterilized. This condensate could consist of a concentration of steam additives that could cause problems with certain operations. For example, in a controlled experiment if animal feeds are being sterilized with poor-quality steam, such steam condensation would contaminate the feed.

For humidification, direct steam injection remains a very popular method. However, when steam contains components that could cause problems to occupant health, a case can be made that such steam additives should not be used in direct-steam-injection-type humidification systems. Several steam-to-steam generator systems are available where building steam can be used as a heat source to evaporate city water (or in some cases D.I. water) to make "clean" steam. Other cold-mist humidification systems are also available that do not depend on steam at all.

Careful study of the steam additives proposed and of building steam should be done. A recent study done by Battel Institute on their own steam ["Determination of Amines in Indoor Air from Steam Humidification" (Edgerton, 1989)] provides a good discussion.

The Battel Study concluded that concentration of amines measured in indoor room air during normal operation of boiler and humidifier can be very low compared with any established health standards. On the other hand, a NIOSH case study by Hills, Lushniak, and Sinks (Hills, 1990) showed that overtreatment of boilers with such water treatment compound can cause health hazard to the occupants. Studies by Fannick, Lipscomb, et al. (NIOSH, 1983) show effects of such compounds in a museum and report problems. Other NIOSH reports [McManus and Baker (NIOSH, 1981)] provide a good background. The workplace amine concentration will depend upon the boiler treatment compound control systems. If excessive chemicals can be introduced, it could result in problems.

At this time, use of control steam for humidification should not be prohibited. Careful review of the current literature on boiler systems is needed before a final decision can be made. Chapter 28 should be reviewed for alternative humidification systems.

1.5.7 Pure Water System

Pure water is required in various research activities. The purity measurement is specific resistance in ohm-cm and is expressed in conductivity (micro-ohm at 25°C). Water used for pharmaceutical laboratories, for example, is defined by United States Pharmacopeia XXI (USP, 1990).

In *Handbook of Facilities Planning*, (Ruys, 1990) provides a good description of various other types of pure water standards. Standards have been established by the National Committee for Clinical Laboratory Standards (NCCLS), the College of American Pathologists (CAP), the United States Pharmacopeia (USP), the American Society for Testing and Materials (ASTM), and the American Chemical Society (ACS).

Classification	Resistivity (megohm/cm)
Absolute purity	18.3
Ultrapure	>1.0
High purity	1.0
Low purity	<1.0
Biopure	Pathogen-free, sterile >1 ppm/TDS

1.5.7.1 Production Methods. There are several methods for producing pure water. The most common are:

Deionization. Impurities are removed by passing water through synthetic resins beds with affinity for dissolved ionized salts and gases. The process will not remove bacteria, pathogens, particulates, or dissolved organic compounds. This process can provide water of 15- to 18-megohm-cm purity. Resin beds require regeneration with sulfuric acid and caustic.

Distillation. Impurities are removed from water by converting the liquid to vapor phase and then recondensing it as distilled water. Distilled water is free of all pathogens except dissolved ionized gases. Distillation can provide water of 800,000-ohm-cm to 1-megohm-cm purity if the feedwater has been pretreated.

Reverse Osmosis. Impurities are removed by utilizing hydraulic pressure to force pure water through a membrane. This process removes some pathogens. It will not remove dissolved ionized gases. Good for water with high total dissolved solids (TDS).

Filtration. Solid particulate impurities are removed by passing the water through a porous membrane or medium. Types include sand filters, diatomaceous earth, cartridge filters, etc.

Other Systems. Combination of the four systems described above are used, or in certain cases special processes may be employed.

1.5.7.2   *Pure Water Piping.* Pure water is very aggressive and corrosive. The impurities in the pipe material in contact with the water can leach out into the pure water. The end-product water then may be unacceptable to the user. Common pure water piping materials are:

Aluminum (type 3003)
Glass
Polyethylene
Polypropylene
Stainless steel
Tin-lined copper

Cost of material, joining methodology, pipe hanging detail, and, most important, possible water contamination described above must be considered before making the final selection.

A recirculated system provides the best assurance of an ongoing clean system. Dead legs in piping systems should be avoided because they could be a source of bacterial growth.

1.5.7.3   *Central Pure Water Supply Versus Onsite System.* In a small project, an onsite system will be most cost-effective. In a large building it is sometimes not cost-effective to produce and pipe the highest grade of pure water throughout the building. A reasonable grade, centrally produced, and "polished" locally in specific laboratories to obtain the final quality, may be more cost-effective.

CHAPTER 2

LABORATORY CONSIDERATIONS

2.1 GUIDING CONCEPTS

All laboratories, regardless of their specific use, have many similar health and safety requirements that should be considered in all design stages. This chapter reviews the requirements that are common to all types of laboratories discussed in this text but does not deal with specific laboratory types. In Part II, specific laboratory types will be discussed and the distinguishing features of a number of commonly encountered laboratory types will be described and illustrated. In some cases, specialized laboratories may not require one or more of the items discussed in this section; in other cases, it will be necessary to describe special facilities not discussed in this chapter; and sometimes it will require a combination of both. Unless otherwise noted, items discussed here will be referred to by section number when specialized laboratory types are covered in Part II.

2.2 LABORATORY LAYOUT

The laboratory layout is critical for the efficient use of space and the safety of laboratory personnel. The laboratory design must be consistent with the building design philosophy described in Chapter 1. This includes provisions for entry and egress, furniture and equipment locations, and access for

Guidelines for Laboratory Design: Health and Safety Considerations, Second Edition
By Louis DiBerardinis, Janet S. Baum, Gari T. Gatwood, Anand K. Seth, Melvin W. First, and Edward F. Groden.
ISBN 0-471-55463-4   Copyright © 1993 by John Wiley & Sons, Inc.

disabled persons. Consideration of safety issues in the planning and schematic design phases of a project saves owners and users costs for corrective modifications of the plans and materials during construction or living with the continued liability of built-in safety hazards. Neglect of safety considerations in design phases can lead to laboratories containing needless hazards to health and safety.

Alternative designs for a standard double-module laboratory, with notations on health and safety considerations, are summarized in Figures 2-1 through 2-8. These figures illustrate the preferred locations for chemical fume hoods, primary access and secondary emergency exits, and recommended module dimensions for all of the laboratory building options on vertical shaft locations, which were discussed in Chapter 1.

2.2.1 Personnel Entry and Egress

There are many specific requirements for laboratory entry and egress. As with building design, major references pertaining to egress and entry for laboratories are published by the National Fire Protection Association (NFPA 101, 1991), U.S. Occupational Safety and Health Administration (OSHA, 1990), and Building Officials and Code Administrators, International Inc. (BOCA, 1990). They set policies on the following items: the number and width of exits, direction of door swing, and permissible door swings into egress pathways. The essential features are summarized below.

The NFPA (NFPA 45, 1991) requires a minimum of two exits from each laboratory under the following conditions as does OSHA (OSHA 1910.37, 1990):

1. A laboratory work area contains an explosion hazard that could block egress from or access to the laboratory.
2. A Class A (high fire hazard) laboratory unit work area exceeds 500 ft^2.
3. A Class B (moderate fire hazard) or Class C (low fire hazard) laboratory unit work area exceeds 1000 ft^2.
4. A laboratory fume hood is located next to a primary exit of a laboratory.
5. A laboratory uses a compressed gas cylinder larger than lecture bottle size containing "a flammable gas or [a gas that] has a health hazard rating of 3 or 4, that could prevent safe egress in the event of accidental release of cylinder contents."
6. A laboratory uses a cryogenic container with "a flammable gas or [a gas that] has a health hazard rating of 3 or 4 that could prevent safe egress in the event of accidental release of cyliner contents."

When laboratories are designed, there is no assurance that they will meet all the conditions listed in the NFPA standards at some future time. Therefore this book recommends that laboratory designers provide two exits from each laboratory work area whenever it is over 500 ft^2, when there is a fume hood in the laboratory, and when NFPA conditions pertaining to the health hazard rating of compressed gas cylinders or cryogenic containers apply. The safest arrangement for a laboratory is for each required exit to open into a separate fire zone and for each to be located so that the internal pathways to it are separated. Thus, when a hazard makes one laboratory escape pathway impassable, the second will provide an alternative safe route out to a fire-rated building egress passage or stairway. This building layout is illustrated in Figure 1-4 and discussed in Chapter 1. The laboratory layout is shown in Figure 2-1. Exit doors should swing in the direction of egress because an outswinging door cannot be blocked by persons being pushed against it by those in a panic behind them. Also, in an emergency, it takes less time to push open an outswinging door than to pull open a door that swings inward. Finally, a solvent fire in a tightly constructed laboratory room might raise room pressure sufficiently to make it difficult to open an inswinging door. On the requirement for outswinging doors, the NFPA distinguishes between Class A and B (high and moderate fire hazard, respectively) and Class C (low fire hazard) laboratory units. Because the laboratory designer has no control over how an individual laboratory may be used in the future, it is recommended that all doors swing in the direction of egress.

To make certain that egress corridors will not be blocked by open doors, outswinging doors should be recessed sufficiently so that the door does not protrude more than 7 in. into the clear width of the corridor when fully open, as shown in Figure 2-2. This makes it unlikely that persons passing in the corridor could block the opening of the laboratory exit door, or be hit as doors open.

Glass panels of 9 ft^2 or less are permissible in B-labeled fire-rated laboratory exit doors, normally used in 1-h fire-rated corridor partitions. Glass panels in doors help prevent collisions of persons entering and exiting. The glass should be placed low enough that persons of less than standard height or who are in wheelchairs can be seen from the other side of the door.

Many codes concerned with removal of architectural barriers in buildings specify lever action handles for workplace doors because levers are easier to activate than knobs in emergencies. Furthermore, the firm downward motion to release the latch does not require the use of hands. Architectural barrier codes also limit the pressure required to open doors containing automatic door closer mechanisms. For interior hinged, sliding, or folding doors, the code pressure limit is 5 lb of force (ANSI 117.1-1980).

Minimum dimensions for exit doors are 32 in. wide by 80 in. high, but standard practice and architectural barrier codes such as ANSI Standard 117.1-1986, "Providing Accessibility and Usability for Physically Handicapped People", recommend a minimum of 36 in. by 80 in. to facilitate the movement in and out of laboratories of disabled persons in wheelchairs. Primary laboratory entry doors (i.e., those used routinely) that are 42 in. by 84 in. in size are even better for moving equipment and carts in and out. In addition, clearance between the side of the door opening and any obstruction, such as a wall or lab bench, is required on both the push and pull sides of the

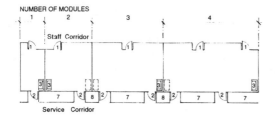

NUMBER OF MODULES

MULTIPLE MODULE LABORATORIES

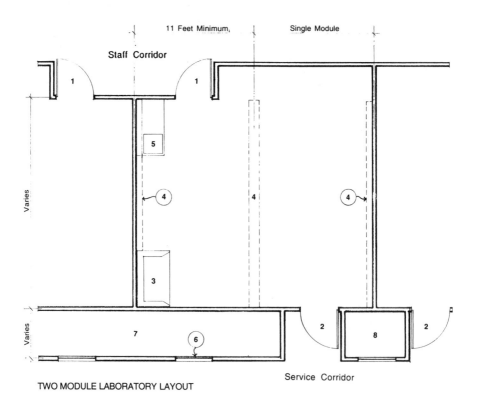

TWO MODULE LABORATORY LAYOUT

KEY

1 Primary Access/Egress
2 Emergency Second Egress
3 Chemical Fume Hood
4 Utility Piping Zone
5 Sink
6 Mechanical Access Door
7 Mechanical Shafts
8 Electrical Closets

FIGURE 2-1. Laboratory layout with utility shaft at service corridor.

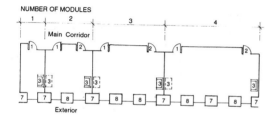

NUMBER OF MODULES

MULTIPLE MODULE LABORATORIES

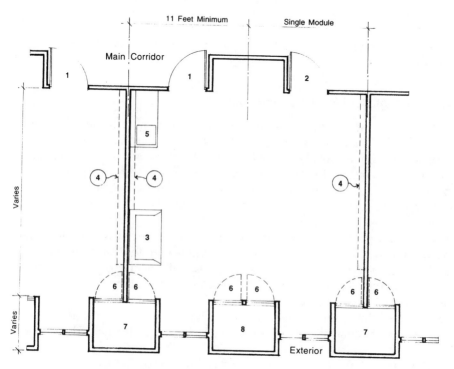

11 Feet Minimum   Single Module

Main Corridor

TWO MODULE LABORATORY LAYOUT

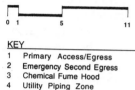

0 1   5   11

KEY

1 Primary Access/Egress
2 Emergency Second Egress
3 Chemical Fume Hood
4 Utility Piping Zone
5 Sink
6 Mechanical Access Door
7 Mechanical Shafts
8 Electrical Closets

FIGURE 2-2. Laboratory layout with utility shaft at exterior wall.

door as shown in Figure 2-3. On the pull side, 18 in. is the minimum when there is no other obstruction within 5 ft of the doorway, but 24 in. is preferred. On the push side, 12 in. is the minimum clearance when the door has both a closer and a latch. Only 4 ft of unobstructed area is needed in front of the door on the push side. Openings such as windows, trap doors, and knock-out panels between laboratories do not qualify as exits or as secondary emergency exits. Doors onto the balconies of multistory buildings are not considered to be legal exits, although in some jurisdictions they may be considered areas of fire refuge when balcony floors can be reached by fire truck ladders. Balconies must be secured with railings that meet height and structural requirements of local codes and national building standards. Exits must be marked in accordance with local and national standards, including Americans with Disabilities Act of 1991, and must remain lit even during power failures.

2.2.2 Laboratory Furniture Locations

Laboratory benches, desks, and other furnishings must be designed and located to facilitate ease of egress and ease of travel within the laboratory. The design team establishes zones in a laboratory by the way the benches, utilities, and major fixed equipment are located. This is the concept of hazard zoning. For example, in laboratories that have fume hoods for conducting potentially hazardous operations, the hoods should be located away from the primary access door as in the case of a hood alcove shown in Figure 2-4. Seated work stations for activities such as microscopy, writing, and computer data entry, where laboratory workers concentrate their attention on a limited field of vision and may not be fully aware of other lab activities, should be near the exit. As shown in Figure 2-5 the laboratory length, from the public access corridor to the exterior wall or service corridor, is divided into zones representing diminishing levels of hazard as one moves toward the primary access door. Secondary egress doors may be located in the hazardous zone.

2.2.2.1 Benches. Center room benches may be of the island type with aisles on all sides making it possible for personnel to move around the bench quickly to reach emergency equipment or an exit. An island type bench is required in the layouts shown in Figure 2-6 and Figure 2-8. When there is an approved exit on both sides or at both ends of a laboratory room a peninsula type bench is acceptable. Single-module laboratories are usually equipped with benches along both long walls and have a central aisle between them as shown in Figure 2-7.

Laboratory benches should be made of sturdy materials that have finishes that can be repaired easily. Benchtops should be level and stable under the heavy equipment and instruments that may be placed on them. Base storage units that are secured to the bench on cantilevered or suspended frames should be secured so that they will not drop off when carts or equipment crash

NUMBER OF MODULES

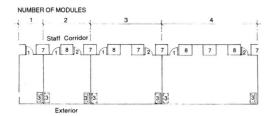

MULTIPLE MODULE LABORATORIES

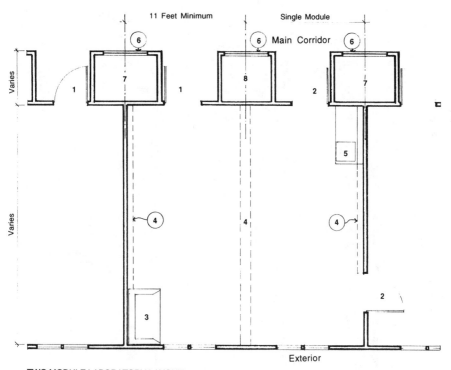

TWO MODULE LABORATORY LAYOUT

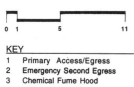

KEY

1   Primary Access/Egress
2   Emergency Second Egress
3   Chemical Fume Hood
4   Utility Piping Zone
5   Sink
6   Mechanical Access Door
7   Mechanical Shafts
8   Electrical Closets

FIGURE 2-3. Laboratory layout with utility shaft at interior wall.

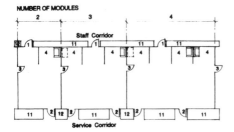

NUMBER OF MODULES

MULTIPLE MODULE LABORATORIES

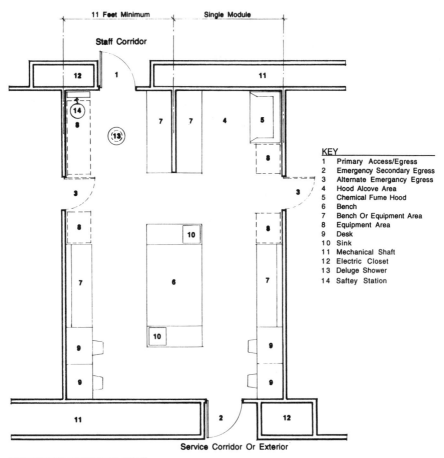

11 Feet Minimum | Single Module

Staff Corridor

KEY

1	Primary Access/Egress
2	Emergency Secondary Egress
3	Alternate Emergancy Egress
4	Hood Alcove Area
5	Chemical Fume Hood
6	Bench
7	Bench Or Equipment Area
8	Equipment Area
9	Desk
10	Sink
11	Mechanical Shaft
12	Electric Closet
13	Deluge Shower
14	Saftey Station

Service Corridor Or Exterior

TWO MODULE LABORATORY LAYOUT

FIGURE 2-4. Laboratory module with hood alcove.

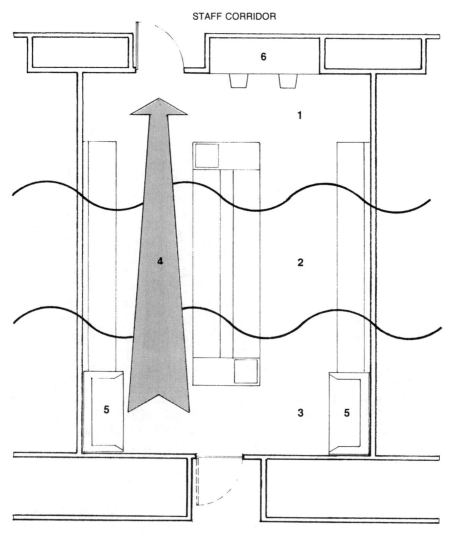

STAFF CORRIDOR

SERVICE CORRIDOR

0 1   5   11

KEY

1   Diminishing Hazard Zone
2   Moderate Hazard Zone
3   Increasing Hazard Zone
4   Emergency Exit Through Diminishing Hazard Zones.
5   Fume Hood
6   Desk/Study Area

FIGURE 2-5. Hazardous zoning diagram.

against them. Base storage units that are mounted directly on the floor are more stable and do not amplify floor vibrations to the benchtop. Knee spaces should be provided along laboratory benches to permit workers to sit comfortably and safely for long periods. With this arrangement, they can put their feet on the floor or on the foot-ring of a stool and maintain a balanced, upright posture while working. A person seated at a knee space provides more area behind for people moving in the aisle, and, when empty, the chair or stool can be pushed all the way into the knee space to further reduce aisle obstruction. If disabled persons are employed in a laboratory, some accommodations in the working height and storage components may be required at those individuals' work stations. Hand wash sinks that meet accessibility standards should also be made available in those laboratories.

2.2.2.2   Aisles.   Major aisles between benches or equipment should have a minimum clearance of 5 ft to allow for safe passage of persons behind those working at a bench (Nuffield, 1961). (See also Section 1.2.2.3.1.) Major aisles between equipment or benches should be aligned in the direction of egress. Because laboratory emergencies are often accompanied by dense smoke and fumes, egress under these obscured conditions may have to be done by crawling out of the lab along the floor. Therefore, when the arrangement of aisles is not regular and aisles do not lead logically to an exit, a person can be overcome by smoke and fumes before figuring how to get out. Laboratory work areas are inappropriate places for exotic circulation geometries.

2.2.2.3   Desks.   Desks can be used safely when they are needed in the laboratory, but they should be located away from potentially dangerous operations. The path from a desk to any egress should not require movement in a direction of or past a zone of increasing hazard. (See also Section 1.2.2.4.)

2.2.2.4   Work Surfaces.   Work surfaces and base storage units should be dimensioned to permit safe access to utility outlets and easy reach to the storage units above the work surface. The standard work surface depth is 2 ft, not including the depth of the utility strip. When deeper surfaces are needed to safely accommodate large pieces of bench-mounted equipment, safe access to utilities must be maintained. When the clearance between the horizontal work surface and the underside of a wall-mounted storage unit is less than 2 ft, projection of these units over the work surface should not be greater than 1 ft. With this geometry, it is not likely that intense heat sources or open flames will be placed beneath the shelf and cause a fire either to the unit or to its contents.

Work surface heights in the laboratory are designed for two body positions: Seated work surface height is between 30 and 32 in., and standing work surface height is between 35 and 37 in. At least one seated work station is recommended in every laboratory where personnel in wheelchairs work and should be considered in laboratories where microanalytical techniques may be used.

NUMBER OF MODULES

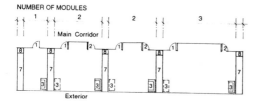

MULTIPLE MODULE LABORATORIES

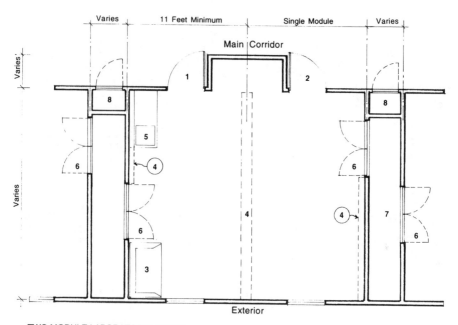

TWO MODULE LABORATORY LAYOUT

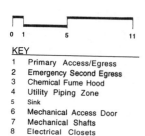

KEY

1 Primary Access/Egress
2 Emergency Second Egress
3 Chemical Fume Hood
4 Utility Piping Zone
5 Sink
6 Mechanical Access Door
7 Mechanical Shafts
8 Electrical Closets

FIGURE 2-6. Laboratory layout with utility shaft between modules.

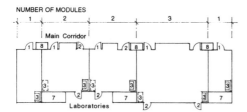

MULTIPLE MODULE LABORATORIES

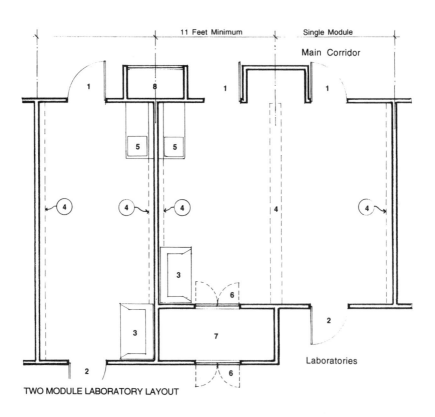

TWO MODULE LABORATORY LAYOUT

KEY
1   Primary Access/Egress
2   Emergency Second Egress
3   Chemical Fume Hood
4   Utility Piping Zone
5   Sink
6   Mechanical Access Door
7   Mechanical Shafts
8   Electrical Closets

FIGURE 2-7. Laboratory layout with utility shaft at central locations.

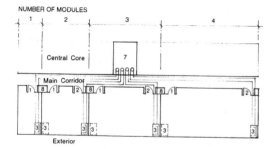

NUMBER OF MODULES

Central Core

7

Main Corridor

Exterior

MULTIPLE MODULE LABORATORIES

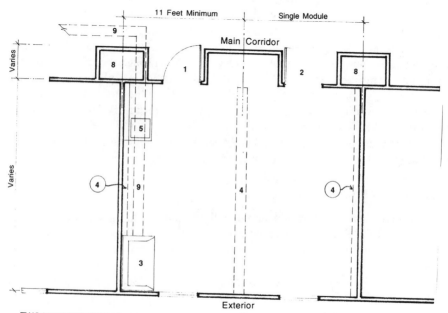

11 Feet Minimum

Single Module

Main Corridor

Varies

Varies

Exterior

TWO MODULE LABORATORY LAYOUT

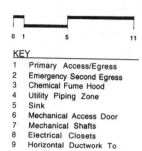

0 1   5   11

KEY

1	Primary Access/Egress
2	Emergency Second Egress
3	Chemical Fume Hood
4	Utility Piping Zone
5	Sink
6	Mechanical Access Door
7	Mechanical Shafts
8	Electrical Closets
9	Horizontal Ductwork To Mechanical Shaft In Central Core.

FIGURE 2-8. Laboratory layout with utility shaft at central core.

2.2.2.5 Safety Stations. A laboratory safety station is an arrangement, and installation of safety and emergency response equipment which is designated for each laboratory. The emergency eyewash fountain, deluge shower, fire extinguisher, fire blanket, safety goggle and protective glove dispensers, shoe cover and protective garment supply, chemical spill kit, first aid kit, and assisted respiratory apparatus are some of the items that a laboratory safety station may accommodate. The laboratory safety station also has a bulletin board or other place dedicated for display of safety regulations and announcements. A safety station location that is consistent from laboratory to laboratory offers greater security to laboratory staff and improves emergency response time and effectiveness. The primary laboratory entry is an excellent location for the safety station because of the high visibility of that location where laboratory staff pass by it daily. In an emergency, staff do not have to waste time recalling where emergency response equipment is kept. When something is missing from or empty in the safety station, staff will notice it more readily and will contact the safety office for replacements. The laboratory entry should be kept unobstructed by equipment, supplies, and personal belongings, such as coats, boots, and handbags, so that the safety station is always accessible. In order to reduce clutter at laboratory entries and elsewhere, lockers or coat rooms should be provided for staff; in addition, adequate storage facilities should be provided within the laboratory and in the building for bulk storage of supplies and unused equipment.

2.2.3 Location of Fume Hoods

Because hazardous operations are usually conducted in the laboratory hood, chemical fume hoods, as discussed in Section 2.2.2, must be located away from the primary laboratory exit for safety reasons. An added advantage of placing a hood at the far end of a laboratory aisle is that it reduces traffic past it, a source of adverse effects on hood performance. (See also Section 1.2.2.4 and Chapter 31.)

2.2.4 Location of Equipment

Layout of fixed and movable equipment is determined by the hazards presented by the equipment, requirements for utility connections, and the area the equipment occupies. The most hazardous processes and associated equipment should be placed away from the primary access door. Equipment posing little or no hazard can be placed closer to the primary egress route. Considerations of area requirements and utility connections are listed in Table 2-1. (See also Section 1.2.2.4.)

2.2.5 Access for Disabled Persons

If a disabled person works, learns or teaches in a laboratory all parts of the laboratory and its emergency equipment should be accessible. Building

TABLE 2-1. Checklist for Selection and Location of Laboratory Furnishings and Equipment

1. List all major pieces of equipment in each laboratory according to the following characteristics:
   (a) Size and weight
   (b) Whether bench mounted, free standing, or other mounting method
   (c) Fixed to wall or floor with permanent utility connections
   (d) Movable with quick disconnects to utilities
   (e) Required utilities and services
       (1) Water: city, purified, recirculated process
       (2) Drain: floor, piped
       (3) Gas: piped, manifolded tanks
       (4) Compressed air
       (5) Vacuum
       (6) Electricity: voltage, amperage, phases, location and number of standard and special receptacles
       (7) Local exhaust services: type, volume, filters, other treatment
       (8) Air supply for local exhaust services: tempered, filtered, humidified
       (9) Steam: pressure, volume, flow rate
   (f) Maintenance clearances
   (g) Operator's position and clearances
   (h) Equipment attachments
   (i) Heavy rotary components
2. Determine the location of and area required for supplies associated with each piece of equipment:
   (a) Glassware
   (b) Instruments, manipulators
   (c) Chemicals
   (d) Disposables: paper goods, plastic ware

ramps, elevators, stairways, toilets, and laboratory emergency eyewashes, deluge showers, fire alarms, fire extinguishers, electrical outlets, switches, emergency alarms, communication equipment, and controls should be designed in accordance with applicable federal and state architectural barriers codes. An example of an essential laboratory safety device, a chemical fume hood, which is adapted for use by a person in a wheelchair is shown in Figure 2-9. Another is shown in Appendix I on emergency deluge showers.

2.3 LABORATORY HEATING, VENTILATING, AND AIR CONDITIONING (HVAC)

Laboratory ventilation is closely related to correct overall laboratory building function in the sense that laboratory HVAC services will be enhanced or constrained by the design choices made for the building as a whole. Prudent

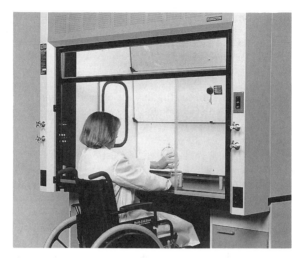

FIGURE 2-9. Chemical fume hood adapted for wheelchair use. Reprinted with permission from Hamilton Industries, Two-Rivers, Wisconsin.

design will take into account the unusually rapid changes that are presently occurring in most research areas. Often they require a greater number of instruments and other equipment per program or per project to support new methods of experimentation. By the very nature of research, old restraints on pressure, chemical composition, toxicity, etc., are constantly being cast off and new demands placed on the provisions of essential facilities. Therefore, reserves should be designed into all laboratory HVAC systems to retard obsolescence. A reserve of 10–25% of new requirements is recommended in all cases, although for industrial systems 35–50% excess is more usual.

2.3.1 Temperature Control

Heating and air-conditioning systems must provide the uniform temperature that is needed for correct operation of many analytic devices. Although close control of humidity will not be necessary in most instances, when close humidity control is required, simultaneous heating and cooling may be needed. Many kinds of HVAC systems have been used for laboratories with success: local cooling units, central systems, or a combination of both. It is important that all elements of the temperature control systems be interlocked so that a uniform temperature can be maintained throughout the year. Often it is advantageous to install a separate air-conditioning and heating system for a laboratory when it is not the major occupant, rather than connecting it to the building system, because laboratories have HVAC requirements that are different from those of a normal office or hospital building, and, if not separated, a large system must operate to maintain preset conditions in a small space.

2.3.2 Laboratory Pressure Relationships

Good laboratory ventilation requires a careful balance between discharge and supply air volumes as well as careful placement of discharge and supply air locations. Even within the same type of laboratory, requirements may vary depending on the hazards of the materials being used, the quantity of hazardous materials being handled, and the nature of the operations involved. Communication with laboratory users at an early stage in planning helps to identify potential hazards before the final design stage. Laboratories using hazardous materials must be maintained at a negative pressure relative to hallways and other adjacent public access areas. Special purpose laboratories requiring positive pressure with respect to hallways and other adjacent areas will be discussed later. All offices, conference rooms, lunch rooms, and other public areas must be maintained at a slightly positive pressure relative to the laboratories to ensure safety.

2.3.3 Laboratory Ventilation Systems

There are three main types of ventilation systems for laboratories based on function:

1. *Comfort ventilation* is a means of supplying measured amounts of outdoor air for breathing and to maintain design temperature and humidity. ASHRAE (ANSI/ASHRAE, 1989) has recommended ventilation rates that are needed to provide an adequate level of indoor air quality. They are being adopted by many states and agencies for health and safety protection of people in workplaces. The document has been accepted by ANSI as a consensus standard. It recommends the outdoor air exchange rates for all types of facilities.

2. *Exhaust ventilation* is designed specifically for health and safety protection.

3. *Makeup air* replaces the volume discharged to the atmosphere through the health and safety exhaust ventilation systems.

These three systems are described in detail in the following paragraphs.

2.3.3.1 *Comfort Ventilation Supply Air for Laboratory Modules.* All comfort ventilation supply air exhausted from laboratories must be replaced with mechanically supplied air or by infiltration from adjacent areas. An adequate supply of air is essential to provide the correct number of room air changes needed to establish safe working conditions. See section 2.3.4.2 on ventilation rate terminology.

2.3.3.1.1 Supply Air Velocity and Entry Locations Inside Laboratories. Makeup air in equal quantities must be supplied to laboratories when air is exhausted through health and safety protection systems. Most of the makeup air will come in as air supplied directly into the laboratory, but a small amount will infiltrate from adjacent spaces when they are at higher pressure relative to the laboratory space. The location of makeup air outlets and the temperature of the air supplied are important. High-velocity supply air jets create sufficient turbulence at a hood face to disrupt exhaust ventilation system performance and be a source of discomfort to occupants of the laboratory. It is recommended that high-velocity outlets not be used in laboratory design. Low-velocity supply air grilles and diffusers should be selected and located so that the air velocity at the occupant's level does not exceed 50 ft/min.

2.3.3.1.2 Air Distribution. Air supplied to a laboratory space must keep temperature gradients and air turbulence to a minimum, especially near the face of the laboratory fume hoods and biological safety cabinets. Air outlets must not discharge into the face of fume hoods. Cross-flows that impinge on the side of a hood alter airflow more seriously than do cross-flows in front of the hood (Schuyler and Waechter, 1987). Large quantities of supply air can best be introduced through perforated plate air outlets or diffusers designed for larger air volumes (ANSI, 1992). The air supply should not discharge on a fire detector, because this slows its response.

Some general air distribution guidelines are as follows (Caplan, 1978):

1. Terminal velocity of supply air jets (near hoods) is at least as important as hood face velocity when the face velocity is in the range of 50 to 100 fpm.
2. The terminal throw velocity of supply air jets (near hoods) should be less than the hood face velocity, preferably no more than one-half to two-thirds the face velocity. Such terminal throw velocities are far less than those for conventional room air supply.
3. Perforated ceiling panels provide a better supply system than do grilles or ceiling diffusers because the system design criteria are simpler and easier to apply, and precise adjustment of fixtures is not required. Ceiling panels also permit a greater concentration of hoods than do wall grilles or ceiling diffusers.
4. Wall grilles or registers should have double deflection louvers set for maximum deflection. The terminal velocity (near hoods) should be less than one-half of the face of the hood velocity.
5. If the wall grilles are located on the wall adjacent to the hood, the supply air jet should be above the top of the hood face opening. For

equal terminal throw velocities, grilles on the adjacent wall cause less spillage than will grilles located on the opposite wall.

6. The terminal throw velocity from ceiling diffusers at the hood face should be less than the hood face velocity.

7. Diffusers should be kept away from the front of the hood face. A larger number of smaller diffusers is advantageous, if the necessary low-terminal velocity can be maintained.

8. Blocking the quadrant of the ceiling diffuser blowing at the hood face results in less spillage.

9. Perforated ceiling panels should be sized so that the panel face velocity is less than the hood face velocity, preferably no more than two-thirds of the hood face velocity.

10. Perforated ceiling panels should be placed so that approximately one-third or more of the panel area is remote (more than 4 ft) from the hood.

11. Additional tests are needed to determine laboratory fume hood performance; Peterson et al. (Peterson, 1983) indicate that the only way to determine hood effectiveness is to test the specific hood under actual room conditions.

12. Room air distribution currents may be evaluated to determine if local air currents are high enough to cause discomfort or disturb experiments (ASHRAE, 1990).

2.3.3.2 Recirculation of Laboratory Room Air. All hazardous materials should be used in chemical fume hoods or with some other type of local exhaust ventilation. When the amount of exhaust air is reduced to a minimum by the use of well-designed and well-functioning chemical fume hoods and alternative local exhaust systems, less energy is expended and comfort is equal to what would be experienced were fumes to be dissipated by dilution air exchange involving the entire laboratory room.

In some research laboratories the room air can be recirculated within local cooling and heating devices (e.g., fan coil units), provided that the minimum amount of outdoor air is supplied through the unit (or from another source) to satisfy health and safety requirements when the laboratory is occupied. Recent AIA guidelines prohibit recirculation of room air in clinical laboratories (AIA, 1992). Refer to Chapter 13 for additional details.

All air from local exhaust facilities, as well as from lavatories, sterilization rooms, and simila⸍ ⸍acilities, must be exhausted to the outside with no recirculation allowed. Air from offices, libraries, conference rooms, and similar nonlaboratory facilities can be recirculated with the addition of the minimum amount of outside air required to maintain health and safety and to comply with building codes.

2.3.4   Exhaust Ventilation for Laboratory Modules

Exhaust air systems of three types may be needed in each laboratory module: (1) removal of general room comfort supply air and contaminated dilution ventilation air; (2) health and safety exhaust ventilation air from biosafety and chemical fume hoods contained in the laboratory; and (3) local or spot exhaust ventilation air. The choice of system(s) will depend on the size of the laboratory, the nature and quantity of materials used, and the type of installed laboratory equipment. These systems may have either constant or variable air volume.

2.3.4.1 Exhaust of General Room Ventilation Air from Laboratories.   The manner in which general ventilation air is exhausted from each laboratory room depends on its size and the nature of the activities and equipment present. In some cases, adequate exhaust of general room ventilation air will be provided by a laboratory fume hood or some other local exhaust air system. In other cases, a combination of general room return air facilities, chemical fume hoods, and additional local exhaust air facilities may be used.

2.3.4.2 Ventilation Rate Terminology.   In ventilation consensus and building codes, room and laboratory gross ventilation rates are usually expressed in terms of air changes per hour or cubic feet of air per occupant, or sometimes both.

Air changes per hour (AC/HR) is a ventilation rate expressed as the number of room volumes exchanged in an hour. It is calculated by dividing the larger of the exhaust or supply volume (in cubic feet per hour) by the volume of the room (in cubic feet). This is the *theoretical* air changes per hour. The actual number of air changes per hour depends on how well mixing occurs, and this is critically dependent on the quality of the design of supply and exhaust systems. Mixing can be evaluated by measuring the decay rate of a tracer gas such as sulfur hexafluoride (SF_6). The ratio of the actual to the theoretical air changes is referred to as the *K factor;* it describes the efficiency of air mixing. K factors are assigned in the ACGIH's *Industrial Ventilation Manual* (ACGIH, 1992) for various locations of the room supply and exhaust points.

Another criterion of ventilation rate is the number of cubic feet of air per minute (CFM) supplied per occupant of the room space. Maximum occupancy must be used, and the room ventilation system will be designed based on this number.

In both cases, the room air volume rate used to calculate AC/HR or CFM per person is the outside air fraction, not the total recirculated air.

REGULATIONS.   Few regulations exist which specify air changes per hour. Some consensus recommendations or guidelines (i.e., ASHRAE [62-89], ANSI, or BOCA) may be adopted as local or state regulations.

OSHA requires a minimum of 6 AC/HR in chemical storage rooms [*Design and Construction of Inside Storage Rooms,* General Industry Standard 29 CFR 1910.106, page 144, OSHA 2206, Nov. 7, 1987 (OSHA, 1987)]. Since most laboratories store some quantities of chemicals, this regulation applies; OSHA has cited university chemical storerooms for inadequate ventilation under this regulation.

Most State Building Codes have adopted standards for the minimum supply of outdoor air. For office areas it is 20 CFM per person.

VENTILATION RATES: CONSENSUS RECOMMENDATIONS, CODES AND STANDARDS, GOVERNMENT REGULATIONS. Professional organizations and government agencies provide recommendations for AC/HR or CFM per person for various room types and buildings. Those that pertain to laboratory buildings are summarized in Table 2-2.

RATIONALE FOR EXISTING AC/HR RECOMMENDATIONS—EMPIRICAL APPROACH. Originally, AC/HR recommendations for laboratories were based on heating, general ventilation, and cooling requirements rather than being based on health and safety concerns. Recently, new recommendations for minimum outside air requirements have been addressed to reduce indoor air quality complaints that have followed. The "tightening" of buildings in response to energy conservation requirements have increased the need for careful evaluation of ventilation needs. The AC/HR values in Table 2-2 have been arrived at empirically and, for the most part, are based on anecdotal evidence.

TABLE 2-2. Recommendations for Ventilation Rates for Laboratory Buildings

Organization	Offices	Laboratories	Comments
1. ASHRAE	20 CFM/person	6 AC/HR	CFM per person is based on a maximum of 7 people per 1000-sq. ft average range and depends on mixing factors and special uses (e.g., animals or odoriferous compounds)
2. AIA		4–12 AC/HR	

Some institutions and industries have adopted their own guidelines. These include the ones listed below.

Institution	AC/HR	CFM/ft^2	Space
1. Exxon (Exxon, 1980)	—	0.5–1.0	Laboratories
2. Los Alamos National Laboratories (LANL, 1991)	8–10	—	Laboratories
3. Eastman Kodak (Kodak, 1985)	10	—	Darkrooms

EXPERIMENTAL APPROACH. Los Alamos National Laboratory (LANL, 1991) has begun an experimental program to determine acceptable AC/HR values for radiation laboratories subject to the buildup of hazardous airborne radiation levels from visually undetectable spills. Preliminary results indicate that 8 AC/HR is the minimum to maintain radiation exposure levels below current health standards even should a small spill of radioactive materials occur.

THEORETICAL APPROACH. Another way of estimating ventilation requirements for laboratories is to calculate the concentration reduction on the basis of theoretical air changes and to correct the result by introducing an appropriate K factor. On this basis it takes approximately 7 AC/HR to reduce the initial concentration by 90% assuming perfect mixing in the laboratory. Assigning a K factor of 1.5 for a well-designed laboratory ventilation system (ACGIH, 1992), the required exchange rate is 10 AC/HR. This method has been used by some to establish minimum exchange rates based on an assumed generation rate in a given room volume.

INDUSTRIAL HYGIENE APPROACH. Traditionally, the industrial hygiene approach is to evaluate the source strength of hazardous materials and to control exposure by the use of custom-designed local exhaust ventilation systems (e.g., chemical fume hoods, glove boxes, slot-ventilated work benches) and work practices. Dilution ventilation is seldom used to control chemical hazards. The major source of assigned AC/HR standards for laboratories relates to building code concerns and not to the specific needs of laboratories to reduce odors from people and materials.

However, operation of a chemical fume hood generally provided a laboratory air change rate in excess of 10 AC/HR so that minimum dilution needs are always met and the establishment of a minimum number of AC/HR for laboratories is usually not addressed. The introduction of variable air-volume systems for laboratory hoods and other exhaust ventilated facilities has created a potential to reduce the total volume of health and safety exhaust to levels approaching zero. A minimum amount of comfort air exchange is always necessary; therefore, there is a need to establish minimum air exchange requirements for health and safety. As noted earlier, heating or cooling requirements often dictate the minimum, and it may not be necessary to set minimum air-exchange rates based solely on health and safety considerations.

All occupied and unoccupied laboratories require a minimum mechanical ventilation rate of 0.5 cubic feet per minute (cfm) per square foot of floor area when no fume hood is operational. When a fume hood is operational, removal of a minimum of 1 cfm per square foot, or the equivalent of six complete laboratory air changes per hour, will be needed. It must be kept in mind that these figures represent minimum exhaust air rates for laboratories. In most cases, larger air exhaust rates will be needed to handle all the air discharged to the atmosphere through laboratory hoods and other local exhaust air facilities inside the laboratory. Opportunities for energy conservation are discussed in Chapter 21.

2.3.4.2.1 Velocities for Removing Room Ventilation Air. Exhaust grilles for general room ventilation should be sized so that the inflow face velocity is between 500 and 750 fpm (feet per minute). Wall-mounted grilles should be placed to provide an airflow direction within the laboratory from the entrance door toward the rear of the laboratory to minimize the escape of fumes to the corridor.

2.3.4.3 Air Rates for Laboratory Hoods and Other Local Exhaust Air Facilities.

Because air exhausted from health and safety facilities cannot be recirculated, but must be discharged to the atmosphere, the energy cost is high. Opportunities for energy conservation are outlined in Chapter 26.

2.3.4.4 Chemical Fume Hoods.

Laboratory hoods, sometimes called *chemical fume hoods* or *fume cupboards,* are a form of local exhaust ventilation commonly found in laboratories using toxic, corrosive, flammable, or malodorous substances. The purpose of a laboratory fume hood is to prevent or minimize the escape of contaminants from the hood to the laboratory air and, by this means, to provide containment. Successful performance depends on an adequate and uniform velocity of air moving through the hood face, commonly referred to as the *control velocity.* Hood performance is adversely affected by external high-velocity drafts across the face, large thermal loads inside, bulky equipment in the hood that obstructs the exhaust slots at the rear or creates eddy currents at the opening, and poor operating procedures on the part of personnel using the hood by their failure to work 6 in. or more inside the hood face. With the sash closed, the hood can minimize the effects of small explosions, fires, and similar events that may occur within, but it should not be depended upon to contain fires or explosions other than trivial ones. To function correctly, a chemical fume hood must be designed, installed, and operated according to well-established criteria.

The chemical fume hood is the laboratory worker's all-purpose safety device. It is probably the single item that most definitively characterizes a laboratory and its importance for the safety and health protection of laboratory workers cannot be overstated. This being so, it is essential for the laboratory designer to understand thoroughly the functions that characterize a satisfactory laboratory hood and the several designs that are on the market. Not all are equally effective or efficient in the utilization of airflow.

Well-designed fume hoods have several important characteristics in common: (1) Air velocity will be uniform—that is, ±20% of the average velocity over the entire work access opening. (2) All the hood surfaces surrounding the work access opening will be smooth, rounded, and tapered in the direction of airflow to minimize air turbulence at the perimeter of the hood face.

(3) Average face velocity will be a minimum of 100 fpm for work with any of the chemical, biological, and radioactive materials usually encountered in university, government, and industrial research and in commercial consulting laboratories. When handling substances associated with a somewhat higher hazard level, 120 fpm average face velocity is recommended. In the past OSHA called for, but did not require, average face velocities in excess of 150 fpm, for laboratory hoods used with any of the 13 carcinogens listed in OSHA 1910.1003 et seq. (OSHA, 1992). Under OSHA Laboratory Standard 1910.1450, promulgated in 1990 (OSHA, 1992), OSHA did not specify a hood face velocity, recognizing that face velocity alone may not be a good indicator of protection (DiBerardinis, 1991). Hood face velocities in excess of 120 fpm are not recommended because they cause disruptive air turbulence at the perimeter of the hood opening and in the wake of objects placed inside the work area of the hood (ANSI, 1992; Chamberlin, 1982; Ivany, 1989). (4) The face velocity of hoods with adjustable front sash will be maintained constant (within reasonable limits) by an inflow air bypass that proportions the air volume rate entering the open face to the open area, or by some other method that produces a similar result.

Use of a totally enclosed and ventilated glove box is recommended for handling very hazardous materials. The use of glove boxes minimizes exhaust air volume to the atmosphere and simplifies air treatment for environmental protection.

There are a number of distinctive types of laboratory hoods in widespread use. Each will be identified in the following sections and described more completely in Chapter 31. Example purchase specifications are also provided in Chapter 31.

2.3.4.4.1 Standard Chemical Fume Hoods. The basic chemical fume hood incorporates the four principles enumerated in Section 2.3.4.4 that characterize a well-designed fume hood. Most laboratory furniture suppliers have one or more models that will meet the listed criteria. A model purchasing specification is presented in Chapter 31.

CONVENTIONAL TYPE. The conventional-type fume hood is one of the oldest forms of laboratory fume hoods. It is designed so that all exhaust air is drawn in through the front face opening. As a result, as the sash is lowered, reducing the face opening, the air velocity is increased proportionately. This can result in excessive face velocities and poor containment. This type of hood is not recommended for use in laboratories.

BYPASS TYPE. This type of hood permits laboratory air to enter the hood chamber through a "bypass" when the sash is closed. This bypass is designed so that a constant face velocity through the hood opening is maintained as the sash is lowered. This prevents excessive face velocities when the sash is nearly closed ($<\frac{1}{3}$ open).

AUXILIARY AIR TYPE. Auxiliary air fume hoods are recommended for laboratory use only under the specific conditions described below. The major difference between the bypass chemical fume hood and the auxiliary air hood is the method employed to provide makeup air to the hood. For bypass hoods, all of the makeup air is provided by the room HVAC system, whereas for supply air hoods, part of the air is introduced from a supply air grille just above and exterior to the hood face. Some manufacturers state that up to 70% of the total hood exhaust volume may be supplied in this manner, with the remainder coming from the laboratory HVAC system. When auxiliary supply air is introduced behind the sash (an incorrect hood design), the hood chamber is likely to become pressurized and blow toxic contaminants out the open front into the laboratory. Only auxiliary air hoods that introduce auxiliary air exterior to and above the hood face are acceptable.

Auxiliary air hoods have two advantages over conventional bypass hoods. First, the air supplied to the hood does not have to be cooled in very warm climates, a significant energy saving. Second, auxiliary air hoods can be used in laboratories that contain so many hoods that the volume of supply air required, were they all standard chemical fume hoods, would result in many more room air changes per hour than are desirable. Providing auxiliary supply air, even to only some of the hoods, can reduce room air velocities and supply air quantities to acceptable levels.

There are disadvantages to the use of auxiliary supply air hoods. First, they are more complex in design than the usual chemical fume hood, and their correct installation is more critical to safe and efficient operation. Often, only a small imbalance between room air and auxiliary air supplies can result in unsafe operating conditions. Second, unsafe conditions occur when the velocity of external auxiliary air supplied to the face of the hood is excessive, because high-velocity air sweeping down across an open hood face can produce a vacuum effect and draw toxic contaminants out of the hood. Third, this type of hood must have two mechanical systems (separate exhaust and supply systems) for each hood. Inasmuch as some additional supply air will need to be added to the laboratory anyway, the auxiliary air hoods require two supply air systems instead of the one supply system that standard chemical fume hoods require.

Generally, the use of auxiliary air hoods is discouraged by safety and health professionals (ANSI, 1992). When the advantages outweigh the disadvantages, auxiliary air fume hoods should be used because energy savings through reductions in summer cooling and winter heating can be realized even in moderate temperature climates. Energy savings during the heating season occur because the auxiliary air need not be heated to as high a temperature as the comfort ventilation air. Use of auxiliary air hoods is particularly attractive for renovations when one or more fume hoods must be added to a laboratory with limited amounts of supply air capacity. Because of the complexity of this type of hood, the performance specifications are more stringent than for standard chemical fume hoods. Specifications for both types are contained in Chapter 31. It is extremely important to choose the particular auxiliary air hood that will be purchased before the supply air system is designed because not all auxiliary air hoods have the same requirements for room and auxiliary air volumes and some meet performance requirements at less than 70% auxiliary air.

2.3.4.4.2 Horizontal Sliding Sash Hoods. Economy in the utilization of conditioned air for laboratory hoods can be achieved most satisfactorily by maintaining the required safe face velocity but restricting the open area of the hood face. A transparent horizontal sliding sash arrangement can cut overall air requirements by 50% if two half-width panels are used on two tracks. Similarly, if three panels are used, the minimum open area reduction is 67% for a two-track setup and 33% for a three-track arrangement. This design also has an advantage over the conventional vertical sash because the full height of the hood opening is always available. When horizontal sliding panels 14 to 16 in. wide are used, they can also serve as safety shields.

Under most conditions, a fume hood with horizontal sliding sashes gives personnel protection equal to that given by a hood with a vertical sash, provided that it is designed and operated to have a minimum average face velocity of 100 fpm at the maximum face opening. All velocities in the plane of the hood face should be greater than 80 but less than 125 fpm under operating conditions. Sometimes, turbulence occurs at sharp panel edges when inflow velocities exceed design values. Visual smoke trails may be used to test for inward airflow across the entire face opening and an absence of turbulence at sash edges.

2.3.4.4.3 Perchloric Acid Hoods. Perchloric acid hoods require special construction, construction materials, and internal water-wash capability. Problems reported with hoods heavily used for perchloric acid digestions are associated with the accumulation of explosive organic perchlorate vapors that condense while passing through the hood exhaust system. They can detonate from percussion during cleaning, modification, or repair. Therefore, use of specially designed fume hoods is required for use with perchloric acid. Perchloric acid hoods should meet the same contaminant fume retention capabilities as the chemical fume hoods described in the preceding sections. In addition, they should be constructed of stainless steel and have welded seams throughout. No taped seams or joints and no putties or sealers can be used in the fabrication of the entire hood and duct system. The perchloric acid hood also requires an internal water-wash system to eliminate the buildup of perchlorates. The water-wash system should consist of a water spray head located in the rear discharge plenum of the hood plus as many more as are needed to ensure a complete wash down of all surfaces of the ductwork from the hood work surface to the discharge stack on the roof. To drain all of the water to the sewer in a satisfactory way during normal wash-down operations, it is important to have the exhaust duct go straight up through the building with no horizontal runs. A straight vertical duct run also facilitates periodic examination for maintenance purposes.

When only small amounts of perchloric acid are used, it is possible to construct an air scrubber in the hood. This can be done when the point of generation of perchloric acid fumes can be identified and a small capture hood designed. The National of Safety Council recommends such a system (NSC, 1985).

2.3.4.4.4 Biological Safety Cabinets. The biological safety cabinet is a special form of containment equipment that will be addressed in more detail in Chapters 12 and 31. It is used for work with cell and tissue cultures and parenteral drugs when the materials being handled must be maintained in a sterile environment and the operator must be protected from toxic chemicals and infective biological agents. The dual functions of protecting the worker and maintaining sterility are achieved by two separate cabinet flows: (1) a turbulence-free downward flow of sterile HEPA-filtered air inside the cabinet for work protection and (2) an inward flow of laboratory air through the work opening to provide worker protection. In addition, all air exhausted from biological safety cabinets is filtered through HEPA filters to provide environmental protection. The inward airflow and the downward airflow are delicately balanced in the biological safety cabinet; great care must be exercised to maintain the design flow rate of each, as well as the ratio between the two. Biological safety cabinets have achieved (a) widespread work protection use for recombinant DNA research and (b) worker protection use for working with infected body fluids and cultures.

2.3.4.4.5 High-Velocity–Low-Volume Spot Exhaust Systems. Effective exhaust ventilation must be provided for all apparatus and procedures used in the laboratory that generate hazardous contaminants or create excessive heat. In some cases it may not be possible or desirable to conduct the operations in a chemical fume hood. For example, when the equipment is large, it may not fit into a hood, or it may fit but occupy too much hood space and affect hood performance adversely. These cases call for special or supplementary ventilation arrangements that often take the form of high-velocity–low-volume local exhaust points consisting of open-ended exhaust hoses or ducts. Flexible exhaust ducts of 4- to 6-in. diameter, often referred to as "sucker hoses" or "elephant trunks," are useful for this service because they can be moved to locations where they are needed. An advantage of local exhaust hoses is their ability to reduce the total amount of air removed from the laboratory as a result of capturing contaminants at the source at high velocities, thereby using less total air volume than would be required for a fume hood. An ASHRAE study indicates that local exhaust systems are much more energy conservative than dilution ventilation (DeRoos, 1979).

Examples of local exhaust hoods are presented in Chapter 31, Section 31.9. There are almost an infinite number of hood and systems designs that can be made. The design factors include the type and size of the equipment, the location of the generation source(s), the quantity and physical properties of the material generated, and the access to the equipment needed by laboratory personnel. The design guideline provided in Chapter 3 of the ACGIH *Industrial Ventilation Manual* (ACGIH, 1992) should be followed. In addition, some of the specific hood "types" described in Chapter 10 of the above reference may be applicable to some of the operations performed in the laboratories.

Spot exhaust facilities require a high-static-pressure exhaust system that must be provided to the laboratory from a system separate from the one serving the hoods. This is because hoods are low-static-pressure devices whereas spot exhaust points require negative static pressures from 2 to 5 inches of water gage (in. w.g.), depending on design factors such as the quantity of air and air velocity at the opening. It is essential to design for adequate local exhaust quantities in the planning stages because it is almost impossible to upgrade the system after installation except by total replacement.

It is possible to plan for flexibility by providing the high-static-pressure exhaust system discussed above and a variety of local exhaust hoods that can be selected and installed as necessary. If the process or equipment changes, it would then be possible to disconnect the "hood type" and install a more appropriate one.

2.3.4.4.6 Variable-Air-Volume Hoods. Variable-air-volume hoods can be any one of the standard hoods discussed in Section 2.3.4.4.3. The major difference is that the exhaust quantity is not constant. Generally the exhaust quantity decreases as the hood face opening decreases. This can be accomplished by a variety of techniques that are discussed in Chapter 31 (Section 31.6) and Chapter 33.

The major advantage of this type of system is a reduced operating cost in the form of energy savings. The disadvantages relate to the relative sophistication of the control equipment, needed maintenance, and long-term repeatability of the control systems. Variable-air-volume systems can be either on a single hood or part of a manifold system.

Since this is a relatively new concept in laboratory ventilation design, a more thorough discussion is provided in Chapter 31 (Section 31.6) and Chapter 33.

2.3.4.4.7 Ductless Fume Hoods. A ductless fume hood is one which treats the exhaust air and returns it directly to the laboratory space. The air treatment mechanism is usually an integral part of the hood structure.

Ductless fume hoods have limited use in most laboratories because of the wide variety of chemicals used (ANSI, 1992; NFPA 45, 1991). The potential contaminant concentration is generally unknown, and the appropriateness of the air purification system installed in ductless hoods must be evaluated for each chemical used. In addition, the warning properties (i.e., odor, taste) of the chemical being used must be adequate to provide an early indication that the air purification unit is effectively preventing emission of toxic vapors back to the laboratory. Alternatively a continuous monitor that can detect concentrations below the TLV for *each* chemical used in the hood could be used.

Ductless fume hoods may be safe when the contaminant is particulate, and provisions can be made for changing filters without excessive contamination of the laboratory. Biological safety cabinets (Class II, Type A) are examples of a type of ductless hood used successfully for control of particulate biological hazards.

Activated charcoal is not efficient for fine particles and is only useful for adsorbing certain gases or vapors. Many gases and vapors of low molecular weight can be displaced from activated charcoal by higher-molecular-weight organic molecules and reenter the room air with continuous flow of clean air under these conditions. Ductless hoods, if designed properly, serve to protect workers at the hood face, but may spread the contaminant into the room air over a long time span at lower concentrations (Kermig, 1991).

2.3.4.5 *Airflow Monitors.* For safety, especially when working with hazardous materials that give no sensory warning by odor, visibility, or prompt mucous membrane irritation of their escape into the workroom from the laboratory hood, it is advisable to provide each hood with an airflow monitor capable of giving an easily observed visual display of functional status. This is recommended by ANSI (ANSI 1992) and is inferred by OSHA (OSHA 1910.1420,). Installation of a hood airflow gage at the site of use has the special advantage of placing responsibility for day-by-day monitoring of hood function with the primary users. Devices that may be used to monitor hood function include liquid-filled draft gauges or Magnehelic gauges,* which measure hood static pressure. Other devices measure airflow velocity in the exhaust duct where velocity is high or at some point at the side of the hood opening as a surrogate measurement of total exhaust. In all cases, the type and location of monitors should be evaluated carefully because incorrect positioning of static pressure taps can give false readings and inline airflow devices may become clogged or corroded. A hood monitoring and airflow control system that has important energy conservation aspects is described in Chapter 26.

2.3.5 Exhaust Fans and Blowers

After total air volume and static pressure requirements have been established, fan selection for a laboratory fume hood or spot exhaust system will depend upon the following factors: (1) ability to fulfill application requirements, (2) ease of maintenance, (3) initial cost, (4) life expectancy, and (5) availability of spare parts.

---

* Dwyer Instruments Inc., P.O. Box 373, Michigan City, IN 46360.

Cast-iron fans are ideal for all exhaust air applications, including spot exhausts, because they are of excellent quality and reliability, have an extended lifespan, and require very little maintenance. Whenever long life expectancy and freedom from maintenance are not critical, steel plate fans are acceptable. Whenever the effluent air from fume hoods customarily contains large amounts of severely corrosive gases plus condensing water vapor, exhaust fans constructed of fiberglass-reinforced polyester (FRP) are highly recommended. Fume hood exhaust fans handling contaminants composed of dusts and mists, or containing flammable, toxic, or corrosive materials, should be located outside occupied areas of the building and as close as possible to the point of discharge to the atmosphere, preferably on the roof of the laboratory building.

Exhaust fans should have a discharge velocity at least 2500 fpm and a stack extending at least 10 ft above the roof parapet and other prominent roof structures. Under no circumstances should weather caps be used on local exhaust system stacks. To take care of precipitation into the open stack when the fan is idle, there should be a drain connection at the low point of the scroll casing that can drain directly to the roof. For small blower-motor sets, there is a distinct advantage associated with selecting a direct-drive blower with a totally enclosed, weatherproof motor of the correct rpm. This selection eliminates belts, belt guard, and motor enclosure, and results in a compact, maintenance-free installation. A typical arrangement of a roof-mounted motor blower and stack is shown in Figure 2-10.

Belted fans should be double-belt-driven, the shaft bearings should contain standard grease fittings for lubrication, and the fan should have a drain at the low point of the scroll casing. Vibration isolaters are required to minimize noise transmission through the connecting ducts and building structure. The entire installation exposed to the weather must be built to withstand wind loads of 30 lb/ft^2 applied to any exposed surface of the fan and ducts rigidly attached to it.

Explosion-proof motors and nonsparking wheels are required when it is possible for the effluent air to contain more than 25% of the lower explosive limit (LEL) of any combination of vapors and aerosols. All fan motors must meet the requirements of the National Electrical Code (NFPA 70, 1990) and conform to applicable standards for load, duty, voltage phase frequency service, and location. Motors should be mounted on an adjustable sliding base. Motors of 0.5 horsepower and larger should be of the squirrel-cage induction or wound rotor induction type and should have ball or roller bearings with pressure grease lubrication fittings. Drives for belted motors should be as short as possible and equipped with a matched set of belts rated at 150% of capacity. A weatherproof metal guard with angle iron frame securely fastened to fan housing and fan base should be provided for protection.

Each fume hood should be exhausted by its own exhaust fan. When multiple fume hoods are manifolded to a single exhaust fan, imbalances in the exhaust airflow can occur as hoods are turned on and off. This can be

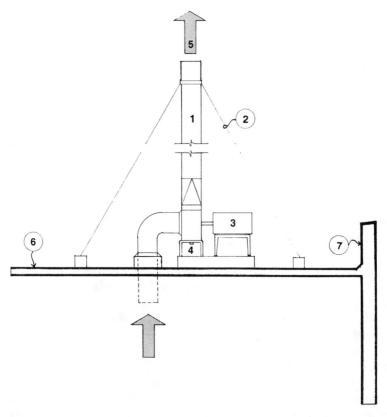

KEY

1 Stack Discharge Directly Upward.
2 Guy Wires To Roof If Required (3 lines/stack).
3 Directly Connected, Totally Enclosed Weather
   Proof Motor.
4 1 Inch Diameter Hole At Low Point In
   Fan Scroll For Drainage To Roof.
5 Exhaust Air.
6 Roof.
7 Parapet.

FIGURE 2-10. Exhaust stack and blower.

overcome by turning all hoods on and off together, but this is no longer considered a satisfactory arrangement from the standpoint of energy efficiency. Following installation of a manifolded hood system, the correct airflow to each hood must be established by adjusting dampers placed in the exhaust duct for that purpose and locking them in position. Balance is difficult to maintain and the system is prone to lose balance, making one or more hoods potentially unsafe. Cross-contamination from various hoods can also occur when the manifolded system is no longer in balance. Because of this and other potential problems, multiple fume hoods on a single exhaust fan are not recommended.

2.3.5.1 Exhaust Air Cleaning for Laboratory Effluent Air. Generally, exhaust air from laboratories is not cleaned prior to release because of the excellent dilution capability of the atmosphere when the air is discharged straight upward, starting 10 ft or more above all roof obstructions. However, certain laboratory procedures may require special effluent gas treatment to avoid polluting the atmosphere. Specific instances will be described in Part II, where detailed descriptions of commonly used laboratories will be found.

2.3.5.2 Exhaust Ducts and Plenums

2.3.5.2.1 Construction of Exhaust Ducts and Plenums. Exhaust ducts for fume hoods and local exhaust systems should, preferably, be of the high-velocity type to avoid settlement of particles in horizontal runs. Additionally, high-velocity systems reduce duct cross section and save space in vertical chases and above-ceiling utility areas. Stainless steel of a high chromium and nickel content is the material of construction for hard service in a corrosive, erosive, and high-temperature environment, but cost is high. Other materials of construction, such as epoxy-coated steel, can be used for less severe applications. Plastic piping has exceptional resistance to many commonly used corrosive chemicals, but physical strength and flame spread ratings are less than for metals. It may be excessive, but California regulations require a 2-h rated separation between fume hood exhaust ducts from different floors within the same chase. All fume hoods and local exhaust system ducts should be constructed of round piping with the interior of all ducts smooth and free of obstructions. All joints should be welded or otherwise sealed airtight.

Flexible ducts used for spot exhaust service should be kept to minimum lengths because flexible duct has an airflow resistance that is as much as two to three times that of metal duct of the same diameter. Poorly formed flexible elbows can increase system resistance to the point where function becomes lost. To reduce energy loss and air noise levels, the open ends of sucker hoses should be tightly capped when not in use. Flexible tubing must be selected from among the noncollapsible types.

2.3.5.2.2 Duct Leakage. All local exhaust system ductwork located inside the laboratory building must be maintained under negative pressure to prevent leakage of contaminated air into occupied spaces. Maximum duct leakage of 2% at design negative pressure should be specified in the building design documents and the installation carefully supervised. To ensure that this standard is achieved, the ducts should be tested, without fail, after installation by capping the ends of duct runs, putting the ducts under the design negative pressure specified in the construction documents, and measuring the volume of inflow air carefully with a sensitive, variable head airflow meter. Inflow rate should not be greater than 2% of the maximum design airflow rate for the duct run under test. It is necessary to point out that balancing measured airflow rates into a system against measured airflow at the discharge end is not an acceptable way of assuring no more than 2% inleakage because field measurements cannot be relied upon to less than ±5%, at best. If condensation inside the ducts is possible, all sections must be pitched to drain to a sewer connection.

Ideally, ductwork should be designed to be self-balancing without the use of trimming dampers. Dampers inside exhaust air systems serving laboratories (with the exception of those required for emergency fire suppression purposes)·are failure-prone from corrosion, vibration, and maladjustment by unknowledgeable service personnel. Maladjusted dampers are frequently observed to seriously compromise the safety objective of the exhaust systems in which they are installed. Therefore, experienced ventilation engineers design systems that will meet building and laboratory design objectives without trimming dampers. To be effective, however, installation contractors must follow directions with sufficient exactitude to accomplish the desired results. Meticulous testing of completed systems is the only way to assure quality installations. However, the future flexibility of any installation may be compromised without dampers.

If condensation can occur inside exhaust ducts due to changes in temperature from a warm building to a cold mechanical or fan room, consideration should be given to draining from the low point of the duct before it leaves the building and trapping this condensation and vapors back into the plumbing system.

2.3.5.2.3 Noise Suppression. Air noise in laboratory ductwork can be minimized by keeping velocities within acceptable ranges (2000–3000 fpm) and by using flexible connections between fans and ducts. Use of sound attenuators in exhaust ductwork is not an acceptable solution to noise problems, although their use is acceptable in supply air systems. Control of noise from sucker hoses (a potential major noise source) was covered in Section 2.3.5.2.1.

2.4 LOSS PREVENTION AND OCCUPATIONAL SAFETY AND HEALTH PROTECTION

2.4.1 Emergency Considerations

A clear objective of laboratory building design is to avoid and prevent unsafe conditions. In spite of the best planning, experience indicates that emergencies sometimes occur from unforeseen events. Therefore, the prudent building designer takes all reasonable steps to contain possible emergencies by the installation of loss control services and equipment. With emergency systems, personnel and physical plant losses can be avoided or at least reduced to a tolerable level. All of the following emergency considerations should be included in the basic design. Rejection of any must be documented and justified. Those not required by law or code represent good practice standards.

2.4.1.1   Emergency Fuel Gas Shutoff.   In facilities where fuel gas is piped throughout the building to laboratories, a method of shutting off the flow of gas at the laboratory must be installed for emergency use. The preferred method is to run the gas supply pipe to a wall just outside the laboratory and locate an emergency shutoff valve station at that point. The station should consist of a simple ball valve located in a box with a breakable glass or easily removable cover and a clear sign announcing its function. Although a major building gas shutoff must be provided along with an excess-flow check valve to serve the entire building, the major building cutoff valve may be too remote from the laboratory room involved in an emergency and the building excess-flow check valve will be too large to sense and to stop the flow of fuel into small laboratory areas. For both reasons, local valve stations are necessary.

Note. The requirements for fuel gases do not apply to nonfuel laboratory gases that are piped into laboratories through small piping systems from a central source. However, nonfuel gas pipes, valves, excess-flow check valves, and other materials and installations must meet applicable codes and standards of the NFPA, the Compressed Gas Association, and the ANSI.

2.4.1.2   Ground Fault Circuit Interrupters.   Ground fault circuit interrupters should be used on all laboratory benches and where portable or nonstationary equipment is used. Ground fault circuit interrupters are devices that compare the current flow in wires feeding and returning from electrical devices, such as mixers, ovens, meters, blenders, pumps, and stirrers. When an imbalance of more than 5 mA occurs, the electrical power to the device will cut off. This is predicated on the assumption that the leakage, or lost current, could pass through the body of an operator. The circuit interrupter functions to limit the amount of "shock" to nonlethal levels. This is in contrast to the electrical fuses and circuit breakers normally found in a laboratory building. They do not open until the power requirements of the equipment online are exceeded and, therefore, fuses and circuit breakers cannot prevent an electrical shock to personnel; they function only to protect equipment and the building against fire.

Ground fault circuit interrupters tend to sum up all the leakage of devices plugged into one circuit. Installation of more interrupters, or fewer electrical appliances per circuit, can help to reduce nuisance tripping. Old electrical devices or appliances, whose insulation is not as good as when they were new, also can trip the interrupters, indicating a need for these appliances to be renovated or replaced. Fuses and circuit breakers, ground fault circuit interrupters, and a comprehensive electrical safety program are all necessary to provide maximum freedom from shock, electrical fire, and adverse electrical equipment problems.

Ground fault circuit interrupters should be used at all laboratory benches, and there should be no more than three duplex outlets per interrupter. Ground fault circuit interrupters should also be used near wet operations such as sinks. Locating the ground fault circuit interrupter unit in a central box at the main electrical panel for the laboratory encourages prompt attention to standing line leaks and other potential problems in the system. The use of no more than three duplex outlets placed close to each other and attached to a ground fault circuit interrupter on the same bench is ideal because this arrangement minimizes long wire runs with many junction boxes that have a leak potential. In addition, when ground faults occur that shut the system down, they can be corrected easily and the system turned back on again quickly.

Stationary electrical equipment, such as refrigerators and ovens, should be equipment-grounded through hard wiring; therefore, these devices need not be ground fault circuit interrupted in addition. One duplex plug outlet on each laboratory bench might be installed without ground fault circuit interrupters to provide a means of temporarily using equipment that would otherwise break the circuit. This outlet should be clearly identified.

2.4.1.3 *Master Electrical Disconnect Switch.*
In each laboratory area, electrical services should be identified and sited in such a manner that all electrical power to the laboratory (except lighting and other life-safety critical items such as exhaust systems and alarms) can be quickly disconnected from one easily accessible location. When such an arrangement is not possible or feasible, a shunt trip breaker system should be installed that will accomplish the same thing upon depression of the master mushroom kill-switch. This switch should be located near the normal emergency exit route, but not in a place where it can be activated inadvertently.

2.4.1.4 *Emergency Showers.*
Emergency deluge showers are used to dilute and wash off chemical spills on the human body. Because many chemicals attack the body rapidly, the location and reliable functioning of the showers is of critical importance. Specifications for emergency showers are given in Appendix I.

Deluge showers, when properly installed, provide a minimum of 30 gallons of water per minute and deliver this volume at low velocity because high-velocity showers can further damage injured tissue. For this reason, only low-velocity deluge shower heads should be used. The valve operating the shower should be a type that requires a positive action to close, such as a ball valve with a non-spring-loaded lever arm. Rigid pull bars of stainless steel stand up better under corrosive conditions than do chains and other metals.

When testing is not performed routinely, valves often become so difficult to operate that a chain can break before the water valve is turned on. The preferred location of a shower is just outside the hazardous area. This could be in a hall not more than 25 to 50 ft from the laboratory. The valve should be located close enough to a wall that normal traffic will not bump into the operating rod. The shower head should be far enough out from the wall to allow a second person to move around it to help an injured person. A large contrasting spot should be painted on, embedded in, or affixed to the floor directly beneath the shower to indicate its location. At least one shower in each area of the building should be tempered to provide water at 70 to 90°F to accommodate an injured person who should remain in the shower for the recommended 15-min period. Normally, cold water temperature is substantially below 70°F and immersion in water at this temperature, for a long period, would be painful for an injured person. For economy, the remainder of the deluge showers can be connected directly to the potable cold water system. Another way of providing a tempered water deluge shower is to use a nearby toilet facility with a shower stall already in place. In this case, a standard emergency shower deluge head should be installed that uses the correct pipe size and an antiscald mixing valve. With this arrangement, the injured person would be able to have a degree of privacy for the 15-min shower, especially important when it becomes necessary to remove contaminated clothing, as it most often does.

Another advantage of using a shower stall in a rest room for a tempered emergency shower is that the runoff water can go through a floor drain. Floor drains are not normally installed in corridors under emergency showers because they are used so infrequently that the traps become dry and allow sewer gases to enter the area. If the tempered water shower is to be in a hall corridor or laboratory, and a drain is deemed necessary, a floor drain capable of handling the entire output of the shower should be installed with it and the occupants alerted to the need to pour water every two weeks, or ethylene glycol every six months, into the drain.

Contrary to popular opinion, emergency deluge showers are not installed to extinguish clothing fires. The best method of extinguishing a clothing fire is to "stop, drop, and roll", and then remove the burned clothing and seek help. Nevertheless, if one is within a step or two of a deluge shower, this device can be used as an effective fire extinguisher for a clothing fire. The hazard associated with moving from the laboratory to a shower with clothes burning is the probability of inhaling burning gases and searing the breathing passages, including the deep parts of the lungs, to a fatal degree. The "stop, drop, and roll" method is preferred.

2.4.1.5 Emergency Eyewash.
The reaction of many chemicals with the human eye is very rapid. For example, the interior chamber of the eye can be reached within 6 to 8 s following a splash of concentrated ammonia. Therefore, most physicians agree that an immediate flush with copious quantities of water is the best first-aid treatment for chemical splashes to the face and eyes.

There are two basic types of emergency eyewash devices available: plumbed and portable. Because many portable units do not have the capacity to deliver 15 min of copious flushing, this discussion is limited to plumbed emergency eyewash units. It is acknowledged that portable units have value when they can be located very close to the user and result in a very quick start to eye flushing that can be completed at the plumbed eyewash station once the initial flush has been accomplished.

There should be at least one eyewash facility per laboratory. They may be located at sinks or at any other readily accessible area in the laboratory. Laboratories using strong acids or bases should have an eyewash within 20 ft of the hazard area. A tempered water unit should be installed for each contiguous group of laboratories. It should be set at 70 ± 5°F. This is necessary because most physicians recommend a 15-min wash prior to transportation to a medical facility. Holding one's eyes open to 35–40°F water for more than a minute or two can become painful and ultimately impossible.

The best units for washing eyes and face are the multistream, cross-flow types. They flush the face and both eyes at the same time with near zero-velocity water. Hand-held units on a hose, such as a kitchen spray, have the advantage of serving as a minishower for splashes of the arms, hands, and other small spills to the body. Specifications for emergency eyewash facilities appear in Apendix II. See also A. Weaver and K. Britt, "Criteria for Effective Eyewashes and Safety Showers" (ASSE, 1977).

2.4.1.6 Chemical Spill Control. A means for effecting prompt spill control should be provided for all laboratories using hazardous materials. They include making provisions for (1) establishment of convenient cleanup stations, (2) one or more storage locations of adequate size to hold the requisite quantity of neutralizing chemicals, adsorbents, disinfectants, and other equipment needed to deal effectively and rapidly with the maximum anticipated spill, and (3) diking materials, consisting of fixed dikes or dike bags. The locations of cleanup stations need to be clearly identified in the building design stages.

2.4.1.7 Emergency Cabinet. Emergency cabinets should be provided for all laboratories and located in an area that will be readily accessible under stress conditions. The cabinet should be sized to hold items specific to the work of the laboratory, as well as a number of general items. General and specific items may include:

General	Specific
Emergency blanket	Medical antidotes
Emergency response information	Chemical spill kits
First-aid kit	Protective clothing
Stretcher	Escape breathing equipment
Resuscitation equipment	

2.4.2   Construction Methods and Materials

The fire-resistive construction requirements of BOCA Article 9 for Class A buildings should be followed (BOCA, 1991). Selection of wall coverings, bench materials, and other furnishings of some laboratories will require special consideration. Examples are: installation of materials resistant to fire; consideration of the potential for electric shock in the use of metal furnishings; and reflectivity of building materials for certain operations involving light, darkness, and lasers.

Adequate electrical outlets for standard line voltage equipment should be provided within each laboratory to facilitate the use of all pieces of anticipated electrical equipment. Older laboratories often have many extension wires draped around the laboratory, resulting in an unsafe condition. This has occurred because laboratory designers of a former period could not envision the enormous increase in the number of electrically driven devices in all modern laboratories. Unsafe extension wire conditions can be avoided in new laboratories by adequate preplanning that includes a systematic and realistic evaluation of current and future electrical outlet requirements. Electrical outlets in laboratories should be equipped with ground fault circuit interrupters (Section 2.4.1.2).

2.4.3   Control Systems

Equipment using hazardous materials, such as constantly flowing toxic or flammable gases, as well as electricity-dependent operations, should have alarm and automatic shutoff circuits capable of rendering the process and equipment safe in case of failure. To build these safeguards into a laboratory requires that potential problems be defined by the expected users of the laboratory during the building and laboratory design stages.

2.4.4   Alarm Systems for Experimental Equipment

A system of electronic communications should be installed in laboratory modules so that, when necessary, monitoring of key laboratory operations can be initiated to signal equipment malfunctions. The system should connect the laboratory to control points selected by the laboratory manager. Selected control points may be the office of the investigator, a hall outside the laboratory, a security guard station, or the fire station. The alarm system is not intended to be a control system; it should be for the transfer of information only. When there will be firm regulations against leaving experiments, reactions, and all other laboratory activities unattended, and when no other need exists, alarm systems for experimental equipment may be omitted. In an operation where hazards are controlled by outside services, such as water for a still, ventilation, or electricity, provisions should be made for shutdown of the operation in emergency situations. Emergency electrical systems are described in more detail in Section 2.4.1.3.

2.4.5 Hazardous Chemical Disposal

Discarded hazardous materials, including flammable liquids, strong reagent chemicals, biological and radioactive wastes, and highly toxic chemicals, should be segregated into specific areas within the laboratory for disposal. The hazardous waste storage area should not be located where an unexpected reaction could immediately involve persons working at normal work stations, and it should not block normal egress routes. In addition, the hazardous waste storage area should be chosen so that, should an undesired reaction occur, it would not affect other areas of the building. Chapter 23 contains detailed information on rooms used for hazardous waste of all kinds.

2.4.5.1 Laboratory Storage of Hazardous Chemical Waste. Hazardous wastes should be stored in appropriate safety containers. Flammable liquid wastes, for example, should be stored in approved waste safety cans that carry UL or FM approval for this use. They should be placed in a location where spills can be caught and retained without damaging the floor or creating slippery work and walking surfaces. Local ventilation may be necessary for especially toxic, malodorous, and volatile waste materials. Because some waste containers are large and bulky, the architect and laboratory designer should consult manufacturer's catalogs to get a good idea of the space that will be needed and to make provision for movement of the accumulated wastes by cart or some other suitable means. For more on chemical storage, see Section 1.4.6.

2.4.5.2 Chemical Waste Treatment Prior to Disposal. Laboratory sinks and floor drains should be tied into a chemical process sewer for the disposal of nonhazardous chemicals requiring neutralization, to make them suitable for ultimate disposal in the sanitary sewer system. Local, city, and state regulations should be examined to determine the extent to which neutralization tanks must be used.

2.4.6 Chemical Storage and Handling

Determining chemical storage and handling locations and methods after a laboratory has been constructed often leads to acceptance of unnecessary risks. These matters should be considered during the design phase of the laboratories. Experience has shown that site inspection of currently used facilities of the potential users of new laboratories is the best way to determine the amounts of chemicals that must be stored in the new laboratory. When an inspection is not possible, the needed information should be obtained from the expected users and from similar laboratory operations. Current needs figures should be increased by a safety factor of 1.5 to 2.0 to allow for growth, and space provided accordingly.

2.4.6.1   *Storage in the Laboratory.* For safety, the quantity of chemicals stored in a laboratory should be kept to a minimum at all times and they should be stored by methods and locations appropriate to their hazard classification. Large supplies of chemicals should be stored in a central storage area serving the entire laboratory complex (see Section 1.4.7). The laboratory director or principal investigator should be consulted to determine maximum storage quantities for each laboratory, and space should be provided in the new facility for adequate chemical stocks based on the information obtained.

2.4.6.2   *Standard Reagents, Acids, and Bases.* Strong acids and bases may be stored in the ventilated base of chemical fume hoods, but separation should be provided to prevent cross-mixing in the event of breakage or leakage. Mild acids and bases such as citric acid and sodium carbonate may be stored with other low-hazard reagents. Open shelves for chemicals should be located out of normally traveled routes and have a $\frac{1}{2}$- to $\frac{3}{4}$-in. lip to prevent movement over the edge due to vibration.

2.4.6.3   *Flammable Liquid Storage Cabinets.* Contrary to a commonly held belief, flammable liquid storage cabinets are intended to protect the contents from the heat and flames of an external fire rather than to confine burning liquids within. Flammable liquids should be kept in UL-approved flammable liquid storage cabinets. The cabinets should be remote from other operations within the laboratory that could become involved in a fire. In addition, they should not be located where they could impede access to an exit in case of fire. Whether or not flammable liquid storage cabinets should be ventilated depends on several factors, and opinions are mixed. When a flammable liquid cabinet must be located under a chemical fume hood (although this is not desirable), it should be provided with minimal ventilation by being connected to the hood exhaust system through a flash arrestor. Typically, an exhaust connection is made into the back of the hood with one or two $1\frac{1}{2}$-in.-diameter pipes that extend from the underhood storage space through the work surface and into the hood plenum. When a storage cabinet for flammable liquids is to be located in some other part of the laboratory, remote from a chemical fume hood, and it is feared that it may generate flammable vapors and malodors from spills, it should be ventilated at a rate of three to five air changes per hour. Another plan is simply to leave the vent ports open with the flash arrestors in place and let the room exhaust handle whatever vapors escape from the cabinet. Each of the methods cited is legally acceptable. The concern over the use of forced ventilation inside the cabinet is a fear of overcoming the effect of the flash arrestors in protecting the contents of the cabinet from an exterior fire.

2.4.6.4   *Special   Chemicals.* Especially hazardous chemicals, and chemicals, biological agents and radioactive materials requiring security control, should be stored in locked cabinets located within the laboratory or in a central chemical storage room. The locked storage cabinets should be adequate in size and contain internal separations to provide for the storage of incompatible chemicals such as perchloric acid and cyanide compounds. Lecture bottles and full-sized cylinders of compressed gases should be stored in a ventilated storage area. Mechanically ventilated hood bases, and other types of vented cabinets, are suitable for this purpose.

2.4.7 Compressed Gas Cylinder Racks

When full-sized compressed gas cylinders are required inside a laboratory, special facilities must be provided for their safe use. They include strapping and anchoring devices, adequate room ventilation to remove leaking gas, and easy accessibility for periodic exchange of cylinders. The area should be large enough to accommodate an extra tank of each gas in use (or at least the most used gases) with adequate room for empties awaiting collection. The direction of ventilation airflow should be out of the room and building immediately after passing the gas storage area. Local piping systems for gas cylinders located in the laboratory should meet the same criteria cited in Section 1.4.8. Local spot exhaust ventilation can also be used to satisfy gas cylinder exhaust ventilation requirements. Special gas cabinets should be used for particularly hazardous gases. See chapters 8 and 19 for more information.

2.4.8 Safety for Equipment

2.4.8.1 Safe Equipment Locations. All equipment installed inside and outside laboratories should be located in such a manner that its failure will not involve other pieces of vital equipment, block egress routes, or create situations that overwhelm the capabilities of the ventilation and sprinkler systems (see also Sections 1.3 and 1.4.4). Very heavy pieces of equipment must not be placed or moved through areas where floor loading capacity has not been determined by informed study to be adequate for the imposed load.

2.4.8.2 Material Locations. Secure locations should be established for materials that could become involved in creating or worsening emergency situations. They include trash and waste baskets, waste chemicals, stored chemicals, and normal combustibles, such as supplies of single-use disposable laboratory materials.

2.5 SPECIAL SERVICES

Special utility services, such as high electrical demand, high-pressure steam, and hydraulic systems, must be installed in accordance with applicable codes and standards when these are available. When experimental use of special services is not covered by standards or codes, a review team of experts is needed to generate guidelines for installation and use of special services to substitute for absence of codes and standards. This is an important safety provision. The more the special laboratory services deviate from those normally provided to all laboratories, the greater the need is to seek review and guidance on the equipment and procedures. The review team should include members of the industrial safety and industrial hygiene professions, building maintenance personnel, the user, and the building design team leaders.

Special support services for laboratories include the following: darkrooms; a variety of shops, such as machine, wood and plastics, glass-blowing, and maintenance shops; chemicals and gas cylinder storage; hazardous, radiation, and biological waste storage; glass-washing and cage-washing facilities; and general laboratory storage. Part III of this book reviews several laboratory support services in the same format as the laboratory types.

PART II

DESIGN GUIDELINES FOR A NUMBER OF COMMONLY USED LABORATORIES

The following chapters address safety and health issues for a number of well-defined laboratory types. Laboratory type is determined largely by the hazardous properties and quantities of the materials and equipment normally used, the work activities performed, and any special requirements of the laboratory that may affect safety and health adversely inside or outside the laboratory. It is of extreme importance that a specific laboratory type be selected in collaboration with laboratory users plus safety and industrial hygiene advisors. After the laboratory type has been selected, all of the issues discussed in the chapter of this manual dealing with that special laboratory type should be evaluated and implemented in the design stages.

The items discussed under Common Elements of Laboratory Design (Part I) apply to all laboratories except when specifically excluded in the chapter on that specific laboratory type, and then an alternative will be given. In each of the specific laboratory types, special requirements unique to that laboratory will be addressed. They may supplement or supersede the requirements of the Common Elements section. It is recognized that for renovation projects it may not be possible to comply with all requirements due to constraints imposed by the existing facility, but the intent of the safety and health recommendations should be observed carefully when making compromises, and safety and industrial hygiene personnel should be consulted for professional advice.

A matrix showing the major safety and health considerations that should be addressed for each laboratory type appears in Part VI (Appendix V). It should be used as a design refresher and as a quick overview. It should not be used to replace the more detailed information contained in the text.

CHAPTER 3

GENERAL CHEMISTRY LABORATORY

3.1 DESCRIPTION OF GENERAL CHEMISTRY LABORATORY

3.1.1 Introduction

A general chemistry laboratory is designed and constructed to provide a safe and efficient workplace for a wide variety of chemical activities associated with analysis, quality control, and general chemical experimentation. Preferably, this type of laboratory will be located in a building containing mainly similar laboratory units, but not necessarily doing general chemistry work. For example, these laboratories do not belong in office buildings.

3.1.2 Work Activities

General chemistry laboratory activities include mixing, blending, heating, cooling, distilling, evaporating, diluting, and reacting chemicals as part of testing, analyzing, and research experiments. Most of this work will be conducted on a bench or in a laboratory chemical fume hood.

3.1.3 Equipment and Materials Used

Stills, extraction apparatus, reactor vessels, heaters, furnaces, evaporators, and crystalizers are standard items of equipment found in general chemistry laboratories. Many analytical devices (e.g., atomic absorption spec-

Guidelines for Laboratory Design: Health and Safety Considerations, Second Edition
By Louis DiBerardinis, Janet S. Baum, Gari T. Gatwood, Anand K. Seth, Melvin W. First, and Edward F. Groden.
ISBN 0-471-55463-4   Copyright © 1993 by John Wiley & Sons, Inc.

trometers, gas chromatographs, spectrophotometers) are commonly used in these laboratories as auxiliary equipment. Typical equipment that should be considered for placement includes gas cylinders, ovens, stills, experiment frames, vibration-sensitive balances, chemicals, glassware, and cryogenics.

Hazardous materials used in general chemistry laboratories include small quantities of chemicals of high toxicity, volatile liquids, dusts, compressed gases, and flammables.

3.1.4 Exclusions

General chemistry laboratories are not designed for handling extremely hazardous chemicals or performing especially hazardous operations. Hazardous operations not recommended for general chemistry laboratories include, but are not limited to, the use of (1) carcinogenic, mutagenic, or teratogenic chemicals, (2) highly explosive materials in greater-than-milligram quantities, (3) high-tension voltage and high-current electrical services, (4) radiofrequency generators and all electrical operations with a high potential for fire, explosion, or electrocution, (5) lasers of over 3-mW output power with unshielded beams, (6) gas pressures exceeding 2500 psig; (7) liquid pressures exceeding 500 psig, and (8) radioactive materials in greater than 1-μCi amounts. The general chemistry laboratory usually needs no special access restrictions except when specialized or highly sensitive equipment and toxic materials are to be used. If such operations are to be carried out routinely, the designation of the laboratory as a general chemistry laboratory should be reevaluated.

3.2 LABORATORY LAYOUT

The most important consideration in the layout of a general chemistry laboratory is safely locating the number and sizes of chemical fume hoods required.

In laboratories where multiple fume hoods are needed, a side-by-side arrangement at the rear of the laboratory should be considered. If only two fume hoods are needed, one hood on each side wall toward the rear of the laboratory would be acceptable.

A suggested layout for a general chemistry laboratory is provided in Figure 3-1. All the items described in Sections 1.2 and 2.2 should be reviewed, and those that are relevant should be implemented.

3.3 HEATING, VENTILATING, AND AIR CONDITIONING

All the items described in Sections 1.3 and 2.3 should be reviewed, and those that are relevant should be implemented.

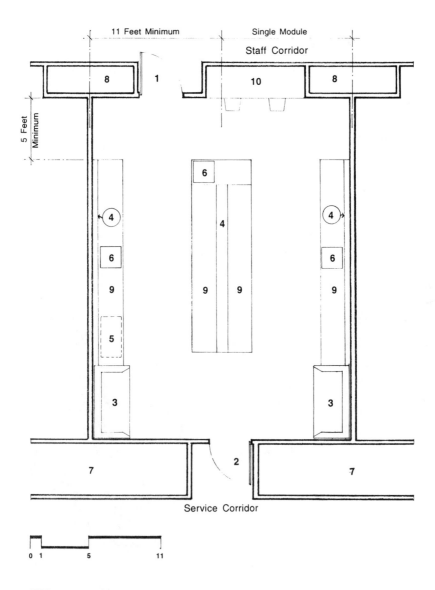

FIGURE 3-1. General chemistry laboratory: Sample layout.

3.4 LOSS PREVENTION, INDUSTRIAL HYGIENE, AND PERSONAL SAFETY

All the provisions in Section 1.4 and 2.4 should be reviewed and implemented.

3.5 SPECIAL REQUIREMENTS

None.

CHAPTER 4

ANALYTICAL CHEMISTRY LABORATORY

4.1 DESCRIPTION OF ANALYTICAL CHEMISTRY LABORATORY

4.1.1 Introduction

An analytical chemistry laboratory is designed, constructed, and operated to provide a safe and healthful work area for activities associated with the analysis of a wide variety of chemicals in amounts ranging from nanograms to kilograms. Many analytical procedures call for handling moderate amounts of hazardous chemicals, including petroleum solvents, explosive gases, and all types of toxic substances.

4.1.2 Work Activities

The activities performed in analytical chemistry laboratories include sample preparation involving mixing, blending, ashing, and digesting. The major activity is to analyze samples, and this requires the operation of a variety of analytical instruments, some of which contain radioactive sources and many of which utilize or produce hazardous radiations such as laser illumination, ultraviolet, infrared, and microwaves.

Guidelines for Laboratory Design: Health and Safety Considerations, Second Edition
By Louis DiBerardinis, Janet S. Baum, Gari T. Gatwood, Anand K. Seth, Melvin W. First, and Edward F. Groden.
ISBN 0-471-55463-4   Copyright © 1993 by John Wiley & Sons, Inc.

4.1.3   Equipment and Materials

The many kinds of analytical equipment utilized in this type of laboratory include spectrophotometers, gas and liquid chromatographs, nuclear magnetic resonance imaging instruments, scintillation counters, mass spectrometer, balances, microscopes, stills, extraction apparatus, ovens, and furnaces. Heavy use of some of this equipment may generate a significant heat load in the laboratory.

Hazardous materials used in analytical chemistry laboratories include small quantities of chemicals of high toxicity, volatile liquids, dusts, compressed gases, and flammables. Toxic materials may be reacted or decomposed into nontoxic compounds during the analytical procedures, but usually they remain in a toxic state during analytical manipulation. Occasionally, nontoxic components may react to produce hazardous reaction products, but this is not usual.

4.2   ANALYTICAL CHEMICAL LABORATORY LAYOUT

The layout of an analytical chemical laboratory is similar to that of a general chemistry laboratory (see Figure 3-1). Each item addressed in Sections 1.2 and 2.2 should be evaluated for its applicability to the specific needs of those who will use the analytical chemical facility, and items that are relevant should be implemented.

Because it is likely that many pieces of analytical equipment of substantial size will be present in an analytical chemistry laboratory at all times, special care should be given to their location relative to egress routes, ventilation patterns, and the interactions of laboratory personnel with chemical handling operations. Safety and industrial hygiene personnel should be consulted for assistance with special problems of hazardous chemical handling, magnetic fields, and laboratory ventilation.

4.3   HEATING, VENTILATING, AND AIR CONDITIONING

4.3.1   Introduction

All the items described in Sections 1.3 and 2.3 should be reviewed, and those that are relevant should be implemented.

4.3.2   Additional HVAC Needs

Special consideration should be given to providing good coverage of the laboratory with local exhaust systems. The effluent gases from some analytical devices, such as gas chromatographs and atomic absorption spectrometers, often contain toxic chemicals that need to be controlled at the source of generation by being vented to the out-of-doors. In some cases, instrument manufacturers provide recommendations specific to their instruments. Otherwise, the Industrial Ventilation Manual (ACGIH, 1992) or a qualified industrial hygienist should be consulted for advice. If perchloric acid use is anticipated, the safety measures described in Section 2.3.4.4.3 should be complied with.

The heat-producing capabilities of each of the analytical instruments and auxiliary equipment should be evaluated when estimating heating and air-conditioning requirements.

4.4   LOSS PREVENTION, INDUSTRIAL HYGIENE, AND PERSONAL SAFETY

4.4.1   Introduction

All of the items described in Sections 1.4 and 2.4 should be evaluated for their applicability to the specific analytical laboratory under consideration, and particular attention should be given to the location of analytical equipment in relation to egress routes and interaction with other equipment or chemical handling operations. Safety and industrial hygiene personnel should be consulted when additional advice is needed.

4.5   SPECIAL REQUIREMENTS

None.

CHAPTER 5

HIGH-TOXICITY LABORATORY

5.1 DESCRIPTION OF HIGH-TOXICITY LABORATORY

5.1.1 Introduction

A high-toxicity laboratory is designed and operated to provide safe use of highly toxic chemicals, including the use of carcinogens, mutagens, and teratogens. Industrial hygiene and safety personnel should be consulted for assistance in determining whether the nature and quantity of chemicals that will be used in a proposed laboratory fall into this category. In January 1990, OSHA promulgated 29 CFR 1910.1450, a standard regulating exposure to chemicals in laboratories (OSHA, 1990). "Particularly hazardous chemicals," defined in this standard, may require special handling procedures or design features, although not all chemicals that meet the definition of particularly hazardous must always be used in a high-toxicity laboratory. The decision depends on amounts and manner of use. Many chemicals used in a high-toxicity laboratory will meet the OSHA definition of particularly hazardous. Sources of assistance to determine whether the quantity of hazardous materials used will require a laboratory designation of high toxicity include local OSHA and NIOSH offices and State Departments of Occupational Health and Industrial Hygiene. The National Institute of Health has published a "Suspected Carcinogens" listing and an internal guideline document entitled "Guideline for the Laboratory Use of Chemical Carcinogens" (NIH,

Guidelines for Laboratory Design: Health and Safety Considerations, Second Edition
By Louis DiBerardinis, Janet S. Baum, Gari T. Gatwood, Anand K. Seth, Melvin W. First, and Edward F. Groden.
ISBN 0-471-55463-4   Copyright © 1993 by John Wiley & Sons, Inc.

1981) that provides specific information on carcinogens. Not all of the design considerations included in this chapter may be needed in every high-toxicity laboratory. Close communication between all involved in the planning process will be necessary to determine specific requirements.

5.1.2   Work Activities

The basic experimental procedures used in high-toxicity laboratories are similar to those conducted in general chemistry and analytical chemistry laboratories, but provisions should be made for the additional safety procedures that will be required when handling highly toxic chemicals in more-than-microgram quantities. Although this section describes a laboratory that is similar to a general chemistry laboratory (Chapter 3), the design and operation of the safety provisions will be much more critical. The safety guidelines outlined here can be applied to other laboratory types such as chemical engineering and physics laboratories, whenever they use highly toxic materials in quantities that do not exclude their use in such laboratories.

5.1.3   Equipment and Materials Used

The equipment used in a high-toxicity laboratory will vary depending on the nature of the work, but, in general, the equipment will be similar to that found in a general chemistry or analytical chemistry laboratory. Many of the chemicals used in this laboratory will fall into the category of "particularly hazardous" as defined by OSHA in 29 CFR 1910.1450 (e)(3)(viii). They include "select carcinogens, reproductive toxins, and substances having a high degree of toxicity." Select carcinogens include (1) chemicals regulated by OSHA as carcinogens, (2) chemicals listed as "known carcinogens" by the National Toxicology Program (latest edition of the Annual Report on Carcinogens), (3) Group 1 chemicals (carcinogenic to humans) listed by the International Agency for Research on Cancer (IARC), and (4) chemicals causing a statistically significant tumor incidence in experimental animals as defined in 29 CFR 1910.1450 (b). There is considerable overlap of chemicals in the four listings, but they are not identical.

5.1.4   Exclusions

Excluded from this chapter are the use of ionizing radiation, biological agents, and animals. Their use with high-toxicity chemicals requires additional design features that are addressed in Chapters 11, 12, and 18, respectively.

5.1.5 Special Requirements

The nature of the materials used makes it necessary for the high-toxicity laboratory to have special access restrictions. OSHA has stringent requirements that must be followed for specific chemicals (OSHA, 1990). More recent OSHA regulations must be consulted to determine if important changes have been promulgated since the cited edition. In some cases, provisions must be made for change rooms and showers. Industrial hygiene and safety personnel should be consulted for advice when high-toxicity laboratories are to be built or when existing laboratories are to be converted to this use.

5.2 LABORATORY LAYOUT

5.2.1 Introduction

Many types of laboratory layouts are possible, depending on the specific nature of the work to be performed and the space available. A layout similar to that of a general chemistry laboratory (Chapter 3) is often adequate. Specific work-space utilization layouts are described in *Safe Handling of Chemical Carcinogens, Mutagens, Teratogens, and Highly Toxic Substances* (Walters, 1980). Using either the NIH, Walters, or OSHA reference source as a starting point, all the recommendations for safety and health contained in Part I should be reviewed and those that are relevant should be implemented. In addition, the following items should be considered for their applicability to the laboratory work to be performed. Industrial hygiene and safety personnel, as well as appropriate state and federal agencies, may have to be consulted early in the planning phase because adoption of some of the recommendations contained in Part I may depend on the specific nature and quantity of the chemicals to be used in relation to applicable regulations for their use and safe disposal.

5.2.2 Change, Decontamination, and Shower Rooms

Adequate facilities must be provided for laboratory workers to change and shower because procedural requirements for working with highly toxic materials necessitate the use of frequent changes of protective clothing. For use of toxic materials, some OSHA requirements specify that a shower must be included as an essential part of a high-toxicity laboratory; use of other toxic materials requires merely the availability of a shower in the building. Traffic flow into change and shower rooms should be designed so that there are separate clean and dirty pathways to and from the facility and there is no way to bypass the shower on the way out of the high-toxicity laboratory. (See Chapter 12 for more information on this topic.)

5.2.3 Work Surfaces

Work surfaces should be constructed from impervious and easily cleanable materials such as stainless steel. Strippable, epoxy-type paint is acceptable. Use of disposable bench coverings during work should be considered as an added safety practice.

5.2.4 Floors and Walls

Floor coverings should be of monolithic (seamless) construction and utilize materials such as vinyl or epoxy, which are impervious to most chemicals and easily formed into seamless sheets. All cracks and construction seams in floors, walls, and ceilings should be sealed with epoxy or another chemically resistant, long-lived sealant. Utility conduits should be epoxy-sealed wherever they penetrate floors, ceilings, and walls. All laboratory lighting should be sealed with similar materials to be vaporproof and waterproof.

5.2.5 Handwashing Facilities

There should be readily accessible handwashing facilities located within the high-toxicity laboratory, as well as in change and shower rooms.

5.2.6 Access Restrictions

All entrances to a high-toxicity laboratory should be posted with permanent signs indicating restricted access due to the use of specific classes of chemicals (e.g., carcinogens, mutagens). The use of special key access should also be considered when a security breach is liable to result in serious illness or dispersal of high-toxicity materials outside the laboratory.

Isolation of the laboratory by tightly sealing doors makes it desirable that there be a viewing window in each door. For the same reason, when flammable solvent usage is heavy, large blowout windows in the exterior walls are recommended. See NFPA 68 for the appropriate sizing calculations of the blowout areas.

5.3 HEATING, VENTILATING, AND AIR CONDITIONING

5.3.1 Introduction

All of the items described in Section 2.3 should be reviewed, and those that are relevant should be implemented. Additional recommendations are given below. Industrial hygiene personnel should be consulted for assistance when safety and health situations not covered in this book are encountered.

5.3.2 Laboratory Fume Hood

An average face velocity of 100 fpm is recommended. We do not recommend that it be increased to 150 fpm as an added safety feature for work with highly toxic materials because several investigators have indicated that the higher face velocity does not necessarily provide added protection (Chamberlin, 1978; Ivany, 1989, DiBerardinis, 1991). However, it should be noted in this connection that in some cases ACGIH and NIOSH recommend a higher exhaust volume. Therefore, this recommendation should be tempered by reference to all current and applicable local, state, and federal codes and regulations.

5.3.3 Glove Box

Some highly toxic materials require the use of a completely enclosed, exhaust-ventilated work space rather than a conventional laboratory fume hood. In these instances a glove box is required. It should meet the specifications defined by the American Conference of Governmental Industrial Hygienists (ACGIH, 1992), as outlined in Chapter 31.

Glove boxes should be maintained as a closed system at all times and kept under a negative pressure of 0.25 in. w.g. They should be thoroughly decontaminated prior to exhaust airflow being shut down, to avoid loss of toxic contents to the laboratory.

5.3.4 Spot Exhaust for Instruments

Instruments used to weigh, manipulate, and analyze highly toxic chemicals should have spot exhaust ventilation at each potential source of contaminant release, or be completely enclosed in an exhaust-ventilated enclosure. Specific design requirements will vary with each type of equipment and chemical used. Consultation with the manufacturer of the equipment and an industrial hygiene engineer is recommended when the design and application of exhaust ventilation facilities is not obvious. See Section 31.8 (Chapter 31) for more details.

5.3.5 Storage Facilities

All facilities used for storage of highly toxic materials, such as cabinets and refrigerators, should be provided with exhaust ventilation to maintain airflow in an inward direction and prevent buildup of toxic contaminants within the storage space. Slot ventilation around refrigerator doors can be very effective. Specially designed storage cabinets are available that require little airflow (DiBerardinis, 1983). These can be used for smaller quantities of volatile materials.

5.3.6 Filtration of Exhaust Air

The air exhausted from fume hoods, glove boxes, and spot exhaust hoses should be decontaminated before release to the environment. The first cleaning stage should be a HEPA filter with a minimum efficiency of 99.97% for 0.3-μm particles when toxic aerosols are present. The second stage should be an activated charcoal adsorber when toxic vapors are present. The size will depend on the total quantity of airflow. All replaceable components should be capable of being changed without exposure of maintenance personnel (e.g., bag-in, bag-out procedures). For some chemicals, an adsorbent other than activated charcoal may be necessary or more desirable. In case of doubt, an industrial hygienist should be consulted.

5.3.7 Directional Airflow

Air infiltration should always flow from uncontaminated to contaminated areas, that is, from corridors to change and decontamination rooms, and, finally, to the high-toxicity laboratory itself. Flow direction should be monitored by appropriate devices consisting of differential pressure sensors equipped with audible and visual alarms to warn of upset conditions.

5.4 LOSS PREVENTION, INDUSTRIAL HYGIENE, AND PERSONAL SAFETY

5.4.1 Introduction

All the items described in Section 2.4 should be reviewed, and those that are relevant should be implemented.

5.4.2 Protection of Laboratory Vacuum Systems

Laboratory vacuum systems should be protected from contamination by installation of traps containing disposable HEPA filter and/or activated charcoal adsorbent systems.

CHAPTER 6

PILOT PLANT (CHEMICAL ENGINEERING LABORATORY)

6.1 DESCRIPTION

6.1.1 Introduction

A pilot plant, or chemical engineering laboratory, is designed, constructed, and operated to provide a safe and healthful work area for activities associated with the handling of substantial quantities of toxic chemicals, petroleum fuels, compressed gases, and other hazardous materials for chemical processing experimentation. A special characteristic of a pilot plant, in addition to large floor area and multistory height, is that materials are usually handled in large quantities (gallons and pounds) as opposed to the small quantities (milliliters and grams) used in most other types of laboratories. Because of the frequent use of flammable and explosive chemicals, pilot plants should be isolated from public areas, other laboratories, and office spaces by distance, special fire protection, and/or explosion-resistant construction (Jones, 1966). Frequently, special access restrictions must be imposed when inherent hazards are associated with the specialized materials and equipment being used.

Guidelines for Laboratory Design: Health and Safety Considerations, Second Edition
By Louis DiBerardinis, Janet S. Baum, Gari T. Gatwood, Anand K. Seth, Melvin W. First, and Edward F. Groden.
ISBN 0-471-55463-4   Copyright © 1993 by John Wiley & Sons, Inc.

6.1.2 Work Activities

The activities performed include mixing, blending, heating, cooling, distilling, filtering, absorbing, crystallizing, evaporating, grinding, size separating, and chemical reacting as a part of production or purification of a product. Some of the procedures and materials require the special ventilation capabilities that will be discussed in Section 6.3. A pilot plant usually requires a more or less permanent crew with special training that includes instruction in safety and health protection. Protective clothing and respirators must be available for all personnel working in the pilot plant.

6.1.3 Equipment and Materials Used

Analytical instruments and a full range of sensors and automatic process controllers will usually be present, in addition to large items of process equipment. Extremely hazardous materials that are sometimes used in pilot plants include chemicals of high toxicity, biological agents, volatile liquids, combustible dusts, and highly reactive or explosive materials. Some radioactive materials may also be used in the pilot plant. Operations often involve the use of high voltages, very-high-radiofrequency generators, and other electrical equipment with a high potential for fire, explosion, and electrocution. High-pressure steam, air, and special gases are employed frequently, as are open flames, furnaces, and similar intense heat generators that make the heat load in pilot plants a special ventilation concern.

6.1.4 Special Requirements

Training of all personnel in the processes to be conducted in a pilot plant, including an understanding of first aid, emergency operations, ordinary pathological and hazardous waste disposal, and the use of emergency equipment, should be mandatory before operations begin.

6.2 PILOT PLANT LAYOUT

Because pilot plant operations involve a large variety of equipment and operations arranged in a constantly changing pattern, it is not possible to illustrate a comprehensive layout. All of the items addressed in Sections 1.2 and 2.2 should be evaluated for their applicability, and safety and industrial hygiene personnel should be consulted for further assistance when unusual requirements are encountered. Additional sources of information include OSHA, NIOSH, and State Departments of Occupational Health, Safety and Industrial Hygiene, in addition to pertinent reference materials. Utilities provided in pilot plants, usually from a number of well-distributed locations, include: high-amperage single- and three-phase electrical current of 120, 240, and 440 V; compressed air to 100 psig; vacuum of $\frac{1}{2}$ atm or lower; steam at least up to 15 psig; hot and cold water; and multiple floor drains leading to waste collection and handling facilities.

6.3 HEATING, VENTILATING, AND AIR CONDITIONING

6.3.1 Introduction

Pilot plant ventilation is needed to provide an environment that is within acceptable comfort limits and to provide a reasonable capacity for diluting contaminants released into the work environment. Air change rates for chemical processing should be 5 cfm per square foot of floor area. For petroleum processes, 3 cfm per square foot is adequate. Ventilation must be provided in a manner that will not contaminate other areas of the building either by infiltration through low-pressure pathways or by contamination of air intakes with pilot plant exhaust air.

6.3.2 Additional Requirements

All of the items described in Sections 1.3 and 2.3 should be examined, and all that apply to pilot plants should be implemented. Special provisions for pilot plant local exhaust systems follow.

6.3.2.1 Local Exhaust Systems. Pilot plants should be provided with fixed general laboratory ventilation systems designed to supply air at ceiling level and exhaust it from floor level. In addition, multiple local exhaust outlets should be provided from a perimeter system of main ducts by the use of quick-disconnect types of fittings and flap-type dampers that automatically close the connections when not in use. The use of such a local exhaust system is capable of serving the entire pilot plant area at the same time that it limits the amount of ventilation air exhausted, thereby conserving energy. Care in the design of these systems is necessary because a static pressure of at least 2.5 in. w.g. at each exhaust point will be required to assure adequate airflow capacity.

Typical equipment found in the pilot plant that may need some form of local exhaust includes mixing and reaction vessels, distillation columns, fermentation units, ovens, filtration apparatus, analytical equipment, and centrifuges. The exhaust air quantity and hood configuration will depend on the particular process and equipment and the generation points of the contaminant. The exhaust hoods may range from an enclosure such as a laboratory fume hood or glove box to a capture or exterior hood such as a canopy or slot hood. Some examples are provided in Chapter 31. General design guidelines in Chapter 3 of the ACGIH *Industrial Ventilation Manual* (ACGIH, 1992) should be followed.

6.3.2.2 *Temperature and Humidity Control.* When necessary, the pilot plant laboratory should be heated, cooled, humidified, or dehumidified. Low humidity levels that may lead to high static electrical charge are not desirable.

Air conditioning of the entire space may not be necessary. Local cooling systems may be adequate for personal comfort.

There may be process equipment that requires close temperature control for operation. For this purpose, a separate system should be provided; usually water-cooled D-X equipment is used. Water-to-waste condenser cooling systems are wasteful of water, suggesting that air-cooled D-X systems be used when appropriate.

6.4 LOSS PREVENTION, INDUSTRIAL HYGIENE, AND PERSONAL SAFETY

All the items described in Sections 1.4 and 2.4 should be evaluated for relevance, and all that apply to pilot plants should be implemented.

6.4.1 Spill Containment

Due to the size of operations in this laboratory, larger quantities of chemicals and other materials may be necessary. Storage containers and locations for them should be carefully selected to provide hazard separation and isolation. Spill dikes may be necessary, and large quantities of spill control materials may need to be stored. Storage of these materials should be outside of the laboratory yet not isolated from it in time of need. Floor drains should have closures that are normally in place to prevent a release of hazardous materials from entering the drain or sewer system.

6.4.2 Electrical Considerations

Consideration should be given to the use of explosion-proof rated electrical wiring and fixtures for all standard room equipment such as lighting, outlets, and switches. This would facilitate the use of highly flammable material in the pilot plant.

6.5 SPECIAL REQUIREMENTS

Unusually high ceilings combined with the use of bulky processing equipment, such as large tanks, in a constantly changing pattern make provision of uniform lighting of good quality and adequate intensity difficult. Insofar as possible, ceiling lighting should be provided by many closely spaced fixtures to avoid the heavy shadows cast by large equipment when widely spaced, high-intensity light sources are used.

Provisions should be made for the installation of an air-supplied respiratory protection system. It should be a separate and protected compressed air system supplied by an oil-less compressor.

CHAPTER 7

PHYSICS LABORATORY

7.1 DESCRIPTION

7.1.1 Introduction

Research carried out in a physics laboratory may include the use of electricity (high current, voltage, and frequency levels), many chemicals (solid, liquid, and gaseous), intense light sources (lasers with greater than 3-mW output power), magnetics, cryogenics, and a variety of high-energy systems including high-temperature steam and compressed air. Research may be carried out with radioactive materials to produce ionizing radiations, but consideration of their use is not included in this chapter. Chapter 11 contains a description of laboratories utilizing more than trivial amounts of radioactive materials.

7.1.2 Work Activities

The basic procedures carried out in physics laboratories include experimental development of mechanical, electrical, hydraulic, and pneumatic systems and examinations of the properties of matter. Operations involve equipment setups and physical measurements for experiments that involve observation, data collection, and analysis.

Guidelines for Laboratory Design: Health and Safety Considerations, Second Edition
By Louis DiBerardinis, Janet S. Baum, Gari T. Gatwood, Anand K. Seth, Melvin W. First, and Edward F. Groden.
ISBN 0-471-55463-4   Copyright © 1993 by John Wiley & Sons, Inc.

7.1.3 Equipment and Materials Used

Equipment used in physics laboratories is varied and depends on the nature of the work. A partial list of physics laboratory experiments and equipment includes the following:

Shock tube studies (air compressors, pressure-relief diaphragms, pressure and gas-flow measuring instruments).

Lasers (electrical circuits, cryogenic liquids).

Spectroscopy (carbon arc source generators, magnets, photography facilities).

Cryogenics (refrigeration equipment, liquified gases, low-temperature measuring instruments).

Electromagnetics (high electrical current services, cryogenic liquids, ion source generators).

High-frequency noise and electricity research (high-current and high-voltage electrical services).

Energy storage systems (rotary machines, heat exchangers, electrical condensers, temperature measuring instruments).

Ionizing radiation systems (x-rays, high-current and high-intensity electrical services, and ionizing radiation measuring instruments).

High vacuum systems.

7.2 LABORATORY LAYOUT

A physics laboratory should be laid out to provide easy access to all areas within, and associated with, the laboratory by wheeled trucks, dollies, cranes, and other materials handling equipment. Services, such as electrical, should have easy access for safe modification of experimental setups. Because a physics laboratory can be used for so many different types of research, equipment configurations are likely to be different from experiment to experiment. For this reason, all services should be flexible enough to serve all parts of the laboratory conveniently. A typical layout is shown in Figure 7-1, within a two-module area. Physics laboratories are often larger than two modules.

All the items reviewed in Sections 1.2 and 2.2 that are applicable to physics laboratories should be implemented. An additional recommendation is noted below.

7.2.1 Egress Routes for Physics Laboratories

In the typical physics laboratory layout illustrated in Figure 7-1, the two required separate egress routes are shown. The doors should be large enough and the route width sufficient to accommodate medical stretchers and other

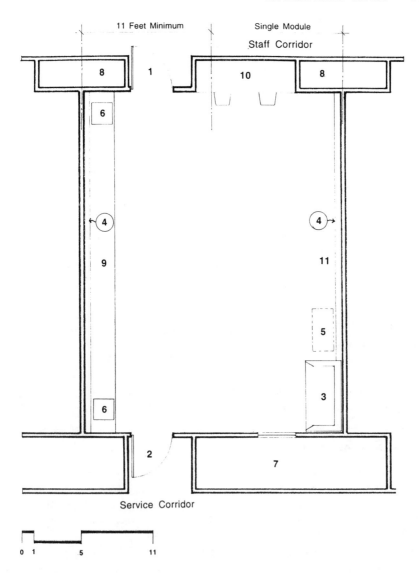

FIGURE 7-1. Physics laboratory: Sample layout.

emergency equipment. They should also be adequate for materials handling equipment used for transporting heavy and bulky items. For this reason, laboratory aisles may need to be wider than the minimum recommended in Chapter 2.

7.2.2 Furniture Location

In laboratories utilizing unshielded laser beams, desk chairs and seated work surfaces (heights of 30–32 in.) are not recommended. Laser beams are often directed and optically transferred at seated eye level. Desks and seated computer stations are better provided in separate rooms.

7.3 HEATING, VENTILATING, AND AIR CONDITIONING

All the items contained in Sections 1.3 and 2.3 that are applicable to physics laboratories should be implemented.

7.4 LOSS PREVENTION, INDUSTRIAL HYGIENE, AND PERSONAL SAFETY

All the items contained in Sections 1.4 and 2.4 that are applicable to physics laboratories should be implemented. Additional recommendations are noted below.

7.4.1 Emergency Eyewash Stations

When chemicals hazardous to face and eyes, such as strong acids, alkalies, and other corrosive materials, are used in a physics laboratory, one or more eyewash stations should be provided according to recommendations contained in Section 2.4.1.5. Because of the large amounts of electrical equipment contained in most physics laboratories, plumbed eyewash stations should be located in a hall or some other nearby area outside the physics laboratory proper. Specifications for such a system are in Appendix II.

7.4.2 Fire Detection, Alarm, and Suppression Systems

Fire detection and suppression systems should be carefully planned from the earliest stages of laboratory design because even under the best of conditions, installation costs are very high for this type of laboratory.

7.4.2.1 Fire and Smoke Detection and Alarm Systems. Physics laboratories need fire and smoke detectors that respond to products of combustion (e.g., photoelectric and ionization detectors) or thermal effects (either rate of temperature rise or a final fixed temperature). A combination of ionization and photoelectric detectors is useful to sense either visible or invisible combustion products generated in a fire in a physics laboratory. When the laboratory uses substantial amounts of flammable gases or flammable volatile liquids, fixed temperature or fixed rate of temperature rise detectors are acceptable, reliable, and less expensive, though slower to alarm. When a physics laboratory is in a part of a building without windows to the outside and when light research, such as with lasers, is not being carried out, flame-sensing detectors are also appropriate. All dectectors should be UL- or FM-approved and be tied into a UL- or FM-approved general building alarm system. Section 1.4.4 contains additional information about fire detection systems.

7.4.2.2 Fire Suppression Systems.

7.4.2.2 Fire Suppression Systems. The basic methods of fire suppression in a physics laboratory should be fixed, automatic systems combined with hand-held, easily accessible portable extinguishers. Although sprinklers are considered to be the best fire control device for most laboratories in a research laboratory building, physics laboratories often contain special electrical hazards that should be protected with fire suppression systems other than sprinklers. These include CO_2, Halon 1301, and total-flooding dry chemical systems. An exception to this rule can be made when the laboratory is electrically deenergized upon sprinkler system activation. Whenever water-sensitive equipment is not used and whenever high-voltage electronics or high-current electrical equipment, such as superconducting magnets, are not present, physics laboratories can be protected from fire spread with automatic sprinkler systems.

7.4.2.2.1 Fixed Automatic Extinguishers. Fixed automatic fire extinguisher systems used in physics laboratories should be consistent with the operations anticipated for that laboratory as explained in Section 7.4.2.2. When operations prohibit the use of a water sprinkler system, fixed automatic Halon or similar extinguisher types should be used. Of the various choices, Halon 1301 offers the following advantages: It is not life-threatening because airborne concentrations are usually between 5 and 7%, and it is a clean extinguishing agent that leaves no residue. Such systems must be designed, constructed, and installed in accordance with "Halon 1301 Fire Extinguishing Systems" (NFPA 12A, 1989). Halon 1301 is considered an ozone-depleting chlorofluorocarbon (CFC), and as the world's use of CFC materials is reduced or eliminated altogether, it may no longer be available. Total flooding CO_2 systems should be avoided whenever the enclosure to be protected will be occupied. Dry chemical systems leave a powdery residue that may harm equipment. Although a powder residue is not the primary concern when selecting a fire suppression system, when there are alternatives of equal efficacy, it becomes an important criterion in the selection process.

7.4.2.2.2 Portable Extinguishers. Hand-portable fire extinguishers of adequate size that contain appropriate extinguishing agents for anticipated fires should be located in the laboratory. Appropriate hand-portable units for a physics laboratory are 15-lb CO_2 extinguishers and 2A-40 BC multipurpose dry chemical extinguishers. Places for these extinguishers should be provided within the laboratory where they can be picked up to assist personnel in making an exit from the laboratory. Sizes of multipurpose extinguishers should always be in the range of 2A to 4A and 40 BC to 60 BC. Portable extinguishers should also meet all the requirements of "Portable Fire Extinguishers" (NFPA 10, 1990).

7.4.2.2.3 Special Systems. Wherever a potential for an electrical fire exists within equipment cabinets that can be protected without involving the entire building system, this type of protection should be provided with Halon systems or CO_2 total flooding systems. Such systems may be wired into the building alarm and annunciation system. All of the ventilation shutdown requirements for these special systems should be met, as outlined in "Halon 1301 Fire Extinguishing Systems" (NFPA 12A, 1989) or "Carbon Dioxide Extinguishing Systems" (NFPA 12, 1989).

7.4.3 Special Equipment Requirements

Careful consideration should be given to the potential contribution of cryogenic and special gases, high electrical energy demands, strong laser beams, and the like, to accident, emergency, and stress situations. Design features should be considered to provide adequate barriers against damage from any of these high-intensity energy forms.

7.4.3.1 Equipment Operation with Hazardous Materials and in Hazardous Modes. Equipment that operates on materials that are hazardous (toxic or flammable) should be provided with all of the building design features (such as ventilation and emergency services) that will make it possible to control the hazards in case of unexpected events. Potential problems may be discovered by discussions between users of the laboratory and safety professionals.

7.4.4 Special Safety Requirements

When screen rooms are required to block radiofrequency energy from entering the laboratory area, entrances to these rooms should be interlocked with the electrical power source, while always permitting easy egress for personnel in the event of an emergency.

A grounding grid system should be installed in a physics laboratory to enable grounding of all pieces of equipment that are electrical in nature or that come in contact with equipment that is electrical. The grid system should be extensive enough and of sufficient size to result in only a small potential difference between the two farthest points.

Because water supply and drainage requirements may be high for water cooling of magnets or other devices, design for such systems should be evaluated early in the building planning phase. For economy of operation, a closed-loop-condenser water-cooling system should be considered for all such facilities.

7.5 SPECIAL REQUIREMENTS

None.

CHAPTER 8

CLEAN ROOM LABORATORY

8.1 DESCRIPTION

8.1.1 Introduction

A clean room laboratory is a specially constructed and tightly enclosed work space with modulating HVAC control systems to maintain design standards for (a) a low concentration of airborne particulate matter, (b) constant temperature, humidity, and air pressure, and (c) well-defined airflow patterns. Clean rooms are classified by particle count per cubic foot of air. Six levels of cleanliness (referred to as "classes") are recognized in clean room practice. It is important that the level of cleanliness that is to be maintained be determined from the beginning of the design process because selection of structural materials, HVAC services, air filters, and clean room layout and furnishings will be dictated by the cleanliness class selected. The cleanliness designations derived from Federal Standard 209D (GSA, 1988) are shown in Figure 8-1, where it may be seen that the six class designations relate to the number of 0.5-μm-diameter particles permitted per cubic foot of air in the vacant clean room. In Table 8-1, it may be seen that clean room classes 100,000, 10,000 and 1,000 designate permissible counts for 5.0- and 0.5-μm particles, clean room class 100 designates permissible counts for 0.5, 0.3, and 0.2 μm, and clean room classes 10 and 1 designate permissible counts for 0.5-, 0.3-, 0.2-, and 0.1-μm particles. Although untreated atmospheric dust usually

Guidelines for Laboratory Design: Health and Safety Considerations, Second Edition
By Louis DiBerardinis, Janet S. Baum, Gari T. Gatwood, Anand K. Seth, Melvin W. First, and Edward F. Groden.
ISBN 0-471-55463-4   Copyright © 1993 by John Wiley & Sons, Inc.

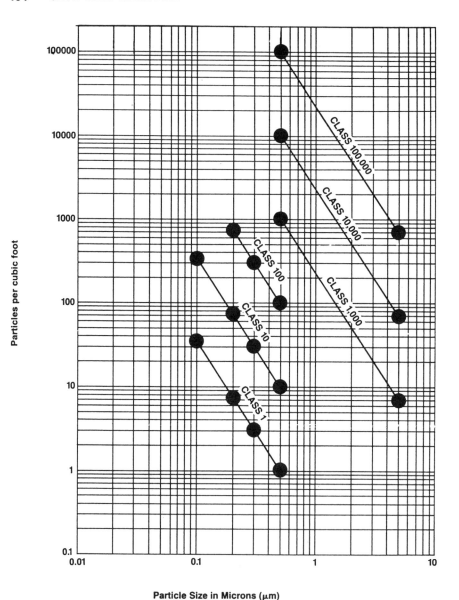

Particle Size in Microns (μm)

FIGURE 8-1. Cleanliness classification.

has a log-normal size distribution, it is not likely that it will remain so after several stages of filtration. Therefore, although Figure 8-1 shows log-normal size distributions for each class, the data points mean that the designated sizes should not be exceeded regardless of the shape of the observed particle size curve. The cleanliness class may be verified by measuring one or more of the sizes designated by a data point in Figure 8-1 or in Table 8-1. Federal

TABLE 8-1. Class Limits in Particles per Cubic Foot of Size Equal to or Greater than Particle Sizes Shown (Micrometers)[a]

Class	Measured Particle Size (Micrometers)				
	0.1	0.2	0.3	0.5	5.0
1	35	7.5	3	1	NA
10	350	75	30	10	NA
100	NA[a]	750	300	100	NA
1,000	NA	NA	NA	1,000	7
10,000	NA	NA	NA	10,000	70
100,000	NA	NA	NA	100,000	700

[a] The class limit particle concentrations shown in this table and Figure 8-1 are defined for class purposes only and do not necessarily represent the size distribution to be found in any particular situation.
[b] NA, not applicable.

Standard 209D contains statistical criteria for dust counting. *Procedural Standards for Certified Testing of Cleanrooms* has been published by the National Environmental Balancing Bureau (NEBB, 1988). Clean rooms designed for particle counts of no more than 10 per cubic foot are becoming common in the microchip industry, and clean rooms designed for no more than one particle per cubic foot of air are considered to be the leading edge of clean room construction. This chapter will deal only with Class 100 clean room laboratories that permit no more than 100 0.5-μm particles per cubic foot of laboratory air.

8.1.2   Work Activities

The activities performed in a clean room laboratory are characterized by a need for extreme cleanliness rather than by the nature of the activities.

Experimentation with, and development of, electronic microchips, minature gyroscopes and switches for guidance systems, pharmaceuticals, and photographic films and film processing techniques are samples of the kinds of work that require ultraclean laboratories for successful operations. Chapter 19 is concerned with clean rooms used as microelectronics laboratories. They differ from this type of clean room in the quantity and toxicity of the materials used. Chemical treatment, precision machining, solvent cleaning of parts and mechanisms, and use of a wide spectrum of precision measuring devices are activities characteristic of clean room laboratories as a class. A clean room laboratory requires all of the special ventilation and air filtration facilities that are described in Section 8.3. The clean room laboratory may also be used for storage of materials that require a high degree of cleanliness as well as temperature and humidity control in accordance with a variety of experimental operating procedures. The laboratory will have access restrictions, and donning of special lint-free clothing will be required for entrance into the clean room laboratory as an aid to maintaining the required low dust level.

8.1.3 Special Requirements

Clean room laboratories may contain multiple rooms, each with different requirements for contamination control. Each room within the clean room laboratory should be maintained at a static pressure higher than atmospheric and higher than that in adjacent indoor spaces to prevent air infiltration from less-well-controlled areas. Differential pressures should be maintained between adjacent rooms of the multi-compartmented clean room laboratory to assure airflow outward from the cleanest spaces to those maintained at a lesser standard of air dustiness. See all of Section 8.3 on this subject. Rooms of simple rectangular solid shape without projections enhance desired airflow patterns and are easier to keep clean.

8.1.3.1 Finish Materials. Clean room laboratories should be constructed of smooth, monolithic, easily cleanable materials that are resistant to chipping and flaking. The interior surfaces should have a minimum of seams and be devoid of crevices and moldings. Walls should be faced with stainless steel or plastic sheeting, or they should be covered with baked enamel, epoxy, or polyester coatings with a minimum of projections. Ceilings should be covered with metal pans, plastic-faced panels, or plastic-finished acoustical ceiling tiles. Floors should be covered with seamless sheet vinyl or with a poured epoxy application that will form a monolithic surface with gently rounded cove base. All of these structural materials should be fire-resistant or treated to acquire this characteristic. Walls and doors should at least be fire-rated 1-h assemblies.

8.1.3.2 Personal Cleanliness. Personnel practices are very important in clean room laboratory operations. To ensure cleanliness, personnel should be provided with lint-free smocks, gloves, shoe covers, and head covers and should have available wash areas with soap or lotions containing lanolin to tighten the skin and thereby reduce sloughing of skin fragments. All equipment and materials should be thoroughly cleaned before being brought into a clean room laboratory.

8.1.3.3 Laminar-Flow Work Stations. It often happens that one or only a few operations require more rigorous dust control than would be provided by the selected cleanliness class of the clean room. Rather than upgrading the entire clean room to accommodate a minor but critical operation, it is possible to use individual work stations located inside the clean room that will provide the required class of dust control locally.

Laminar-flow work stations are self-contained units consisting of a canopied workbench equipped with a fan and HEPA or ULPA (ultralow-penetration aerosol) filters. Filters are placed above the work surface (vertical flow) or in the rear wall of the work space (horizontal flow) and serve to envelop the entire work area in a flow of particle-free air. The interior of the cabinet will be at positive pressure relative to the rest of the clean room, and all of the air introduced into the cabinet will exit through the work opening into the worker's breathing zone. This means that toxic materials must not be used or generated in the horizontal or vertical flow workbench. If toxic products will be present, work stations that provide worker as well as work protection (e.g., class II biological safety cabinets) must be substituted for laminar-flow clean bench work stations. When clean room needs are greater than can be accommodated by a laminar-flow cabinet but not large enough to call for a normal room-size facility, it is sometimes satisfactory to construct a self-contained ''small room'' clean room inside a standard laboratory.

Another method used to upgrade work stations above the clean room rating is to install packaged fan and filter units at strategic locations inside the clean room. Usually, these units are mounted in the ceiling directly over a workbench that requires special protection from contamination. A typical fan and filter ceiling unit is shown in Figure 8-2. Often clear plastic curtains extending from filter to bench top will be used on sides and back to provide additional confinement of the particle-free zone.

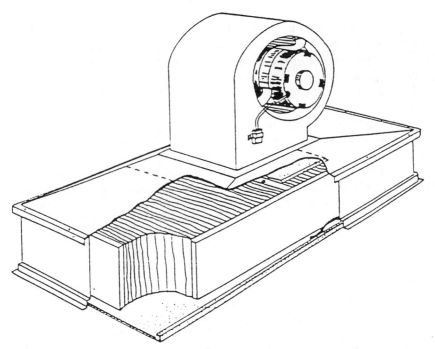

FIGURE 8-2. Fan–filter unit. *Source:* Weber Technical Products Hampstead, New Hampshire.

8.1.3.4 Auxiliary Ventilation Facilities. The high ventilation rates required to maintain air cleanliness may dictate a need for auxiliary fan rooms, and special attention must be given to their design, cleanliness, and maintenance requirements. See ASHRAE (1991) for additional information.

8.2 LABORATORY LAYOUT

8.2.1 Introduction

The layout of a clean room laboratory may be a single room, with a dust-free air-pressurized vestibule, or it may be a laboratory suite of several separate rooms interconnected by an internal, pressurized corridor as shown in Figure 8-3. For emergency evacuation, the internal corridor should connect to a building egress corridor through pressurized or nonpressurized vestibules. A clean-air-pressurized robing room is an essential adjunct of a clean room laboratory. A pressurized vestibule may be used for this purpose if there is adequate space for storage of clothing and installation of a handwashing sink, a shoe cleaner, and/or a sticky shoe mat. Otherwise, a separate robing room must be provided for in the plan. When a robing room separate from a pressurized vestibule is used, it should be at a higher air pressure than the access corridor, but be at a lower air pressure than the vestibule and clean laboratory rooms. Access to the clean laboratory room will usually be through an air-blast chamber placed between the robing room and the laboratory rooms. Its purpose is to blow lint, skin fragments, hair, and so forth, from personnel prior to their entry into the cleanest areas of the suite. The minimum dimension of a clean laboratory room work-space module should be 7 ft 6 in. A typical clean room is shown in Figure 8-3. A vertical section view through a class 100 clean room laboratory is shown in Figure 8-4. It shows a multifilter ceiling, floor-level return air openings, and the HVAC components needed to service the clean room laboratory facility. Very frequently, a clean room laboratory suite will be constructed completely inside another structure such as manufacturing building or a chemistry laboratory building and will draw its makeup air supply from the space surrounding it. Makeup air may also be drawn in from the outside through a prefiltering and tempering unit. The items contained in Sections 1.2 and 2.2 apply to clean room laboratories except when modified in the following sections.

8.2.2 Ingress and Egress

8.2.2.1 Egress Routes. A minimum of two separate emergency egress routes from each clean room laboratory unit is recommended. One of these may be the normal entrance through the robing lock. The exits should be as far apart as possible and lead to different fire zones as a safety measure. In addition, there should be no dead-end corridor longer than 20 ft when mea-

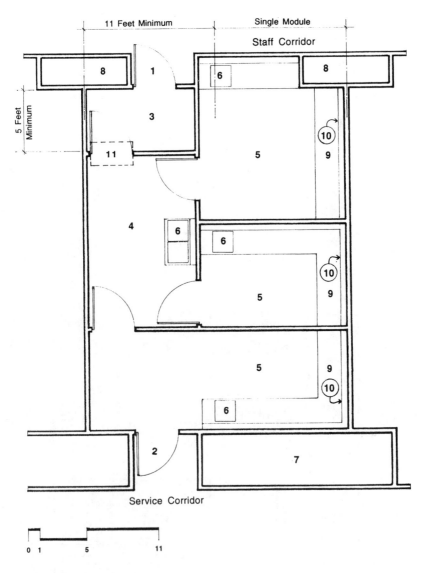

FIGURE 8-3. Clean room laboratory suite: Sample layout.

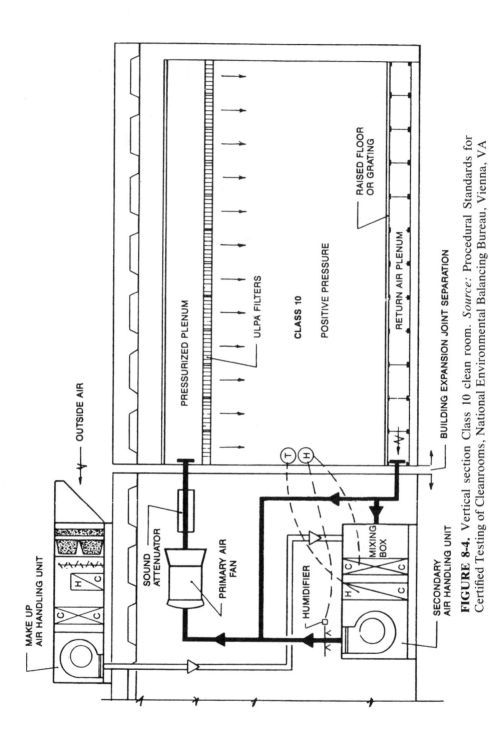

FIGURE 8-4. Vertical section Class 10 clean room. *Source:* Procedural Standards for Certified Testing of Cleanrooms, National Environmental Balancing Bureau, Vienna, VA

sured from the centerline of the door of the farthest room to the door that opens onto a building egress corridor. Check applicable codes thoroughly before deviating from this recommendation. Emergency exits may be so labeled and audibly alarmed to deter use except under emergency conditions. An approved crash bar or "California paddle firelock" hardware will allow emergency egress but limit unauthorized entry. In large clean room complexes, emergency egresses may have to be supplemented with crash panels in walls that abut safe passages to the outside.

8.2.2.2 Traffic Flow. When particulate contamination from outside the clean room laboratory unit is controlled by pressurized anterooms and robing rooms, a one-way flow of traffic from a separate entrance to a separate exit is not essential. However, in situations where contamination generated or contained within the clean room must be prevented from leaving the area, a restricted one-way flow of traffic may be needed. This calls for a place where personnel may change into protective garments, or put them over their street clothes, before they enter a pressurized vestibule giving access to the clean laboratories. A similar space at the exit end of the clean room facility that includes wash basins or showers, and an autoclave for sterilizing outgoing materials and protective garments, may be needed as well. When showers are required, the one-way internal corridor should loop back to the changing room in order for personnel to gain access to their street clothes. The project engineer should discuss with industrial hygiene and safety consultants the degree of cross-contamination to be allowed between the building air supply and the clean room. There are many possible arrangements whereby a series of controlled spaces can be made to meet the various air cleanliness and interlocking air pressure gradient specifications. Biological safety laboratories are discussed in Chapter 12.

8.2.2.3 Interlocking Doors in Ingress and Egress Pathways. It is frequently proposed that access corridors, pressurized vestibules, and robing rooms associated with clean rooms and clean room laboratories be provided with interlocking doors so that personnel will be prevented from opening both doors simultaneously and, by this means, inadvertently introduce contamination into clean areas. This is a very dangerous arrangement because under the stress of a fire, explosion, or other emergency in a clean room laboratory, it is unlikely that an orderly opening and closing of doors will take place. People can become trapped between locked doors in a vestibule or robing room under panic-producing situations. It is acceptable to place visual and audible alarms on double-access doors to alert supervisory personnel to failures in dust-control discipline so they may take prompt corrective measures, but no barriers to emergency egress should be permitted.

8.3   HEATING, VENTILATING, AND AIR CONDITIONING

8.3.1   Introduction

The HVAC equipment that will be needed for a clean room laboratory depends on the clean room class, temperature and humidity requirements, the need for fume hoods and spot exhaust hoses, and the presence of major heat-generating equipment and activities. The items contained in Sections 1.3 and 2.3 apply to clean room laboratories and should be implemented. In addition, the following provisions should be considered for inclusion in building and laboratory plans.

8.3.2   Environmental Control

8.3.2.1   Ventilation. Good ventilation is critical in a clean room. Large volumes of dust-free air and excellent air velocity and directional control are needed to provide the required cleanliness level. The ventilation system for the clean room should be independent of the general building supply system. This is shown in Figure 8-4. When toxic chemicals or infectious microorganisms are to be used, local exhaust or chemical fume hoods will be needed to meet the provisions of Section 2.3. Information on ventilation of toxic gas cylinder storage cabinets has been published (Burgess, 1985) and is included in Chapter 19.

8.3.2.2   Filtration. Filtration is required for all outside supply air (called *primary air*) and for the combined prefiltered outside air plus recirculated air (called *secondary air*) delivered to a clean room laboratory. Primary air should be filtered: first, by a roughing filter capable of removing coarse particles and fibers; second, by a prefilter of 85–95% atmospheric dust efficiency; and third, by a HEPA filter, efficiency rated at 99.97% or higher. Secondary air is usually filtered only through HEPA filters. In a clean room laboratory, the supply air may be delivered to a ceiling plenum containing a continuous bank of HEPA filters that provide filtered air circulation downward to the floor where return air grilles will be located. A fraction of the air will be exhausted to the outside or to the building housing the clean room laboratory, and the remainder will be recirculated through the HEPA filters. When horizontal airflow is desired, a similar arrangement is possible by building a wall of HEPA filters at one end of the room and locating return grilles in the opposite wall. These ventilation plans are illustrated in Figure 8-5.

8.3.2.3   Room Pressure Balance. Pressure control within the clean room should be maintained by static pressure controllers that operate dampers, fan inlet vanes, or a combination of both to maintain the correct ratio of supply to return and exhaust air. To provide close control of room pressure, distur-

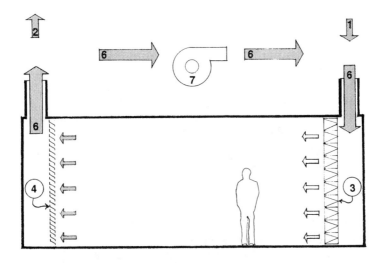

HORIZONTAL LAMINAR FLOW CLEAN ROOM

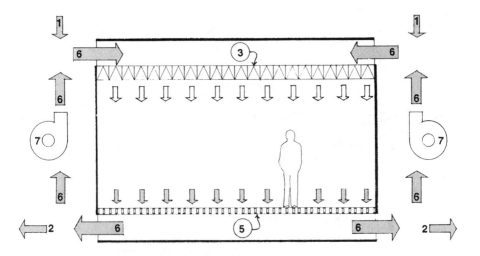

VERTICAL LAMINAR FLOW CLEAN ROOM

KEY

1 Air Supply
2 Air Exhaust/Return
3 HEPA Filter
4 Exhaust Grille
5 Grated Floor
6 Recirculated Air Loop
7 Fan
☐ Clean Air
▨ Contaminated Air

FIGURE 8-5. Clean room airflow.

bances to airflow should be minimized by operating fume hoods and local exhaust points continuously and providing air locks between adjoining areas. The specifications for maintaining room pressure balance were set forth by the General Services Administration in Federal Standard 209b for clean rooms processing government contracts. Generally, the pressure differential between contiguous pressure control zones will be 0.05 in.w.g. Higher differentials between zones tend to make large doors difficult to open and close, whereas lower differentials lack effectiveness in separating the zones. In certain applications, simplified systems without the use of separate primary/secondary fan systems may be sufficient. See ASHRAE (1991), Chapter 16 for an additional reference.

8.3.2.4 *Malfunction Alarms.*

Alarms should be provided to indicate a malfunction of the directional airflow arrangement. Additional alarms should be provided for the filter banks to notify service personnel that the filters are becoming loaded with dust to a point where airflow delivery is being affected adversely. It should be kept in mind that it is necessary to measure total airflow rate as well as filter bank pressure drop to evaluate the condition of the filters correctly.

8.3.2.5 *Humidity Control.*

It is important that humidity and temperature conditions be controlled. Humidity control is required for corrosion and condensation control, static electricity elimination, and personal comfort. Temperature control provides stable conditions of operation for instruments and personnel.

ASHRAE 1991, Chapter 16 discusses this issue in more detail. In general, conditions to be met are as follows:

68°F–73°F ± 1°F temperature
30–50% ± 5% relative humidity

In critical applications, tolerances of ±0.1 to 0.5°F and ±0.5 to 5% RH may be required. If a relative humidity of less than 35% is to be maintained, special precautions for static electricity control must be taken.

The actual design conditions will depend on the specific laboratory process.

8.4 LOSS PREVENTION, INDUSTRIAL HYGIENE, AND PERSONAL SAFETY

The information contained in Sections 1.4 and 2.4 applies to clean room laboratories and should be implemented, with the following addition.

8.4.1   Portable Fire Extinguishers

In addition to the fire extinguisher recommendations contained in Section 1.4.4.2.2, space should be provided during the planning stage for hand-portable fire extinguishers to be used in the event of a filter fire. The size and type of extinguishers should be determined with the aid of a safety professional. Carbon-dioxide-filled extinguishers are favored by clean room operators because they leave no particulate residues. In the presence of the high airflow rates that are characteristic of clean rooms, however, an inert blanket of carbon dioxide has only limited persistence and will prove inadequate under these special circumstances. It should be remembered that when there is a fire in a clean room, the facility is already severely contaminated and the extinguishing agent should be selected solely on the basis of effectiveness.

8.5   SPECIAL REQUIREMENTS

8.5.1   Lighting

Clean room laboratory lighting should be flush-mounted and sealed from the room. Illumination levels of 100 ft-c at 30 in. above the floor will be adequate to provide a reasonable illumination level for fine, precision work that will avoid eye fatigue. Because of the smooth, hard nature of the room's surfaces, special care must be exercised to provide soft, even lighting and an absence of direct or reflected glare.

CHAPTER 9

CONTROLLED ENVIRONMENT (HOT OR COLD) ROOM

9.1 DESCRIPTION

9.1.1 Introduction

A controlled environment room is a laboratory or a laboratory adjunct in which temperature and humidity are maintained within a specified range so that laboratory activities can be conducted and laboratory products maintained under controlled conditions. A controlled environment room can be maintained to within 1°F at an elevated temperature up to 120°F, or at a reduced temperature down to 35°F. In addition, relative humidity can be controlled to within 0.5% of the full humidity span.

9.1.2 Work Activities

Although controlled environment rooms are primarily designed and used for storage of sensitive materials that require maintenance within a specified temperature and relative humidity range, they are frequently used, in addition, for conducting activities normally performed in a general chemistry laboratory or in biology, bacteriology, or cell culture laboratories.

Guidelines for Laboratory Design: Health and Safety Considerations, Second Edition
By Louis DiBerardinis, Janet S. Baum, Gari T. Gatwood, Anand K. Seth, Melvin W. First, and Edward F. Groden.
ISBN 0-471-55463-4   Copyright © 1993 by John Wiley & Sons, Inc.

9.1.3 Equipment and Materials Used

It is expected that few pieces of analytical equipment will be located permanently in controlled environment rooms because of the unfavorable storage environment. Specialized storage containers will usually be found there. Controlled environment rooms usually have no special access restrictions unless sensitive equipment or dangerous materials are being used therein.

9.1.4 Exclusions

Controlled environmental rooms are not designed for handling extremely hazardous chemicals or performing especially hazardous operations. Hazardous operations not recommended for controlled environmental rooms include, but are not limited to, the use of (1) carcinogenic, mutagenic, or teratogenic chemicals, (2) highly explosive materials in greater-than-milligram quantities, (3) high-tension voltage and high-current electrical services, (4) radiofrequency generators and all electrical operations with a high potential for fire, explosion, or electrocution, (5) lasers of over 3-mW output power with unshielded beams, (6) gas pressures exceeding 2500 psig, (7) liquid pressures exceeding 500 psig, and (8) radioactive materials in greater than 1-μCi amounts. The controlled environment rooms (hot or cold) usually need no special access restrictions.

9.1.5 Special Requirements

Ideally, environmental rooms should be the product of a single manufacturer and completely furnished and installed by the same manufacturer to avoid division of responsibility. If possible, rooms should be prebuilt at the manufacturer's plant and pretested prior to shipment. Pretesting conditions should simulate, or even exaggerate, the environmental conditions that will be found in actual service. Tests should include a thorough check of the mechanical, electrical, and temperature-control systems. It is advisable to have an owner representative present during the manufacturer's preinstallation testing program.

9.2 LABORATORY LAYOUT

9.2.1 Introduction

A controlled environment room is a laboratory type that need not always conform to the dimensional guidelines set out in Section 1.2.2.3. The interior area of manufacturers' standard controlled environment rooms can vary from closet size [20 net square feet (nsf)] to double-laboratory module size (400 nsf) or even larger. The minimum dimension may be less than 7 ft 6 in. because a controlled environment room is not usually classified as a habitable room. If a controlled environment room larger than double laboratory size is needed, safety personnel should be consulted for assistance in the preparation of design specifications.

Temperature conditions within controlled environment rooms are usually outside the comfort zone for personnel, so work efficiency must be carefully considered during the layout of work surfaces and storage units. When work surfaces and sinks are present inside the controlled environment room, they should be located close to the door so that the zone of greatest activity will be near the exit. Chromatography separations normally run for many hours and must be visually checked frequently. To increase efficiency, a flexi-frame chromatography support grid should also be located near the exit. Because space is usually limited in this type of laboratory facility, storage units should be placed to the rear so that materials on them will be less likely to get bumped and spilled. Examples of average and small environmental rooms are shown in Figure 9-1.

9.2.2 Egress

9.2.2.1 Doors. Two doors are not practical or required in controlled environment rooms less than 200 nsf, because of limited wall perimeter. However, controlled environment rooms of 400 nsf and over should have a second egress. For rooms between 200 and 400 nsf, a second egress should be considered on the basis of (a) individual need and (b) understanding of the materials stored, processed, and conducted therein.

9.2.3 Furniture Location

Work surfaces and storage shelving should be designed and located to facilitate ease of egress. Other furnishings should be kept to a minimum, so that aisles are not obstructed. In controlled environment rooms having a minimum inside width of 15 ft, an island-type work surface, bench, sink, or storage unit may be located inside the room, provided a minimum egress aisle width of 60 in. is maintained to the exit. Aisles between parallel rows of work surfaces or storage units, as well as aisles between an interior wall and a work surface or storage unit, should be not less than 36 in. Desks should not be placed in controlled environment rooms.

9.2.4 Access for Disabled Persons

To make controlled environment rooms accessible to persons in wheelchairs, a clear floor area of at least 5 × 5 ft will be required for turning the wheelchair around, and a work surface 32 in. high should be provided. A ramp is commonly used to roll in laboratory carts. However, ramps with a slop not greater than 1 : 12 may be used for access by a wheelchair and should be capable of safely supporting that load.

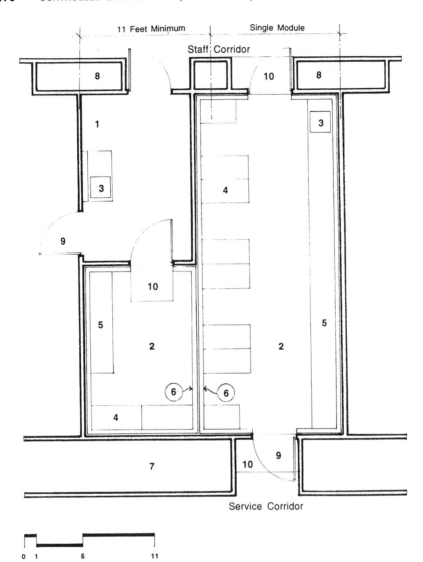

0 1   5   11

KEY
1   Equipment Area
2   Controlled Environment Room
3   Sink
4   Storage Shelving
5   Work Surface
6   Manufacturer's Insulated Wall
7   Mechanical Shaft
8   Electrical Closet
9   Emergency Second Egress
10 Ramp

FIGURE 9-1. Controlled environment rooms: Sample layout.

9.3 HEATING, VENTILATING, AND AIR CONDITIONING

9.3.1 Introduction

HVAC requirements for a controlled environment room depend on the temperature and humidity conditions to be maintained, as well as on the activities to be performed. Outside air-volume rates should be kept to a minimum, particularly when close humidity control is required. Most cooling systems are of the direct expansion refrigeration type with separate condensing unit and evaporator coil section.

Unless there are special requirements, such as a need for a fume hood or numerous local exhaust points, outside air exchange requirements for controlled environment rooms will be minimal. A minimum of 50 cfm of outside conditioned air is recommended when people must work inside the controlled environment room regularly for prolonged periods of time. To provide fresh filtered air for the people using the room, a supply air blower with full-modulating control should be connected to a ceiling plenum located at the entrance to the work area and the supply air should be discharged directly through the evaporator or heating coil. Air should be exhausted to the building exhaust air system through an adjustable damper. Supply and exhaust air volumes will depend on room size and the activities to be conducted inside the controlled temperature room. When toxic chemicals will be used, spot exhaust points or chemical fume hoods will be needed, and they should conform to the recommendations in Section 2.3.4.4.

9.3.2 Temperature and Humidity

9.3.2.1 Room Temperature. For low-temperature control, the refrigeration system should contain a direct expansion unit of industrial quality, designed to operate continuously with an integral evaporator coil. Room temperature controller and other instrumentation should be designed to control coil temperature over the full temperature range on a precise demand basis. The control mode should be fully modulating with proportional action from 0 to 100% of total condensing unit capacity over the full-rated temperature range. The main controller should include a means for direct setting of the control point, input and output meters to display the proportioning action of the control unit, a temperature indicator to permit monitoring room temperature conditions with a set-point accuracy of not less than 0.5% of the full-rated temperature span, and a proportioning band of not less than 1% above or below the control point. The temperature-sensing unit must possess adequate inherent stability, accuracy, and sensitivity to provide the degree of temperature control required for the operations that will be conducted in the controlled environment room. A recorder with a 12-in. circular 7-day or 24-h chart should be installed in a central control panel to assist in monitoring the stability of the set conditions.

9.3.2.2 Room Humidity. When humidity control is required in controlled environment rooms, extreme care must be exercised in the selection of the humidifier. For humidification, steam injection remains the most popular choice. However, steam from a central boiler plant or a local boiler in a facility is treated with chemicals which, some studies have shown, can cause health hazards. (See Section 1.5.6).

There are many humidifiers available such as cold mist types, steam generators which use building steam to evaporate city water or pure water to make clean steam. Self contained humidifiers make clean, dry, low pressure steam electrically and collect the minerals in a replaceable cartridge. Control should be in response to a pneumatic or electronic proportioning control system. The controller should be fully calibrated and include an electronic sensing unit, an integrated recorder with a 12 in. recording chart, and a set-point accuracy of not less than 0.5% for the full humidity span.

9.3.3 Emergency Alarm and Control System

A safety control and alarm system, provided by the manufacturer, should be mounted on an outside wall of the cold or warm room adjacent to the entrance door. It should consist of an independent electrical low and high temperature control system that will take over operation in the event of a main control failure and contain an alarm buzzer to give audible warning in the event of temperature deviations. The safety control and alarm should be equipped with the following components: a main on/off switch for the entire system, a silencing switch for the buzzer, and a reset switch to return the system to normal operation. Terminals should be provided to connect the alarm system into a remote central location.

9.4 LOSS PREVENTION

Information contained in Sections 1.4 and 2.4 applies to all controlled environment rooms, and all relevant items should be implemented. In addition, when an experiment could give rise to a hazardous situation, either by depleting oxygen or by releasing toxic contaminants, provisions should be made during the building design phase to install facilities for supplied air or self-contained breathing apparatus, plus one or more atmospheric monitors to identify the hazardous situation and provide an appropriate alarm.

9.5 SPECIAL REQUIREMENTS

9.5.1 Materials of Construction

Controlled environment rooms may contain a rapidly degrading environment for some of the services and equipment installed in it. For example, hot rooms can reduce the normal expected life of electrical wiring through deterioration of the insulation. The National Electrical Code should be consulted to determine whether an over-design in wire size and insulation type will be required due to the specific temperatures and current loads that will be encountered.

Frequently, cold rooms become wet with moisture that condenses from building air that enters the room when people go in and out. The moisture affects materials of construction such as steel shelves, electrical conduits and fixtures, and moisture-absorbing materials. The use of a vapor-tight electrical system can help prevent early deterioration and possible shock hazards. A careful selection of materials should result in nonrusting shelves, structural components, and finishes of the cold room.

In designing for humidities above 60%, there is a potential for microbiological growth. In this case the materials of construction should be easily cleanable and be of a nonorganic nature to the extent possible. If there is to be human occupancy under these conditions, the use of HEPA filtration of the recirculated air should be considered.

9.5.2 Lighting

Interior lighting should be high-output fluorescent designed to provide 70 ft-c evenly distributed when measured at 40 in. above the floor. Ballasts should be mounted externally, and fluorescent lamps should have moisture-proof covered socket ends. Lens panels should be the diffuser type made from acrylic.

CHAPTER 10

HIGH-PRESSURE LABORATORY

10.1 DESCRIPTION

10.1.1 Introduction

A high-pressure laboratory is designed, constructed, and operated to permit safe experiments at gas pressures over 250 psig and liquid pressures over 5000 psig.

Because of the high-energy potential of high-pressure fluid systems, special consideration must be given to the location of the laboratory within the building structure and to its materials of construction. For example, a 10-ft^3 volume of dry nitrogen at 6000 psig has the energy equivalent of approximately 300 lb of TNT. Designing a laboratory to handle a potential explosion of this magnitude requires great care. Ideally, the laboratory should be a free-standing barricaded building, but when it is located inside a laboratory building, the ultimate in control of the qualities and quantities of materials that will be used in its construction and in the management of laboratory procedures will be required.

10.1.2 Work Activities

The investigative procedures used in a high-pressure laboratory include those used in a general chemistry laboratory except that they will be conducted on gases and liquids at higher pressures. Pressure testing of vessels, high-

Guidelines for Laboratory Design: Health and Safety Considerations, Second Edition
By Louis DiBerardinis, Janet S. Baum, Gari T. Gatwood, Anand K. Seth, Melvin W. First, and Edward F. Groden.
ISBN 0-471-55463-4   Copyright © 1993 by John Wiley & Sons, Inc.

pressure reactions, and some handling of very high temperature, as well as cryogenic, liquids may take place in the high-pressure laboratory. Small quantities of high explosives may also be used.

10.1.3 Equipment and Materials Used

High-pressure oil pumps, high-pressure gas compressors, high-pressure piping and valves, accumulators, barricades, and compressed gas cylinders are used in high-pressure laboratories. There is likely to be a need for some chemical storage. The location of each of these items should be considered carefully from the safety standpoint. Because much of the equipment is necessarily heavy, floor loading requirements must be evaluated early in the building planning process. Access restrictions will apply to this laboratory during many operating conditions.

10.1.4 Exclusions

High-pressure energy systems that exceed the deisgn capabilities of the facilities must be carried out in other locations. Prior to construction, the designer/architect/engineer must determine the maximum size of experiments and pressure conditions that will be allowed in the facility, and management controls must be instituted to make certain these limits are never exceeded.

10.2 LABORATORY LAYOUT

10.2.1 Introduction

All provisions of Section 2.2 apply to high-pressure laboratories and should be implemented. In addition, wall, floor, and ceiling construction should be designed to contain the explosive violence of a maximum design energy release. One method of construction is to contain the released energy; an alternative method is to channel the release of energy into safe pathways designed to protect persons and property outside the laboratory.

Full containment of an explosion requires the construction of high-pressure areas, or "cells," along outside walls. The walls and ceiling of cells are constructed of reinforced 18-in.-thick concrete, or of steel, or of combinations of materials of adequate strength to contain the maximum anticipated explosion. All exterior approaches to the cells must be secured. The cell is sealed with a door assembly of required strength and fire rating. This strength should be consistent with that of the walls and ceiling. Personnel conducting experiments in which explosions are anticipated are restricted from the active cell during the actual experiment. An adjacent room, adequately protected from the maximum design energy release, may be needed for remote monitoring and control equipment.

Venting panels may be installed in a wall or, if the high-pressure laboratory is in a single-story building, in the roof to direct the safe release of explosive energy. The panels are designed to open outward by the force of a sudden pressure rise before structural damage occurs. The free area required and release pressure are specified in Venting of Deflagrations (NFPA 68, 1988) for laboratories according to the amounts and nature of the explosive or volatile materials used. High-order explosions or detonations have pressure rise rates so rapid that venting panels may not open in time to offer adequate building protection and should not be relied upon for this type of explosive research. The area outside the blowout or pressure relief panels should be protected as described in the following paragraph.

Locations for blowout panels and cell walls should be selected in locations that prevent passersby from close approach. Blowout panels should be securely tethered to immovable elements of the building structure with loose ropes of strong steel cable to prevent their release as free projectiles following a forceful explosion. In addition, heavy mesh screening or a similar barrier should be installed to contain the debris that may be ejected through the opening. Earth berms of sufficient height and breadth are the best means for containment, but if there is not enough area outside the high-pressure laboratory for a protective berm, a solid blank wall with the strength to withstand the blast should be constructed. As a minimum, a high, heavy-duty chain-link fence should be erected far enough away to prevent any dangerous approach to the panels or cells.

When the high-pressure laboratory is within a multistory building, it should be located on the ground floor. Windows or other building openings above blowout panels should be protected from possible flame spread and explosion debris by a solid canopy above the blowout panel of sufficient strength and fire rating to withstand the hazards.

All laboratories and storage rooms involving a significant explosion hazard must comply with NFPA, local building codes, and fire department regulations.

The construction of a fully equipped high-pressure laboratory can be avoided by substituting a specially designed, heavy-walled vessel that is large enough to house the experiment plus auxiliary equipment and strong enough to contain the energy resulting from any forseeable accident. For specific details of construction of heavy-walled pressure vessels, the ASME Unfired Pressure Vessel Code, Section VIII (ASME, 1989) should be consulted.

10.3 HEATING, VENTILATING, AND AIR CONDITIONING

10.3.1 Special Requirements

A high-pressure laboratory presents a number of unique problems for the designer. High-pressure laboratories are seldom air-conditioned, but when they are, the ducts and fans may have to be specially constructed and mounted to withstand explosions. Also, special venting arrangements must be provided to relieve excessive pressure safely under unexpected circumstances.

When a high-pressure laboratory functions at ordinary pressures, all the provisions of Sections 1.3 and 2.3 that are applicable to its new use should be implemented, plus those that apply to whatever type of laboratory it most closely resembles.

Ventilation rates of 30–45 air changes per hour, characteristic of pressure cells, are considered more than adequate for most alternative operations, but the use of more than trace amounts of highly toxic chemicals will require reevaluation of ventilation rates.

10.4   LOSS PREVENTION, INDUSTRIAL HYGIENE, AND PERSONAL SAFETY

All the items contained in Sections 1.4 and 2.4 apply to high-pressure laboratories and should be implemented. In addition, the following recommendations should be given careful consideration.

10.4.1   Portable Fire Extinguishers

Space should be provided for hand-portable fire extinguishers of larger capacity than those used in other laboratory types. They should be sized and typed in consultation with a safety specialist to make certain they will provide adequate protection for the planned operations.

10.4.2   Compressed Gas Cylinder Racks

Racks similar to those described in Section 2.4.7 should be located so that an explosive incident within the high-pressure laboratory will not spray the storage area with shrapnel that may puncture other tanks and cause the explosive release of additional toxic or flammable materials.

10.5   SPECIAL REQUIREMENTS

None.

CHAPTER 11

RADIATION LABORATORY

11.1 DESCRIPTION

11.1.1 Introduction

A radiation laboratory is designed and constructed to provide a safe and efficient workplace for a wide variety of activities associated with materials that emit ionizing radiation that can be harmful either by direct radiation in the electromagnetic spectrum (such as neutron, gamma, or x-ray energy), by ingestion, or by inhalation of particulate materials that emit ionizing radiation (e.g., alpha, beta). Ordinary chemicals labeled with radioactive isotopes may be in liquid, gaseous, or solid form. They retain the same inherent properties to produce toxic chemical exposures as nonradiolabeled forms and have the added hazard of being radioactive. General chemistry activities may also be performed in this laboratory. It is preferable that this type of laboratory be located in a building containing similar laboratory units but not necessarily all conducting work with radioactive materials. Not all of the design considerations presented here may be needed in every radiation laboratory. Close communication between users and designers will be necessary to determine when certain items can be omitted with safety. Consultation with a health physicist who may be the laboratory's radiation protection officer is recommended.

Guidelines for Laboratory Design: Health and Safety Considerations, Second Edition
By Louis DiBerardinis, Janet S. Baum, Gari T. Gatwood, Anand K. Seth, Melvin W. First, and Edward F. Groden.
ISBN 0-471-55463-4   Copyright © 1993 by John Wiley & Sons, Inc.

11.1.2 Work Activities

The basic experimental procedures used in this laboratory will be similar to those conducted in general chemistry and analytical laboratories, but special provisions must be made for the additional safety procedures that will be required when handling radioactive materials or radiation-producing equipment. Although this chapter will describe a facility similar to a general chemistry laboratory, the safety requirements that will be outlined can be applied to other laboratory types, such as chemical engineering, physics, or high-toxicity laboratories, in which radioactive materials will be used. As a design guideline, any laboratory that uses radioactive materials with a total activity greater than 1 μCi should be considered a radiation laboratory. See Appendix C in 10 CFR 20 for a list of radioactive materials and their activities (NRC, 1991).

11.1.3 Equipment and Materials Used

The equipment used in a radiation laboratory will depend on the specific nature of the work, but, in general, it will be similar to that found in a general or analytical chemistry laboratory. X-ray producing equipment may be included.

11.1.4 Exclusions

This section does not cover the use of ionizing radiation in connection with animal studies or biological agents. It does not cover activities that produce nonionizing radiation.

11.1.5 Special Requirements

A radiation laboratory should have special access restrictions. In addition to permits and licenses, there are stringent EPA, OSHA, and Nuclear Regulatory Commission (NRC) requirements pertaining to acquisition, storage, use, release, and disposal of radioactive materials that must be followed. In some cases, provisions must be made for change rooms and showers. Health and safety professionals, including health physics specialists, should be consulted when the total quantities of radioactive materials are likely to equal or exceed a regulatory threshold and whenever estimates of radiation levels exceed 25% of the maximum allowable human exposure levels. (See 10 CFR 20, Appendix C.)

Energy-emitting devices should be directed away from laboratory entrances and primary egress aisles.

11.2 LABORATORY LAYOUT

11.2.1 Introduction

Many laboratory layouts are acceptable. Final choice will depend largely on the type of work to be performed and available space. A radiation laboratory layout similar to that of a general chemistry laboratory, as well as a number of specialized work utilization layouts, are described in *Safe Handling of Chemical Carcinogens, Mutagens, Teratogens, and Highly Toxic Substances* (Walters, 1980). Whatever laboratory design is selected, all the provisions of common elements of laboratory design contained in Sections 1.2 and 2.2 apply and should be followed. Some thought should be given to keeping the size of the laboratory to a practical minimum. This will facilitate decontamination should it become necessary. For large radiation-producing equipment, such as is found in an x-ray laboratory, limiting size will reduce the number of potentially exposed individuals.

In addition, the items that follow should be reviewed for their applicability to the particular laboratory work being planned. Because the need for some of the special items will depend on the nature of the work and the quantity of radioactive materials that will be used, health and safety professionals should be consulted for assistance.

11.2.2 Change, Decomtamination, and Shower Rooms

Adequate facilities for laboratory workers to change and shower must be provided when procedural requirements call for the use of full-body protective clothing. For the use of some substances, OSHA and NRC requirements specify that a shower be available as a part of the laboratory, whereas the use of other radioactive chemicals necessitates only that there be a shower available in the building. Operations and exposures that only require the use of gloves and a laboratory coat may not need shower facilities.

11.2.3 Work Surfaces

Work surfaces should be smooth, easily cleanable, and constructed from impervious materials such as stainless steel with integral cove joints. Strippable epoxy-type paint can be used as a substitute. The use of an impervious and disposable bench covering during work should be considered as an aid in the cleanup of radioactive spills.

11.2.4 Floors and Walls

In areas using liquid or particulate radioactive or radiolabeled chemicals, floor coverings should be of monolithic materials, such as seamless vinyl or epoxy. Cracks in floors, walls, and ceilings may need to be sealed with epoxy or a similar nonhardening material. All penetrations of walls, floors, and ceilings by utility conduits may need to be similarly sealed with an appropriate sealant such as epoxy. Lighting may need to be provided by sealed vapor- and waterproof units, and lighting fixtures should be flush with the ceiling to eliminate dust collection. The decision to require the above construction methods will depend on the volatility, flammability, toxicity, and quantity of radiolabeled and other chemicals used in the facility.

11.2.5 Handwashing Facilities

Handwashing facilities should be located within the radiation laboratory and in change and shower rooms.

11.2.6 Access Restrictions

Entrances to the laboratory should be posted with permanent signs indicating restricted access due to the use of radioactive materials or radiation-producing equipment (e.g., an x-ray unit). Laboratories in which a person may receive in excess of 100 mrem in 1 h at 30 cm from the radiation source or from any surface that radiation penetrates must post a sign on the entrance door labeled "Caution: High Radiation Area." Laboratories in which radiation levels could reach 500 or more rads in 1 h at 1 m from a radiation source or any surface through which radiation penetrates must post a sign on the entrance door labeled "Grave Danger: Very High Radiation Area" and must be equipped with entry control devices. The use of special key access and alarm should be considered for these areas. Refer to 10 CFR 20 for exact requirements for "high" and "very high" radiation areas. Because of access restrictions and isolation of the laboratory, it is recommended that there be a viewing window in the doors. Such windows should not be of such size and construction that they compromise design fire ratings. Radiation producing equipment should be placed so that radiation beams are directed away from entry or exit doorways.

11.3 HEATING, VENTILATING, AND AIR CONDITIONING

11.3.1 Introduction

The items contained in Section 2.3 apply to radiation laboratories, and all that are relevant should be implemented. Additional considerations are given below. Industrial hygiene or radiation safety professionals should be consulted for assistance when large amounts of radioactive materials or ionizing radiation-producing devices will be used.

11.3.2   Fume Hood

The NRC recommends an exhaust volume of 150 cfm/ft^2 of maximum open-face area for fume hoods handling radioactive materials. Where not required by regulation we recommend 100 cfm/ft^2 of maximum open-face area. An isotope fume hood with cleanable surfaces (Chapter 31, Section 31.5) should be used.

The use of variable-volume hoods is not recommended. This is because of the potential need to have air cleaners in the system and because allowable emissions are based on concentrations which will increase with decreased airflow rates.

11.3.3   Glove Box

Work with volatile and powdery radioactive materials, or the use of amounts that may produce concentrations in air above values which define an airborne radioactive exposure of concern, calls for the use of a completely enclosed, ventilated work space, called a *glove box* or *gloved box,* rather than an isotope fume hood. The glove box should meet the specifications of the American Conference of Governmental Industrial Hygienists (ACGIH, 1992) and those outlined in Chapter 31, Section 8. Local ventilation at the entrance port may be required. The glove box is a closed system and should be kept under a negative pressure of 0.25 in. w.g. static pressure relative to the laboratory. The laboratory itself may be under a negative pressure of 0.05 in. w.g. relative to the atmosphere to prevent exfiltration of potentially contaminated air, making the box interior 0.3 in. w.g. lower than atmospheric pressure.

11.3.4   Spot Exhaust for Instruments

Analytical instruments that are used to weigh or manipulate radioactive chemicals should be equipped with spot exhaust ventilation at potential points of contaminant release, or be completely enclosed in a ventilated enclosure such as a glove box or an isotope fume hood. Specific design requirements will vary with each type of equipment and chemical used. The design principles in Chapter 3 of the ACGIH *Industrial Ventilation Manual* should be followed (ACGIH, 1992). Consultation with industrial hygienists and health physicists is highly recommended when the use of radioactive materials may generate radiation levels greater than 25% of the recommended maximum exposure levels.

11.3.5   Storage Facilities

Cabinets, refrigerators, and all other equipment items and areas used for storage of volatile radiolabeled materials should be provided with local or general exhaust ventilation sufficient to maintain a directional airflow and prevent buildup of radioactive contaminants within the storage space. Laboratory users need to be consulted to determine whether any radiolabeled materials that may release vapors or gases will be stored in the facility. Sections 1.4.7 and 2.4.6 contain design information for chemical storage facilities.

11.3.6   Filtration of Exhaust Air

Sometimes air exhausted from fume hoods, glove boxes, and spot exhaust systems must be filtered before release to the environment to avoid atmospheric pollution and to conform with NRC regulations. Industrial hygienists and health physicists should be consulted for an evaluation of this need. The first stage of filtration should be a HEPA filter with not less than 99.97% efficiency for 0.3-μm particles. A second stage containing an adsorbent such as activated charcoal may be required when the radioactive effluents are in the gaseous or vapor form. The components of the air-cleaning train and the size of elements will depend on the materials to be used and the total quantity of air to be exhausted. All air-cleaning elements should be capable of being changed without exposure of maintenance personnel (e.g., bag-in, bag-out procedures).

11.3.7   Exhaust Stream Monitors

The NRC requires continuous air monitoring of exhaust air streams for some radioisotopes when the concentrations exceed regulated levels. Access for sampling locations must be provided in the initial design when it is anticipated that effluent air monitoring may be needed. Monitoring pumps should be hard-wired to prevent accidental outage.

11.4   LOSS PREVENTION, INDUSTRIAL HYGIENE, AND PERSONAL SAFETY

The recommendations contained in Sections 1.4 and 2.4 apply to laboratories using radioactive materials and should be followed.

11.4.1   Radioactive Waste

Where special procedures for handling radioactive materials in laboratory waste are required, provisions for separate radioactive waste storage areas and easy access to shipping areas must be made in the building plans. All provisions of Section 1.4.6 apply to radiation waste storage facilities plus the use of radioactive shielding when required. See Chapter 23 for more details.

11.5 SPECIAL CONSIDERATIONS

11.5.1 Shielding

Some radiation-producing equipment and radioactive materials require the use of special shielding. Generally, cobalt and cesium irradiators should be placed in separate, small rooms that are lead-lined or constructed with dense concrete walls, floors, and ceilings to shield the source. These facilities add many tons of extra weight to the structure and require special consideration during structural design.

11.5.2 Iodinization Facility

It may be desireable to designate one laboratory per floor or an appropriate number of floors where iodinizations can be performed. This will allow for better control of this activity.

CHAPTER 12

BIOSAFETY LABORATORY

12.1 DESCRIPTION

12.1.1 Introduction

Accidental contact with infectious and toxic biological agents is an occupational hazard for persons who work in laboratories handling oncogenic viruses, infectious agents, and similar harmful biological substances. Often, research activities with these substances call for additions of radioactive tracers and chemicals that are known to be mutagenic, teratogenic, or carcinogenic, thereby substantially increasing the hazardous nature of the work. As of 1978, the registry of laboratory-acquired infections listed in excess of 4000 cases; it has been estimated that this number represents only a fraction of the total, because reporting of these cases is not compulsory.

The Center for Disease Control, an agency of the U.S. Public Health Service, has classified laboratories handling hazardous biological agents into the following four biosafety categories (HHS, 1988):

Biosafety Level 1. Suitable for experiments involving agents of no known, or of minimal, potential hazard to laboratory personnel and the environment. The laboratory need not be separated from the general traffic patterns of the building, and work is generally conducted on open benchtops. Special containment equipment is not required or generally used. However, hand-

Guidelines for Laboratory Design: Health and Safety Considerations, Second Edition
By Louis DiBerardinis, Janet S. Baum, Gari T. Gatwood, Anand K. Seth, Melvin W. First, and Edward F. Groden.
ISBN 0-471-55463-4 Copyright © 1993 by John Wiley & Sons, Inc.

washing facilities are required in the laboratory, and provision for storage of laboratory coats, gowns, or uniforms is recommended.

Most general-purpose laboratories, such as a general chemistry laboratory (Chapter 3), are suitable for work at biosafety level 1 provided that work surfaces are easily decontaminated and can resist frequent application of decontamination chemicals.

Biosafety Level 2. Suitable for work involving agents of moderate potential hazard to personnel and the environment. A Level 2 facility differs from a Level 1 facility in the following ways: (1) There must be provision for limiting access to the laboratory when experiments are being conducted, and (2) all procedures involving large volumes (or high concentrations) of agents, or involving the likelihood of creating infectious aerosols, must be conducted in biological safety cabinets or other physical containment equipment. In addition, the entire laboratory must be designed for easy cleaning, and provision must be made for installation of an autoclave for decontaminating all laboratory infectious wastes. If windows can be opened they must be fitted with screens.

Biosafety Level 3. Required for work in "clinical, diagnostic, teaching, research, or production facilities in which work is done with indigenous or exotic agents which may cause serious or potentially lethal disease as a result of exposure by the inhalation route" (HHS, 1988). It is suitable for experiments involving agents of high potential risk to personnel and the environment. Provisions must be made to control access to the laboratory at all times, and the Level 3 laboratory must be equipped with special engineering and design features plus physical containment equipment and devices. All procedures involving the manipulation of infectious material must be conducted within biological safety cabinets or similar physical containment devices, or by personnel wearing appropriate personal protective clothing and devices. It also differs from a Level 2 facility by requirements that (1) all vacuum lines must be protected with HEPA filters and liquid disinfectant traps, (2) the Level 3 laboratory must be separated from other building areas by a double-doored passage that may also serve as a clothes change and shower room, and (3) all windows must be sealed in the closed position.

Biosafety Level 4. Required for work involving dangerous and exotic agents that pose a high individual risk of life-threatening disease to laboratory personnel. The facility is either in a separate building or in a controlled area within a building that is completely isolated from all other areas of the building, and access to the laboratory must be strictly controlled at all times. Within work areas of the facility, "all activities [must be] confined to Class III biological safety cabinets or Class I or Class II biological safety cabinets used along with one-piece positive pressure personnel suits ventilated by a life support system" (HHS, 1988). In addition to the facilities needed for a Level 3 laboratory, provision must be made to remove double-packaged biological materials that are to remain in a viable state through a disinfectant dunk tank or fumigation chamber contained in a double-doored airlock designed for this purpose. Except for such viable materials, everything else leaving the laboratory must be passed through a double-doored autoclave or fumigation chamber for decontamination. Materials and supplies that are not carried into the facility through the change room will be brought in through a double-doored lock and the lock will be decontaminated each time the inner door is opened. The physical security of a Level 4 facility must be maintained with the aid of secure, locked doors. An access control system connected to a computer that monitors entry and egress is highly recommended. Facilities must be provided for a complete change of clothing on entering and for disrobing and showering when leaving.

The publication *Biosafety in Microbiological and Biomedical Laboratories* (HHS, 1988) should be consulted to determine the biosafety laboratory level required for specific etiological (infectious) agents, oncogenic viruses, and recombinant DNA strains. Although this chapter deals only with laboratory design for safe handling of microbiological agents requiring a biosafety level 3 facility, possible future conversion to a Level 4 facility should be kept in mind as a factor in the design. A biosafety level 1 laboratory requires no special provisions, and a biosafety level 4 laboratory requires facilities that should be designed and constructed only by specialists in this technology. The special construction and facilities of a biosafety level 2 laboratory are all covered in the biosafety level 3 laboratory description.

Level 3 biological safety laboratories are designed as secondary containment facilities by the creation of physical enclosures and negative air pressures and include within them primary containment facilities in the form of biological safety cabinets in which hazardous operations are conducted.

12.1.2 Work Activities

The activities performed in a biosafety level 3 laboratory include work with low- to moderate-risk biological agents. Often, microgram quantities of radionuclides, carcinogens, teratogens, and mutagens are employed as adjuncts to work with biological agents, but in no case is it intended that the biosafety level 3 laboratory be used for work with large quantities of gaseous and liquid chemicals that are hazardous by virtue of their toxicity, radioactivity, or flammability unless additional safety features are provided such as a chemical fume hood designed for this class of service.

12.1.3 Equipment and Materials Used

The unique equipment item in biosafety laboratories is the biosafety cabinet, designed to provide personnel, product, and environmental protection when working with hazardous biological agents. Biological safety cabinets are described and illustrated in Section 31.8. The most widely used biological safety cabinet is officially designated "Class II (Laminar Flow) Biohazard Cabinetry," and its construction, performance, and testing are covered by National Sanitation Foundation Standard No. 49 (NSF, 1992). Class I cabinets provide personnel and environmental protection but no work protection, and for that reason they are seldom used for work with biological agents that must be protected from contamination. Class III cabinets are total-enclosure, negative-pressure glove boxes equipped with at least one double-door lock containing a facility for decontaminating items withdrawn from the cabinet. Class III cabinets provide the highest level of personnel, work, and environmental protection, and they are used in biosafety level 4 laboratories.

Additional equipment items found in biological safety laboratories include centrifuges, high-speed blenders, sonicators, and lyophilizers. Each of these equipment items has an ability to generate large numbers of respirable aerosolized particles that represent a potential infective dose when working with hazardous biological agents. Incubators, sterilizers, and refrigerators are also commonly found in biohazard laboratories. Specialized analytical devices that are likely to be present include gas and liquid chromatographs, mass spectrometers, and liquid scintillation counting devices. One or more conventional fume hoods may be present when hazardous chemical procedures that do not require a sterile air environment are carried out in connection with the biological work.

12.1.4 Exclusions

Biosafety laboratories are not usually designed for work with greater than trace amounts of radioisotopes, carcinogens, mutagens, teratogens, or highly toxic systemic poisons. Restrictions on quantities of flammable solvents follow very closely the safe handling practices applicable to analytical chemistry and similar laboratories. (See Section 2.4.6.3). There is rarely, if ever, a need for high-voltage and high-current electrical services, or for radiofrequency generators and other electrical operations with a high potential for fire, explosion, or electrocution. Other high-energy-releasing sources, such as lasers, x-rays, and gamma energy emitters that are lethal to biological agents, are unlikely to be present in the biosafety laboratory.

12.2 LABORATORY LAYOUT

12.2.1 Introduction

A biosafety laboratory may be a single room amid other laboratories of divergent uses, or it may be a suite of many rooms interconnected through pressurized central corridors with building entrances and essential common service facilities, for example, animal quarters, supply rooms, and washing and sterilizing services. The minimum dimensions of a biosafety laboratory containing one Class II biosafety cabinet with auxiliary equipment and suitable for a single worker is 7.5 × 10 ft. In addition to a 4- or 6-ft-wide cabinet, it should include washing facilities, autoclave, and a space for donning and discarding protective garments at entry. Cloakroom and lockers should be provided for personal articles not required in the laboratory. A typical layout of a biosafety laboratory is shown in Figure 12-1. The recommendations contained in Sections 1.2 and 2.2 apply to biosafety laboratories and should be followed except as supplemented or modified in the following sections.

12.2.2 Floors and Walls

Biosafety laboratories require impervious surfaces and structural joints that are vermin-proof and easily cleaned and decontaminated. Walls and floors should be monolithic and made of washable and chemically resistant plastic, baked enamel, epoxy, or polyester coatings. The monolithic floor covering should be carried up the wall base with a smooth cove joint.

12.2.3 Access Restrictions

Access to biosafety level 3 laboratories should be limited by providing self-closing and self-locking secure doors all around, plus a double-doored clothes change room (a shower facility may be included) at the entry of the laboratory. All self-closing and self-locking doors must present no barrier to egress in the event of an accident inside the laboratory. The biosafety laboratory should be maintained under negative pressure relative to corridors, offices, and so on (in all cases with nonrecirculated air), as an aid in preventing release of biological agents to areas outside the laboratory. All handling of hazardous biological agents should take place in biological safety cabinets.

Because access to biosafety level 3 laboratories is severely restricted, it is helpful to install internal viewing windows consistent with the fire rating of the wall. Ease of surveillance of laboratory operations promotes safety. Systems for voice communication between laboratory personnel and others outside the laboratory are recommended.

12.2.4 Handwashing Facilities

Each laboratory room should contain one or more foot- or elbow-operated handwashing sinks. The new infrared-actuated automatic faucets would be a good choice for a biosafety level 3 laboratory.

12.2.5 Decontamination

Each laboratory or laboratory suite must contain an autoclave for decontaminating wastes.

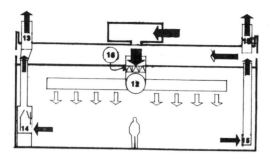

AIR SUPPLY/EXHAUST

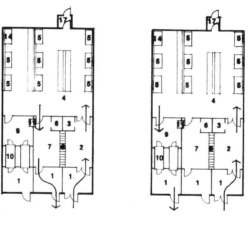

PERSONNEL CIRCULATION   MATERIAL FLOW

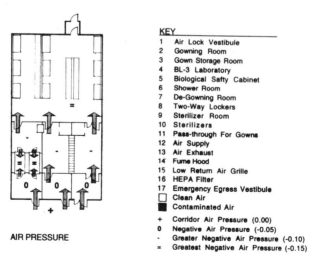

KEY

1   Air Lock Vestibule
2   Gowning Room
3   Gown Storage Room
4   BL-3 Laboratory
5   Biological Safty Cabinet
6   Shower Room
7   De-Gowning Room
8   Two-Way Lockers
9   Sterilizer Room
10 Sterilizers
11 Pass-through For Gowns
12 Air Supply
13 Air Exhaust
14 Fume Hood
15 Low Return Air Grille
16 HEPA Filter
17 Emergency Egress Vestibule
□ Clean Air
■ Contaminated Air

+ Corridor Air Pressure (0.00)
0 Negative Air Pressure (-0.05)
- Greater Negative Air Pressure (-0.10)
= Greatest Negative Air Pressure (-0.15)

AIR PRESSURE

FIGURE 12-1. BL-3 biological safety laboratory layout.

12.3 HEATING, VENTILATING, AND AIR CONDITIONING

12.3.1 Introduction

Biosafety laboratories containing several biological safety cabinets that must be vented to the roof often require complex HVAC facilities and sophisticated controls to maintain adequate air supplies and the correct pressure relationship among adjacent spaces.

The HVAC comfort requirements for a biosafety laboratory with regard to temperature, humidity, and minimum air exchange rates are those outlined in Sections 1.3 and 2.3.

12.3.2 Ventilation

Class II, Type A biosafety cabinets are designed for exclusive use with biological hazards and may discharge all the exhaust air to the laboratory after filtration through a HEPA filter. A Type A cabinet may also discharge exhaust air through a building exhaust system or a dedicated exhaust stack, but it should not be hard ducted to any system containing an exhaust fan because the cabinet is only designed to operate correctly as a self-contained unit. Therefore, when a Type A biological safety cabinet is to be vented through an exhaust system, there must be at least a 1-in. opening between the end of the cabinet discharge port and the start of the exhaust system duct to prevent detrimental interactions between the cabinet fan and the exhaust system fan (NSF, 1992). Earlier editions of Standard Number 49 [Class II (Laminar Flow) Biohazard Cabinetry (NSF, 1983)] contained a description of a cabinet designated as "convertible." It could be used as an unducted Type A cabinet in one configuration or as an ducted Type B cabinet in a different configuration. This cabinet designation is no longer authorized; cabinets must be configured and sold as either Type A or Type B. Class II, Type B biosafety cabinets are designed for use when treating hazardous biological agents with small quantities of radioactive tracers and carcinogenic chemicals. They are designed only to discharge all exhaust air to the atmosphere through a dedicated exhaust system that terminates above the roof of the laboratory building. Therefore, the number and type of all fume hoods and Class II biosafety cabinets that will be vented to the atmosphere must be identified early in the design stage so that adequate supply air can be provided to satisfy exhaust requirements and maintain predetermined pressure gradients.

Practice varies with regard to shutting down airflow in vented biosafety cabinets. It is common practice to run them continuously. If, however, they are shut down during nonworking hours, the HVAC system must be carefully modulated to reduce supply airflow in order to maintain the design room air pressure gradients. This requires sensitive pressure sensors and rapid-response feedback mechanisms, as well as visible gauges for monitoring correct functioning.

Other items commonly found in clean room laboratories that need to be vented are autoclaves and fermenters. Canopy-type hoods or high-velocity–low-volume slot-type local exhaust points (ACGIH, 1992) are suitable for this service. It may only be necessary to utilize the exhaust system when the autoclave or reactor is opened. However, a canopy-type hood can be used continuously to draw off the high heat load associated with autoclaves and fermenters when they are being sterilized between runs.

12.3.3 Filtration

For some classes of agents, all exhaust air from biological safety cabinets, even when discharged directly to the atmosphere, must be filtered through HEPA filters for environmental protection. When some of the laboratory air is discharged to the atmosphere through a separate system, it also should pass through a HEPA filter before discharge. When using trace quantities of especially toxic volatile chemicals in conjunction with biological experiments, it may be prudent to add an efficient adsorber (generally activated carbon) as a second effluent air cleaning stage.

12.3.4 Controls

Pressure control can be maintained within a biosafety laboratory by providing a constant ratio of supply to return and exhaust air with the aid of differential pressure controllers, modulating dampers, fan inlet vanes, variable speed devices, or combinations of all of these. Airflow variations should be minimized as an aid in providing control over room pressure. Continuous operation of hoods, cabinets, and local exhaust facilities is an important aid in maintaining reliable pressure control at all times. Air locks to adjacent areas should be provided. The specifications for air locks should be the same as those required by the General Services Administration in Federal Standard 209b.

12.3.5 Alarms

Malfunction alarms should be provided for the HVAC systems in a biosafety level 3 laboratory. An additional alarm should be provided to notify personnel when HEPA filters are becoming dust-loaded to a critical point. The information on air cleaning system monitoring instruments and alarms contained in Section 8.3.2.2 and 8.3.2.4 for the clean room laboratory also apply to biosafety laboratories.

12.4   LOSS PREVENTION, INDUSTRIAL HYGIENE, AND PERSONAL SAFETY

All provisions in Sections 1.4 and 2.4 that apply to biosafety laboratories should be reviewed and implemented when applicable.

12.5   SPECIAL REQUIREMENTS

12.5.1   Warning Signs

Proper identification of hazardous biological agents is necessary to restrict traffic into hazardous areas and to alert all who enter the area to take precautionary measures. A standardized, easily recognized sign is customarily used for this purpose. It is displayed at each entry to the restricted area at a place where it can be seen easily and is displayed *only* for the purpose of signifying the presence of actual or potential biological hazards. Because entry control is very important for biosafety laboratories, an internationally recognized biohazard warning sign, colored magenta, has been adopted. It is shown as Figure 12-2.

12.5.2   Personal Protective Equipment

Personal protective equipment, such as laboratory coats, gloves, safety glasses and disposable masks, should be issued to and worn by all who enter a biosafety level 3 laboratory. In some cases, shoe covers, head covers, and coveralls will also be required. Selection of adequate personal protective equipment is the responsibility of the laboratory director, but provision for storage of clean garments and safe disposal of those that become soiled must be delineated at the laboratory design stage.

12.5.3   Decontamination

If is frequently necessary to decontaminate biological safety cabinets with gaseous formaldehyde and it is sometimes necessary to decontaminate entire laboratories by this method. To decontaminate an entire laboratory with gaseous formaldehyde, it is necessary to isolate the space in a gas-tight condition. Although temporary plastic curtains and caulking compounds may have to be resorted to in all cases to achieve ultimate leak tightness, forethought in the design of biosafety laboratories can simplify the job enormously with an equivalent saving in expense and down time. An example of constructions to be avoided in biosafety laboratories is the conventional suspended ceiling with interconnections to other laboratories and offices in the space above the laid-in ceiling panels. Room partitions should extend to

FIGURE 12-2. Biohazard sign.

the underside of the structure and be completely sealed. It must be remembered that at the conclusion of the decontamination period, it will be necessary to ventilate the space directly to the atmosphere to purge it of all formaldehyde before it can be reentered.

Formaldehyde is a suspected human carcinogen, but no satisfactory substitute for decontaminating cabinets and laboratory rooms is known. Scrupulous attention to correct safety practices and the faithful use of personal protective equipment are essential when decontaminating with formaldehyde.

12.5.4 Waste Disposal

Some of the wastes from biosafety laboratories will be similar to the infectious wastes generated in hospitals and must be handled by identical methods. All infectious wastes must be sterilized before disposal, and sterilized and noninfectious hospital-like wastes must be packaged and disposed of as "red bag waste" to comply with regulations.

Hypodermic needles are frequently used in biosafety laboratories, and provisions must be made for approved needle disposal. This calls for needle boxes in the laboratory and disposal by approved methods such as incineration. See Chapter 23 for more information on waste disposal.

CHAPTER 13

CLINICAL LABORATORY

13.1 DESCRIPTION

13.1.1 Introduction

A hospital laboratory provides all of the routine clinical testing required for patient care. Inpatient and outpatient specimens are collected, tests are conducted, residual specimens and completed test materials of a chemical and biological nature are disposed of in a safe manner, and reports are provided.

13.1.2 Work Activities

Clinical laboratory activities include common procedures associated with hematology, bacteriology, and pharmacology. They involve mixing, blending, centrifuging, heating, cooling, distilling, evaporating, diluting, plating-out pathogens, examining specimens under the microscope, making radiochemical measurements, plus many similar operations. The use of automated, electronically controlled instruments to perform routine tests in busy hospital laboratories has become prevalent, thereby reducing to a minimum the need for handling chemicals. Clinical microbiological techniques have not advanced in a similar way and they remain largely manual operations.

Guidelines for Laboratory Design: Health and Safety Considerations, Second Edition
By Louis DiBerardinis, Janet S. Baum, Gari T. Gatwood, Anand K. Seth, Melvin W. First, and Edward F. Groden.
ISBN 0-471-55463-4   Copyright © 1993 by John Wiley & Sons, Inc.

13.1.2.1 Detailed Information. Degenhardt and Pfost (ASHRAE, 1983) list a number of common laboratories found in hospitals and clinics. Some of these laboratories use systems or processes that put them in different laboratory classifications. A partial list is shown below.

1. Histopathology. Three significant activities are performed in this area.
   a. Gross specimen examination (dissecting): Considerable quantities of formaldehyde (formalin) are used in this area as a fixer.
   b. Automatic tissue processing: Alcohols, benzene, toluene, and xylene are used as drying and clearing agents in processing specimens.
   c. Slide preparation: Xylene is a commonly used clearing agent.
2. Cytology. Smears made from body fluids are prepared for examination. Staining solutions that contain ethyl ether, ethyl alcohol, and xylene may be used.
3. Microbiology. Infectious materials that require the use of laminar flow hoods are contained in this area. Additionally, radioactive isotopes may be used, requiring a special hood. These special venting provisions are required to ensure the safety of the staff. Chapter 31 should be consulted for additional details.
4. Virology. This area is used for viral isolation and requires local capture hoods to protect workers from infection and to protect culture specimens from contamination.
5. Chemistry. Perchloric acid, a strong reducing agent, may be used in this area, requiring a special exhaust hood. Section 31.4 should be reviewed for additional details.
6. Immunology. Special exhaust requirements will depend on the size of the laboratory and what activities take place. A laboratory of minimal activity may require only a fume hood; a laboratory with maximum activity may have radioactive isotope procedures and animal quarters, each requiring special exhaust systems.
7. Urinalysis. The primary consideration in this area is the control of specimen odors.
8. Nuclear Medicine. Radioactive isotopes (e.g., radioactive xenon) are used in diagnostic procedures in this area. Section 31.5 should be reviewed for additional details.
9. Central Supply. Ethylene oxide may be used in some sterilizing procedures and will require a fume-hood or local exhaust system. Care must be taken in designing ethylene oxide exhaust systems to ensure that the TLV is not exceeded.

Clinical laboratories are not usually located in medical schools; they are associated with hospitals, clinics, or independent firms. They either specialize in certain specific types of tests or attempt to perform them all. An example of the former type is a hematology laboratory (where only human blood tests are processed) or a large analysis laboratory (where blood, urine, stool, and other bodily fluids are analyzed).

In clinical laboratories, the trend is toward computerization, miniaturization, and automation. In the past, even when using automatic systems, some samples were required to be prepared and then processed on a manual basis.

Today's most advanced equipment, however, can perform almost every called-for test on a single sample by programming the testing system. A large number of tests can be completed very rapidly by automation.

In the hospital, rapid delivery of specimens to the clinical laboratory is often necessary; pneumatic tube systems are sometimes employed for this service.

13.1.3 Materials and Equipment Used

The amount and variety of materials and equipment that will be present in a clinical laboratory depend upon the type of facility it serves—that is, whether it is a large general teaching hospital, a satellite medical center, a clinic, or a doctor's office. Typical equipment includes microscopes, hotplates, mixers, autoclaves, balances, centrifuges, and such special instruments as blood cell counters, atomic absorption spectrometers, gas and liquid chromatographs, and mechanized, automatic specimen-analyzing devices of a chemical nature. Because many manual operations are still performed in the microbiology laboratory, handling of presumptive infectious specimens and examination of bacterial, viral, and fungal cultures derived from infectious specimens should be conducted in biological safety cabinets. See Section 12.1.3 and Chapter 31 for information on biological safety cabinets.

13.1.4 Exclusions

Hospital clinical laboratories are not designed for handling large quantities of hazardous materials or for conducting dangerous procedures; in fact, neither should occur in a medical care facility. The procedures used in hospital laboratories are largely standard and routine. Clinical laboratories seldom engage in research for its own sake or conduct unusual test procedures, although some types of medical research utilize routine clinical test results as essential elements of data. There are no unusual access restrictions for a clinical laboratory.

13.2 LABORATORY LAYOUT

13.2.1 Introduction

The layout of a clinical laboratory will be determined by its size and by the nature and number of clinical tests that will be performed. It usually resembles a combination analytical chemistry and biological laboratory. Good practice standards for laboratory layouts, as outlined in Sections 1.2 and 2.2, should be followed. Special requirements may be imposed by the Joint Commission on Accreditation of Health Care Organizations (JCAHO, 1991).

13.3 HEATING, VENTILATING, AND AIR CONDITIONING

The HVAC system for a hospital clinical laboratory will be required to maintain reasonable temperature control to ensure correct operation of the various electronic and other testing devices that will normally be present. The recommendations contained in Sections 1.3 and 2.3 are generally applicable to hospital clinical laboratories and should be considered for implementation with the following additions and comments.

13.3.1 Chemical Fume Hood

Chemical fume hoods are seldom needed in clinical laboratories, but when they are used, they should conform to the recommendations contained in Section 2.3.4.4 and Chapter 31. Perchloric acid hoods are required in this type of laboratory only when perchloric acid digestions of samples are conducted routinely. When perchloric acid hoods are needed, the recommendations contained in Section 2.3.4.4.3 and Chapter 31 should be followed.

13.3.2 Local Exhaust Air Facilities

When there are discrete systems or processes in the hospital clinical laboratory that emit dangerous or obnoxious fumes (e.g., some hematology procedures) or large amounts of heat (e.g., autoclaves), it is advantageous to reduce the total ventilation air requirements for the laboratory by providing local exhaust air facilities for each device rather than to depend upon the general ventilation system for this purpose. Local exhaust facilities may consist of a canopy hood directly over the process or equipment, an engineered slot-type capture hood especially designed and built for the application, or a simple flexible exhaust hose to handle a small emission source such as an atomic absorption instrument flame. See Chapter 31 for more information.

13.3.3 Biological Safety

Biological safety cabinets are needed in many hospital clinical laboratories to handle specimens from infectious patients. Biosafety cabinets should be ventilated in accordance with requirements contained in Section 12.1.3. Most Class II biological safety cabinets used in hospital clinical laboratories will be Type A models that permit recirculation of air inside the laboratory, rather than requiring a direct exhaust connection to the roof. When volatile chemotherapy drugs will be associated with clinical specimens, it is advisable to use a Type B cabinet that is exhausted to the outdoors.

13.3.4 System Components

Hospital clinical laboratories are generally served by a central building HVAC system. Accepted practices of ASHRAE (ASHRAE, 1989, 1990, 1991 and 1992) and SMACNA (SMACNA, 1990) are satisfactory, and no special treatment of system components is required.

13.3.4.1 Supply Air Systems. Hospital clinical laboratories are acquiring increasing amounts of large, bulky, automatic electronic analytical equipment and, as a consequence, distribution of supply air is becoming more difficult, not only to maintain a uniform temperature in the room, but to prevent drafts and unsatisfactory air distribution patterns. For best results, supply air should be provided by a ducted distribution system located above the ceiling that discharges to the laboratory through multiple outlets designed to avoid drafts.

13.3.5 Temperature Control

A uniform temperature ($\pm 1.5°F$) is needed for reliable operation of some analytical devices. Heating and air-conditioning control systems must be provided. When close humidity control is required, although it is seldom necessary, simultaneous operation of heating and cooling systems may be needed. Many kinds of HVAC systems, including local cooling units and central systems, have been used for hospital clinical laboratories with success. An important consideration is that all elements of temperature control systems should be interlocked so that a uniform temperature can be maintained throughout the year. Many times it is advantageous to install a separate air-cooling and heating system for a hospital clinical laboratory rather than connect it to the building system because the laboratories have HVAC requirements that are different from those of the rest of the hospital building. When not separated, a large hospital central system will have to be operated at an inappropriately, and hence an uneconomical, level to maintain desired conditions for a small laboratory suite.

13.3.6 Heat Gain Calculations

To accurately design the cooling needs of a clinical laboratory, information is needed on the heat emission rates of commonly used equipment and instruments. Table 13-1, from Alereza (Alereza, 1984), gives the heat load from commonly used items.

TABLE 13-1. Heat Release Rate of Hospital Clinical Laboratory Equipment that May Be Located in an Air-Conditioned Laboratory

Appliance Type	Size	Maximum Input Rating (BTU/h)	Recommended Rate of Heat Gain (BTU/h)[a]
Autoclave (bench)	0.7 ft^3	4270	480
Bath; hot or cold circulating; small	1.0–9.7 gal, −22–212°F	2560–6140	440–1060 (sensible) 850–2010 (latent)
Blood analyzer	120 samples/h	2510	2510
Blood analyzer with CRT screen	115 samples/h	5120	5120
Centrifuge (large)	8–24 places	3750	3580
Centrifuge (small)	4–12 places	510	480
Chromatograph	—	6820	6820
Cytometer (cell sorter/analyzer)	1000 cells/s	73,230	73,230
Electrophoresis power supply	—	1360	850
Freezer, blood plasma, medium	13 ft^3, down to −40°F	340[b]	136[b]
Hotplate, concentric ring	4 holes, 212°F	3750	2970
Incubator, CO_2	5–10 ft^3, up to 130°F	9660	4810
Incubator, forced draft	10 ft^3, 80–140°F	2460	1230
Incubator, general application	1.4–11 ft^3, up to 160°F	160–220[b]	80–110[b]
Magnetic stirrer	—	2050	2050
Microcomputer	16–256 kbytes[c]	341–2047	300–1800
Minicomputer	—	7500–15,000	7500–15,000
Oven, general purpose, small	1.4–2.8 ft^3, 460°F	2120[b]	290[b]
Refrigerator, laboratory	22–106 ft^3, 39°F	80[d]	34[d]
Refrigerator, blood, small	7–20 ft^3, 39°F	260[b]	102[b]
Spectrophotometer	—	1710	1710
Sterilizer, free-standing	3.9 ft^3, 212–270°F	71,400	8100
Ultrasonic cleaner, small	1.4 ft^3	410	410
Washer, glassware	7.8-ft^3 load area	15,220	10,000
Water still	5–15 gal	14,500[c]	320[c]

[a] For hospital equipment installed under a hood, the heat gain is assumed to be zero.
[b] Heat gain per cubic foot of interior space.
[c] Input is not proportional to memory size.
[d] Heat gain per 10 ft^3 of interior space.
[e] Heat gain per gallon of capacity.

Source: Alereza (1984).

13.4 LOSS PREVENTION, INDUSTRIAL HYGIENE, AND PERSONAL SAFETY

13.4.1 Introduction

The recommendations contained in Sections 1.4 and 2.4 are generally applicable to clinical laboratories and should be considered for implementation with the following additions and comments.

13.4.2 Egress

Because clinical laboratory activities range from simple to very complex and diverse, segregation of activities should be considered in large facilities. Otherwise, maintenance of adequate exit routes and routes to fire extinguishers, deluge showers, and other emergency equipment may become difficult.

13.4.3 Chemical Storage

Chemical storage facilities should be carefully located with exit-access, fire-fighting, spill-control, and nonrelated laboratory operation exposures in mind. Storage within the laboratory should be provided only for small amounts of chemicals, and this restriction should be maintained by careful laboratory management.

13.4.4 Fire Suppression

When placing hand-portable fire extinguishers in the laboratory, the requirements of Section 1.4.4.2.2 should be tempered with the fact that in a hospital facility it may be more critical to get to a fire extinguisher than to be able to make a rapid exit. For this reason, 2A-40 BC dry chemical extinguishers, with their increased extinguishing capacity over CO_2, are considered the most appropriate for use in this type of laboratory. They should be placed in the back of the laboratory as well as at each exit door to enable personnel remote from the exit to get to a unit quickly and safely.

13.4.4.1 Sprinkler. Water-sprinkling systems provide excellent fire suppression for clinical laboratories. In general, semirecessed heads in ceiling tiles provide reasonable protection against accidental breakage of sprinkler heads and resulting water damage. Further protection against accidental release can be obtained by the use of preaction systems whereby two detector heads in parallel must trigger to open a solenoid valve that allows water to pass. For example, in a clinical laboratory with a preaction-type sprinkler system, there has to be smoke as well as fire before sprinklers are actuated.

13.4.5 Codes

NFPA 99, Health Care Facility Code (NFPA 99, 1990), which covers laboratories in health-related institutions, should be consulted for additional regulatory requirements.

13.4.6 Medical Waste

The waste from clinical laboratories may have come in contact with the body fluids of patients and is presumed to be infectious. Until recently, steam autoclaving of such waste prior to municipal waste disposal was considered adequate. With recent concern about HIV infection, a climate has been created where all medical waste is considered to be infectious whether autoclaved or not. EPA and other regulators prescribe strict guidelines on how these wastes must be handled and disposed of. A typical system begins with the collection of medical waste in double red bags. The red bags are packed in cardboard boxes and shipped from the laboratory to a central location where they are burned in approved medical waste incinerators. The incinerators may be onsite, or the red bag waste may be shipped to an offsite incinerator with shipping manifests. A strictly enforced trail is required to track the waste from source to ultimate destruction. See Chapter 23 for more information on waste handling.

CHAPTER 14

TEACHING LABORATORY

14.1 DESCRIPTION

14.1.1 Introduction

A teaching laboratory should be planned, designed, and constructed to provide a safe working and learning environment for groups of students. In many high-school classes, the number of students may not exceed 30, whereas in some undergraduate college and university laboratories, the number is sometimes larger. Graduate-level laboratory instruction is normally conducted in research laboratories, which are covered elsewhere in this book. Teaching laboratories should be designed to demonstrate and encourage safe practices and operations because a disregard or ignorance of safety at the student stage will be carried over into the professional work that follows schooling. For example, laboratories designed for physics that involve electrical apparatus capable of providing serious electrical shock hazards or an ignition source potential should not be combined with chemical laboratories that use flammable liquids and gases.

Experiments carried out in a microscale chemistry teaching laboratory are similar to those performed in a typical general chemistry laboratory, except the scale is one-tenth to one-hundredth. This means that the quantities of hazardous materials that are used will be much less than in conventional chemical experiments. The introduction of microscale organic chemistry in

Guidelines for Laboratory Design: Health and Safety Considerations, Second Edition
By Louis DiBerardinis, Janet S. Baum, Gari T. Gatwood, Anand K. Seth, Melvin W. First, and Edward F. Groden.
ISBN 0-471-55463-4   Copyright © 1993 by John Wiley & Sons, Inc.

instructional laboratories was first implemented in 1983 (Mayo, 1989). By 1990, over 900 colleges and universities in the United States were conducting organic chemistry instructional laboratories at the microscale level. In addition, the introduction of microscale inorganic chemistry instructional laboratories is underway. Although microscale techniques have not become widespread in research and development laboratories, the potential exists.

The activities performed in a microscale chemistry laboratory are similar to those performed in a general chemistry laboratory and teaching laboratory except on a smaller scale. See Chapter 3.

14.1.2 Work Activities

Tasks performed in teaching laboratories will fall into two laboratory types: wet laboratories and dry laboratories. Wet laboratories employ bench experiments that use liquid, solid, and gaseous chemicals, heating devices and open flames, and that discharge both gaseous and liquid effluents.

Dry laboratories use few liquid chemicals. They are characteristic of traditional physics, engineering, and biology laboratories. Experimentation involves the use of electrical components, light generators and optical instruments, mechanical devices, and microscopes, with some use of city gas and water.

14.1.3 Equipment and Materials Used

The materials and equipment found in teaching laboratories are determined by the subjects that are taught. A general chemistry teaching laboratory, for example, will tend to resemble a general chemistry research or analytical chemistry laboratory with respect to equipment and layout, although a teaching laboratory will have its own unique features.

Teaching laboratories for physics, biology, and geology and other experimental sciences will also resemble their research counterpart with respect to equipment and materials. A primary difference is the use of most teaching laboratories by more than one group of students. Materials from one class are generally cleaned up and stored, the benches cleaned, and the area set up for the next class. Therefore, providing adequate secure and safe storage areas is a particularly significant issue for teaching laboratories.

14.1.4 Exclusions

Teaching laboratories, as defined here, are not intended for use for specialized research activities or for conducting hydraulics, civil engineering, materials testing, mechanical engineering, or electronics work. The latter laboratory types have unique requirements and more closely resemble pilot plants.

14.1.5 Special Requirements for Microscale Chemistry Laboratories

Because the chemical quantities required are one-tenth to one-hundredth those used in conventional experiments, the glassware and associated equipment must be similarly reduced in size. However, the analytical equipment (e.g., chromatographs, spectrophotometers) remains unchanged.

14.2 LABORATORY LAYOUT

14.2.1 Introduction

Teaching laboratories, wet or dry, require a maximum number of work stations in a minimum area. In spite of the pressure to maximize use of all available space, benches should be so located that easy, multidirectional movement and egress are maintained. Ease of movement is needed to facilitate getting to and from supply points or rooms, shared instruments, and fume hoods. In addition, instructors must be able to move about freely, to see all areas, and to provide quick response to emergency situations. Peninsula bench arrangements do not permit such movement, but wall benches and island benches do. Island-type benches for teaching laboratories are recommended for classes of 12 or more. Smaller classes do well working at wall benches.

The distances discussed in Section 2.2.2.2 should be maintained between benches and between benches and walls. Experience shows that 32 ft^2 per student is an absolute minimum for a teaching laboratory. This minimum should only be considered when other aspects of the design allow ideal placement of fume hoods, adequate circulation when the room is fully occupied, and rapid and easy egress in case of emergency. Consideration must be given also to adequate areas for storage and cleanup. By the time floor space per student reaches 70 ft^2 there is adequate room for design flexibility.

Distance between benches when students must work back to back must not be less than 6 ft. Otherwise, safe circulation is not possible for students and teachers who might be carrying chemicals, equipment, or other materials.

14.2.2 Laboratory Arrangements

Figure 14-1 shows a typical arrangement for a teaching laboratory. This layout features fume hoods at the rear of the room arrayed side by side and perpendicular to the work benches. Each student works facing a hood. This allows instructors to see students working at the hoods as well as to look directly into the work hood area and quickly recognize when there is an obvious hazard developing, such as a fire or run-away reaction. Good sight

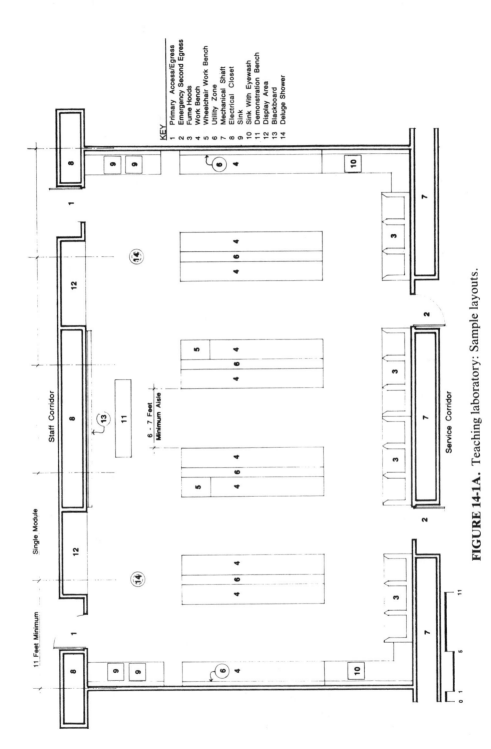

KEY
1   Primary Access/Egress
2   Emergency Second Egress
3   Fume Hoods
4   Work Bench
5   Wheelchair Work Bench
6   Utility Zone
7   Mechanical Shaft
8   Electrical Closet
9   Sink
10 Sink With Eyewash
11 Demonstration Bench
12 Display Area
13 Blackboard
14 Deluge Shower

FIGURE 14-1A. Teaching laboratory: Sample layouts.

FIGURE 14-1B. Teaching laboratory.

lines give alert instructors and students a small but significant headstart to react to an emergency in an appropriate manner. A disadvantage of the arrangement shown in Figure 14-1 is that students may use the aisle in front of the hoods to traverse the laboratory, producing sufficient traffic in front of fume hoods to compromise effective air capture. To reduce traffic in front of the hoods, commonly used instruments, supplies, and less hazardous resources should be located at the ends of benches toward the instructor's demonstration table. This arrangement increases student circulation in the less hazardous sector of the laboratory, an example of the hazard zoning concept discussed in Chapter 2. An understanding of staffing practices is critical when deciding upon the layout of a chemistry or other type of teaching laboratory. For example, when there is a low student/instructor ratio, good laboratory visibility may be less difficult to arrange. In laboratories that do not require many fume hoods, or experimental setups that obstruct sight lines, there will be far fewer restraints on laboratory designs.

14.2.2.1 *Organic Chemistry Teaching Laboratories.* In organic chemistry teaching laboratories, where there may be a requirement for up to one fume hood per student, fume hood density becomes very high and designing a safe laboratory layout is more difficult. Examples of good design solutions are to intersperse fume hoods with individual student benches, on a one-to-one basis, or to have every two students share one fume hood. Either arrangement reduces the potential traffic in front of each hood by limiting access to a few students and an instructor. In addition, the transfer of materials from bench to hood becomes very convenient and encourages students to use the hood. However, it is necessary to weigh the advantages of close proximity between student bench and fume hood against the greater safety margin inherent in having banks of fume hoods separated from the benches shared by students. In addition, arrangements that distribute fume hoods throughout the laboratory tend to restrict good sightlines that cover the entire laboratory because the hood superstructures are so tall. This means that neither instructors nor students are able to observe a safety problem occurring at another work station or in another part of the laboratory. In more traditional laboratory arrangements, a clear view of the entire laboratory is possible.

14.2.2.2 *Microscale Chemistry Teaching Laboratory Arrangement.*

The layout of a microscale chemistry laboratory resembles a General Chemistry Laboratory (Chapter 3) or a conventional Teaching Laboratory (Chapter 14). All the items described in Sections 3.1.2, 3.2.2, 14.1.2, and 14.2.2 should be reviewed, and those that are relevant should be implemented. The major changes will be in the furniture and ventilation requirements.

With the increased use of microchemistry analytical techniques, careful consideration should be given to the individual student work stations. For example, work stations may be designed for the student to be seated rather than standing because greater manual precision is needed and a seated position with both feet on the floor improves balance, postural stability, and manual precision. In this case, the benches should be 29 to 32 in. in height and leg space must be provided by knee holes in the benches used for manipulating microchemical instruments and materials. When work benches are low, to permit work while seated, the aisle should be wide enough for students seated back to back to push back their chairs and still leave room for instructors and other students to pass. A 7-ft aisle width is recommended.

14.2.3 Egress

There should be a minimum of two exits from each teaching laboratory, with each exit opening into a separate fire-safe egress. When teaching laboratories are large, additional exits may be required to be certain that a travel distance of 50 ft to an exit is never exceeded. All exit doors should swing in the direction of exit travel.

All of the remaining egress recommendations in Chapters 1 and 2 should be followed.

14.3   HEATING, VENTILATING, AND AIR CONDITIONING

The HVAC recommendations contained in Chapters 1 and 2 are generally applicable to teaching laboratories and should be considered for implementation. Additional comments regarding HVAC facilities for teaching laboratories follow.

Microscale laboratories usually require lower room ventilation rates than do macroscale chemical laboratories because the potential for contamination is substantially reduced. A rate between 0.5 to 1.0 cfm per square foot or 4 to 6 air changes per hour is generally adequate when fume hoods are not in operation.

14.3.1   Chemical Fume Hoods and Other Local Exhaust Systems

Fume hoods should be located so that they are remote from the main entrances and exits, and do not face exit routes or block them in the event of a fire, explosion, or a violent reaction within the hood. Hoods should be located near the back or outer walls of the laboratory, near the least-traveled pedestrian routes but within easy access to the students.

At least one local exhaust facility is desirable at each bench in a wet laboratory to provide a convenient location where effluents from small fuming, smoking, or noxious experiments can be removed safely. We recommend one individual local exhaust facility per student in organic chemistry teaching laboratories but only one per four work sites for general chemistry teaching laboratories. To comply with the principle of hazard zoning (see Chapter 2), local exhaust facilities should be located on the bench nearest the most hazardous zone of the laboratory. Bench exhaust facilities are not usually needed for physics and similar teaching laboratories, but at least one chemical fume hood should be provided in every laboratory for solvent dispensing and dry activities that should not be conducted on an open bench. Downdraft exhaust ventilation should be considered as a substitute for laboratory hoods where microanalytical techniques are used. Examples of local exhaust ventilation for microscale chemistry are shown in Chapter 31, Section 31.8.

14.3.1.1   Local Exhaust Systems for Microscale Chemistry Laboratories. There will be a reduction in the number of fume hoods needed in microscale teaching laboratories over those required in conventional laboratories because the quantity of chemicals used and the size of the experimental equipment are reduced by one-tenth to one-hundredth. This means that most experiments can be performed conveniently at a lab bench, and consideration should be given to the use of a modified downdraft bench or the installation of slot exhaust ventilation along the rear of the laboratory bench in place of

conventional laboratory hoods. The ACGIH *Industrial Ventilation Manual* provides design guidance.

14.4   LOSS PREVENTION, INDUSTRIAL HYGIENE, AND PERSONAL SAFETY

The loss prevention, industrial hygiene, and personal safety recommendations contained in Chapters 1 and 2 are applicable to teaching laboratories and should be considered for implementation in addition to the recommendations that follow.

14.4.1   Emergency Showers

Emergency showers for large teaching laboratories should be placed inside the laboratory proper and so located that no more than 25 ft of travel distance is required from any point. Showers should not be placed in front of chemical or flammable liquid storage cabinets or shelves, or directly in front of fume hoods. Additional requirements for emergency showers are given in Appendix I.

14.4.2   Emergency Eyewash

For laboratories using chemicals and containing four or more multistudent benches, an eyewash facility should be located at each bench. It can be a hand-held type. For laboratories with fewer than four benches, there should be at least one eyewash facility per laboratory and it should be so located that no more than 3–4 s is required to reach it from the most remote work station. In addition to the hand-held type eyewash mentioned above, at least one tempered full-face eyewash facility should be located in each teaching laboratory. Additional information on eyewash facilities is given in Appendix II and in Section 2.4.1.5.

14.4.3   Chemical Storage and Handling

No highly reactive or flammable chemicals should be stored in a teaching laboratory. An adjacent chemical storage room or a specially constructed protected area should be provided for this purpose. In teaching laboratories, provisions should be made to shelve or otherwise hold only the amount of chemicals necessary for a day's or a single class' experiments. Construction details of safe chemical shelving and storage cabinets are covered in Section 2.4.

Where large quantities of flammable solvents are stored a modified fume

FIGURE 14-2A. Solvent dispensing hood.

hood as seen in Figure 14-2 can be used. The solvent dispensing stations are built into the hood along with a fire supression system.

14.4.3.1 *Microscale Chemistry Laboratories.* Because the quantities of chemicals used are small, storage space and shelving needs will be reduced in proportion.

14.4.4 Hazardous Chemical Disposal

Central points for the collection and temporary storage of chemical waste should be provided in each laboratory. They should be remote from students at their work sites and not located in egress routes. Locations near fume hoods are recommended. Large waste storage areas are unnecessary because wastes should be removed at least daily.

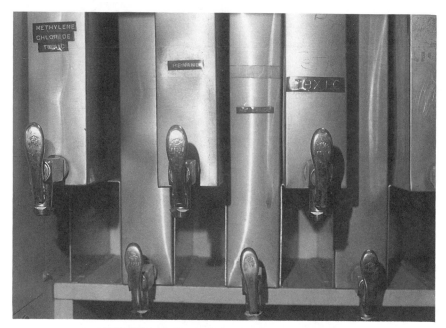

FIGURE 14-2B. Individual solvent dispenser.

14.4.5 Fire Extinguishers

Provisions should be made for locating fire extinguishers within each teaching laboratory. With island benches, one fire extinguisher should be located at each bench. The type of extinguisher is dependent on the use of the laboratory. A clean agent such as CO_2 is appropriate for chemical operations. Size 4A-40 BC or larger ABC-type dry chemical units should be located in the hall to be used as a backup.

14.5 SPECIAL REQUIREMENTS

14.5.1 Preparation Room

There should be a room associated with each teaching laboratory, or group of teaching laboratories, that can be used for the preparation of experimental equipment and materials. If the teaching laboratories do not involve the use of chemicals or hazardous substances, no special facilities are needed. When chemicals are used, the following considerations are important.

   Chemicals stored in the preparation room (when not stored in the areas referred to in Section 14.4.3) will be in the nature of bulk chemical storage. Approved storage cabinets should be provided (see Sections 1.4.7 and 2.4.6) in adequate numbers to handle all flammable liquids. Provisions should be made to store all chemicals according to safe compatibility characteristics. The preparation room should have good general ventilation to dilute released materials below their hazard level—that is, explosivity in the case of flammables and toxicity in the case of toxic materials.

The preparation room should have a well-planned fire-suppression capability that includes fixed automatic fire-suppression facilities and hand-portable fire extinguishers of the ABC dry chemical type with ratings of 4A-40 BC or better. Hand-portable extinguishers should be located strategically to assure that a fire can be attacked quickly and kept from threatening the laboratory itself.

An arrangement such as a pass-through or a Dutch door should be used to eliminate the need for students to enter the preparation room.

CHAPTER 15

GROSS ANATOMY LABORATORY

15.1 DESCRIPTION

15.1.1 Introduction

The gross anatomy laboratory is a group of spaces for the preparation, storage, and dissection of human cadavers, animal carcasses, or portions thereof, for the purpose of teaching gross anatomy or for research. The spaces described in this section include the receiving area, morgue, cold storage room, and dissection room. Emphasis will be given to the gross anatomy teaching laboratory because of its special requirements beyond those of anatomical research. Guidelines in Chapters 1 and 2 should be followed closely for anatomical research laboratories.

15.1.2 Work Activities

15.1.2.1 Morgue. Medical research and teaching institutions receive anatomical gifts from individuals, or their families, who arrange before death for the donation of their bodies after death. Deceased persons are brought to the morgue receiving dock by morticians or family members. Donated bodies are prepared in a manner that preserves all tissues, so that the cadaver can remain unrefrigerated for long periods without harmful deterioration. The cadavers are pickled, then placed either in cold boxes for short-term storage,

Guidelines for Laboratory Design: Health and Safety Considerations, Second Edition
By Louis DiBerardinis, Janet S. Baum, Gari T. Gatwood, Anand K. Seth, Melvin W. First, and Edward F. Groden.
ISBN 0-471-55463-4   Copyright © 1993 by John Wiley & Sons, Inc.

or in freezers for long-term storage. After dissection or use, all of the remains are returned to the morgue to be transported for burial or cremation.

15.1.2.2 Dissection Laboratory. Gross anatomy is generally taught by lectures in conjunction with laboratory sessions. In the laboratory, students inspect and identify all tissues revealed by dissection. Dissections are usually done by students, but may be done by instructors in the case of many allied health science courses. Activities include cutting, sawing, manipulation with fine instruments, and cleaning up. During laboratory sessions, instructors may give demonstrations with models and visual aids. They may also use previously dissected specimens called *prosections*.

In typical medical training, a single cadaver may be dissected by a group of 4 to 6 students during the entire course, which may be as short as two weeks or as long as a semester. Students refer to their anatomy textbooks, adjacent to the dissection table, as they manipulate and view the cadaver. Both visual and tactile information is gathered. Students must be able to inspect the cadaver from all angles and at very close range, 1 in. to 1 ft away. At many schools, students are required to clean up the dissection rooms after each class period, and therefore a number of large sinks are needed.

15.1.2.3 Research Laboratory. The activities in research laboratories that use gross anatomical specimens may include cutting, mixing, mechanical or biochemical procedures, tissue culture, microscopy, and tissue sample preparation, staining, and labeling with radioisotopes.

15.1.3 Equipment and Materials Used

15.1.3.1 Morgue. The morgue contains mortuary equipment for the removal of blood, body cleansing, and perfusion of the bodies with preservatives. Chemicals used include glycerin, glycol, formalin (formaldehyde), phenol (carbolic acid), soap, and water. Organic pathologic materials may be present. Personnel must be carefully trained to recognize and safely handle these specimens. Saws and other instruments are used to collect body parts for separate storage in jars or for prosections. There are large scales for weighing specimens and materials-handling equipment, which may include carts, gurneys, and forklift trucks.

All surfaces within the morgue should be impermeable, nonporous, and washable. Formaldehyde vapor permeates all paper, cloth, plastic, and wood materials stored in the morgue. Permeable supplies should be stored in a separate room. Bulk chemicals should also be stored separately. Floors should be covered with a seamless, smooth material, with an integral cove that extends 8 in. up each wall. Flooring that becomes slippery when wet should not be used. Flooring that may crack under heavy loads of a forklift truck should be avoided. All drains in a morgue should have grease traps, so that organic matter can be cleaned out of the waste water.

Walls should be waterproof, seamless, and smooth. Glazed tile or epoxy paint on block, metal lathe and plaster, or water-resistant sheetrock is recommended.

Plaster or concrete ceilings can be painted with epoxy. Suspended ceilings may be a lay-in type of metal pan, preferably nonperforated. Light fixtures, wall outlets, and switches should be vaporproof because of the moist environment brought about by frequent cleaning procedures.

15.1.3.2 Dissection Laboratory. The materials present in the gross anatomy dissection laboratory may include preserved bodies, fresh unpreserved anatomical specimens, organic waste matter, and chemicals. Chemicals include glycerin, glycol, formalin (formaldehyde), phenol (carbolic acid), soap, and water. Equipment, in addition to dissection instruments, may include saws, drills, and cleaning tools. There will be dissection tables which may be mobile or fixed. Teaching aids may include models, skeletons, charts, and audiovisual equipment.

Formaldehyde, recognizable at concentrations lower than one part per million in air, will be ever-present. To help reduce this problem, all surfaces should be impermeable, nonporous, and smooth, as in the morgue. Even more than in the morgue, surfaces here should be easily cleanable, because students may do the work. The recommended material for all sinks and furnishings in a dissection laboratory is stainless steel. All metal joints should be welded and coved for thorough cleaning of tables, sinks, countertops, and storage cabinets. Wood and plastic materials should be avoided for construction or in furnishings.

For ceilings in gross anatomy dissection laboratories, acoustical dampening qualities may be more important than impermeability. The noise level during classroom hours could make a dissection laboratory with all hard surfaces unbearable.

All floor and sink drains should have grease traps for capture of organic matter, although cleaning of these traps may cause a service personnel problem. A means of mechanizing this chore should receive serious consideration in the design planning phase.

15.1.3.3 Research Laboratory. The materials present in a research laboratory that uses gross anatomical specimens are those listed above for the dissection laboratory. In addition, all the chemicals, equipment, and instruments associated with a general research laboratory may be present.

15.1.4 Exclusions

15.1.4.1 Morgue. An educational or research institution morgue should not be used for clinical autopsies or commercial use. The morgue should be restricted from public access, including donors' family members and students. Authorized, trained mortuary technicians and faculty are charged with operating the morgue to maintain the privacy and dignity of those who make the donations.

15.1.4.2 Dissection Laboratory. The dissection laboratory is not an operating room or laboratory for experimentation with live animals. The dissection laboratory's electrical and ventilating systems are not designed for use in an atmosphere of flammable anesthetics.

15.2 LABORATORY LAYOUT

15.2.1 Introduction

The public may regard the activities pursued in morgues and dissection laboratories with alarm, disgust, or morbid curiosity. To secure these areas from public access and view, as well as to ensure odor containment, the location of the morgue and dissection laboratories should be carefully considered so as to provide for the following:

Access to a loading dock and receiving room.
Private passageway between receiving room and the morgue. Private passageway between morgue and dissection laboratories.
Windows not overlooked by other buildings.
Entries to morgue and dissection laboratories not directly from a main corridor

There should be a covered receiving dock and enclosed room that contains a mortuary-type refrigerator or cold box for deposit and temporary storage of bodies brought by a mortician or others when the morgue is unattended. The cold storage facility should be large enough for donations made over long holiday weekends to avoid stress on family members and discomfort for morgue staff. Access from this room to the morgue should be restricted. Refrigerators or cold boxes that can be opened from either side permit discreet transfer of donations directly into the morgue or the secured passageway to the morgue. The receiving room should be adjacent to a morgue on the same level or connected by a key-restricted elevator to a morgue on a different level. Similar private circulation is desirable between the morgue and the dissection rooms to reduce the chance of accidental contact with noninvolved students and staff or with the public.

15.2.2 Individual Room Arrangements

15.2.2.1 Morgue. The layout of a morgue for a research or medical education institution depends on the projected quantity of anatomical gifts, number of staff, and whether there is a permanent anatomical collection or museum.

The morgue consists of a preparation room, materials and chemical storerooms, pickling rooms, cold rooms, freezer rooms or mortuary refrigerators, sample preparation rooms, and locker rooms. There should be space in the preparation room for as many gurneys as needed to transport all the donations held in the receiving area. Adequate space is needed to make it easy to maneuver the heavily loaded gurneys and carts from one procedure area to the next. Bodies are moved temporarily to racks in ambient room-temperature storage areas immediately adjacent to the preparation area. The alignment of racks and aisles should be arranged so that the staff can safely move the cadavers into cold rooms or freezer rooms adjacent to the morgue or dissection laboratories. In the cold rooms, cadaver storage can be horizontal or vertical. Every area in the morgue should be arranged to facilitate transport of very heavy and stiff specimens.

Support areas for materials and chemical storage, sample preparation, offices, and staff locker and shower facilities should be immediately adjacent to the preparation room. These functions are kept together so that the formaldehyde odor can be contained.

15.2.2.2 Dissection Laboratory.
A gross anatomy teaching laboratory consists of locker/shower rooms for men and women, dissection rooms, and storerooms. These functions should be kept together for security and odor containment. Seminar rooms may also be adjacent to the laboratory suite. Circulation need not be in a one-way loop, as in a surgical area, to maintain sterility. Gross anatomy laboratories generally do not have concerns for sterility or for biohazards; microorganisms do not flourish in properly prepared cadavers. However, if fresh specimens from operating rooms are investigated, universal microbiological precautions should be maintained.

Dissection tables are 2 ft 11 in. high, 6 ft 6 in. long for normal-sized human cadavers, and 2 ft 6 in. wide. They are usually arranged with minimum clearances of 50 to 54 in. head to toe and 40 to 50 in. side to side to accommodate an average size cadaver and four students with their books and instruments. More area per table and an adjustable height table are required if students in wheelchairs attend the class. There should be good light and excellent ventilation at all tables. Large scrub sinks should be provided for handwashing and general cleanup. Countertops and display cases can be distributed along the perimeter of the room for additional teaching materials. Skeletons require hangers or racks, but all should be stored in locked cabinets or storerooms to protect them from theft. Blackboards and projection screens should be positioned so that all students can see them from the vantage point of the dissection tables.

15.2.2.3 Research Laboratory.
The layout recommendations contained in Chapters 1 and 2 are generally applicable to research laboratories that use gross anatomical specimens, and they should be followed closely.

15.2.3 Egress

The egress recommendations contained in Chapters 1 and 2 are generally applicable to gross anatomy laboratories and should be followed closely, in addition to the following recommendation that applies specifically to morgues.

15.2.3.1 Morgue. There should be two exits from the morgue preparation area. Exits from cold and freezer rooms should be designed with the egress recommendations contained in Chapters 1, 2, and 9 firmly in mind. All doors in environmentally controlled rooms should be operable from either side, so no one can be trapped inside.

15.3 HEATING, VENTILATING, AND AIR CONDITIONING

15.3.1 Introduction

HVAC systems for anatomy laboratories are critical because they must provide a safe environment for those who must work in a laboratory with potentially high concentrations of formaldehyde, phenol, and other toxic chemicals. OSHA 29 CFR 1910.1048 limits 8-h exposure to formaldehyde to 0.75 ppm and limits short-term, 15-min exposure to 2 ppm. The HVAC recommendations contained in Chapters 1 and 2 are generally applicable to gross anatomy laboratories and should be followed closely. The basic intent is to maintain formaldehyde, phenol, and other chemical concentrations below objectionable levels. The odor threshold for formaldehyde is <1 ppm. Odor control in the laboratory and surrounding area is also critical. Additional recommendations follow.

An outside air change rate of 15 to 20 per hour is recommended for laboratory and morgue. It is best to introduce the supply air high in the room and to exhaust it from a low room location to draw the contaminated air below the work area and well below the breathing zone. Specially equipped fixed autopsy tables with built-in downdraft exhaust can provide local capture of chemical fumes more effectively than general room exhaust points located near the floor.

Recirculation of air is not recommended. Certain chemicals, such as potassium permanganate, that have good absorption/adsorption properties for formaldehyde have been used in some cases to remove formaldehyde vapors from recirculated air. For such systems to be successful, large recirculation rates must be maintained and careful monitoring of the activity of the air cleaning chemicals must be conducted on a continuing basis.

Local exhaust or capture hoods may not be feasible for gross anatomy teaching laboratories because students tend to work extremely close to the parts being dissected. Therefore, low room air exhaust grilles located on the perimeter walls are likely to be the most effective solution to embalming fluid vapor exposure. Nevertheless, local exhaust opportunities should be studied for each function to determine the feasibility of this method of vapor and odor control. Use of stainless steel ventilation ducts should be considered because of their corrosion resistance. The room should be at a negative pressure with respect to all public areas. No special treatment of exhaust air is needed provided that the room exhaust air can be discharged well above all surrounding buildings and terrain features.

15.3.2 Individual Room Requirements

15.3.2.1 Morgue. Cadavers may be prepared before they are recieved in the morgue or storage area. However, many are prepared in the morgue. The embalming process consumes substantial amounts of formaldehyde solution. Exposure levels are necessarily higher when embalming occurs in the morgue.

Automatic feed systems into closed vented vessels should be used to prepare large volumes of embalming fluid. Small quantities can be prepared in a laboratory fume hood or ventilated enclosure.

15.3.2.2 Dissection Laboratory. In gross anatomy teaching laboratories, students will be dealing with cadavers already prepared and preserved in embalming fluid, an aqueous solution of formaldehyde, methanol, ethanol, phenol, and glycerine. The extent of the chemical exposure depends upon the type of dissections made and the organ being studied. For example, levels of exposure are higher for dissections involving the opening of a body cavity. The greatest exposure appears to be in the abdomen. Quantitative measurements substantiate this experience. Personal exposure concentrations can easily exceed OSHA's permissible exposure level if adequate ventilation is not provided. The exposure levels sometimes do not decrease with time because, as cavities dry out, students intermittently wet the cadaver with additional embalming fluid.

15.4 LOSS PREVENTION, INDUSTRIAL HYGIENE, AND PERSONAL SAFETY

The loss prevention, industrial hygiene, and personal safety recommendations contained in Chapters 1 and 2 are generally applicable to gross anatomy laboratories and should be followed closely.

CHAPTER 16

PATHOLOGY LABORATORY

16.1 DESCRIPTION

16.1.1 Introduction

A pathology laboratory suite may consist of a clinical or research laboratory plus support areas that often include a clinical morgue, autopsy room, organ storage, slide preparation and storage, specimen and slide reading area, photography laboratory, and a small conference room for review of autopsy materials between pathologists and physicians. The support areas are based within a hospital; the research laboratory may be at the same location or at another institution.

16.1.2 Work Activities

16.1.2.1 Clinical Morgue Autopsy Room. The performance of autopsies is the primary activity within a clinical morgue. Autopsy is legally required in many jurisdictions for persons who die of unknown, accidental, or suspicious causes or of certain infectious diseases. Otherwise, permission of the family or next of kin is required. For the deceased patient, autopsy is care that extends beyond death and calls for cooperation between pathologist and clinician. An autopsy determines the cause of death and condition of all organ systems by means of gross and microscopic inspection. The pathologist

Guidelines for Laboratory Design: Health and Safety Considerations, Second Edition
By Louis DiBerardinis, Janet S. Baum, Gari T. Gatwood, Anand K. Seth, Melvin W. First, and Edward F. Groden.
ISBN 0-471-55463-4   Copyright © 1993 by John Wiley & Sons, Inc.

collects samples of all organs to preserve for investigation or research. After the autopsy, the pathologist presents the results to the clinician for review, confirmation of diagnosis and treatment, and discussion. Autopsies provide valuable data on effects of disease on the human body as well as effectiveness of methods of treatment.

An animal morgue is another type of clinical morgue associated with a hospital, research institution, or commercial animal-breeding facility. For further reference read Chapter 15.

16.1.2.2 Sample Preparation Room. Activities conducted in a sample preparation room include fixing whole organs and organ samples by immersion in sinks, tanks, or jars of chemicals and by freeze-drying. In addition, pickled, frozen, and freeze-dried organ tissues may be thinly sliced, fixed in chemical solutions, stained with a variety of dyes and chemicals, and then mounted on slides for microscopic examination. If volume of samples prepared in the pathology laboratory is high, a separate histology laboratory for slide preparation may be warranted. This separation is preferred because the chemicals used for slide preparation may be toxic.

16.1.2.3 Cold Storage Room. In a hospital morgue, bodies remain in cold storage for only a short time before they are autopsied or removed for funeral preparation by morticians. Family members may view the deceased before removal. It is desirable that the hospital provide a separate room adjacent to the cold room for this purpose. It is important that the design of this room is sensitive to the difficulties of persons who have come to view the deceased.

16.1.2.4 Pathology Laboratory. Activities conducted in research or clinical pathology laboratories involve the use of diseased and damaged tissues plus contaminated or infectious materials of living organisms. Operations include tissue cutting, dissection, chemical mixing, mechanical manipulation of organs and tissues, biochemical procedures of widely varying nature, microscopy, microbiological culturing, sample preparation, and staining.

16.1.3 Materials and Equipment Used

16.1.3.1 Clinical Morgue Autopsy Room. Autopsy rooms contain autopsy tables, scales, instrument cabinets, bone saw, electric drills, local exhaust devices, tape recording equipment, tissue cleansing sinks, and handwashing sinks. Many hospitals use the concept of universal precautions; i.e. patients are considered infectious unless proven otherwise. Autopsies therefore are done with use of significant protective gear for staff. An autoclave to sterilize contaminated or infected materials should be within or adjacent to the morgue. The autopsy room may also have an ultracold freezer for immediate preservation of tissues plus other refrigerators for storing donated organs. Chemicals used in an autopsy room include soap, water, alcohol, and decontamination agents, such as chlorine- or phenol-containing solutions.

All surfaces should be impermeable, nonporous, and washable so the morgue can be decontaminated. Flooring should be seamless and nonslippery, with an integral cove extending at least 8 in. up each wall. All drains should have strainers and grease traps so that organic matter can be cleaned out. Walls should be waterproof, seamless, and smooth. Glazed tile or epoxy-painted concrete block, lathe and plaster, or water-resistant sheetrock on metal studs are recommended. All storage cabinets or cupboards should be stainless steel.

Plaster or concrete ceilings can also be painted with epoxy. Suspended ceilings may be lay-in metal pan. High concentrations of formalin- and formaldehyde-containing chemicals are not generally found in high concentrations in the autopsy room itself, so sound-attenuating material may be used above the metal pan ceiling. Due to the presence of moisture from frequent cleaning procedures, light fixtures, wall outlets, and switches should, preferably, be vaporproof.

16.1.3.2 *Sample Preparation Room.*

Equipment in a sample-preparation laboratory may include a cryostat, chemical fume hood, lyophilizer, microtome, and a large sink with gravity feed or a circulating pump for formalin or other chemical fixation. Contaminated materials and infectious organisms may be present. Tissues from patients should be handled in biological safety cabinets specially designed for these agents (see Chapter 12 and Section 31.8). Figure 16-1 shows tissue-trimming work stations designed for organ and tissue dissections. Staff must be trained to handle infectious specimens safely, according to approved microbiological techniques, and they must autoclave all materials before disposal to destroy harmful biological agents.

Chemicals include formalin, formaldehyde, xylene, phenol (carbolic acid), soap, water, and various histological stains and fixatives. Concentrations of some of these chemicals may produce a high explosion potential in the sample preparation room. Install local exhaust devices and work stations for procedures that are potentially hazardous, such as the slide preparation exhaust chamber illustrated in Figure 16-2, which is designed to reduce xylene fumes.

All surfaces should be impermeable, nonporous, and washable, as in the autopsy room. Flooring should be seamless and nonslippery, with an integral cove extending at least 8 in. up each wall. All drains should have strainers and grease traps so that organic matter can be cleaned out.

Ceilings should not have sound attenuating materials because high concentrations of formaldehyde will be present. OSHA 29 CFR 1910.1048 sets the limit for long-term exposure at 0.75 ppm for an 8-h exposure and sets the short-term exposure limit at 2 ppm for a 15-min exposure.

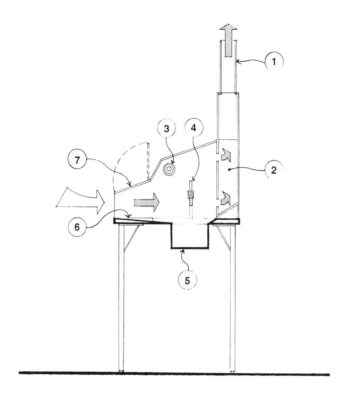

KEY

1 Exhaust Duct
2 Negative Pressure Exhaust Plenum.
3 Acrylic Tube With Removable
  Light Inside.
4 Faucet
5 Double Chamber Sink
6 Flat Cutting Surface
7 Acrylic Face Shield
☐ Clean Air
▨ Contaminated Air

AIR FORMULAS

A = B

FIGURE 16-1A. Tissue-trimming work station: Section.

FIGURE 16-1B. Tissue-trimming work station.

16.1.3.3 Clinical Pathology Laboratory. A clinical pathology laboratory will contain contaminated or infectious materials as well as chemicals, equipment, and analytical instruments associated with a general research laboratory.

16.1.4 Exclusions

16.1.4.1 Morgue. The clinical morgue should be physically isolated and administratively restricted from public access and from hospital staff who do not work in the morgue or have related activities.

16.2 LABORATORY LAYOUT

16.2.1 Introduction

The clinical morgue and pathology laboratory are critical facilities in a hospital, but patients, visitors, and noninvolved personnel should be reasonably shielded from these areas. As with medical education morgues, the location of the clinical morgue is important. There should be a covered dock for receiving bodies for autopsy and for removing bodies for funeral preparation. This dock should not be shared with receiving, materials handling, or trash

FIGURE 16-2. Slide preparation station.

removal activities. If the morgue is not adjacent to the dock, a short, private corridor or key-operated elevator should connect them directly without passage through other public areas. For transport of patients who die in the hospital, personnel must use corridors and elevators used by the public and other staff. Unless specially designed gurneys are used to discretely transport dead bodies, it is better for nonmedical staff morale if the morgue is away from main service corridors so they do not routinely see corpses. The location of clinical pathology laboratories is not as critical. Tissue samples from hospital operating rooms or from autopsy can be transported safely and discretely in closed, opaque containers. If highly infectious materials are frequently used, the laboratory should be adjacent to operating rooms and the morgue.

16.2.2 Individual Room Arrangements

16.2.2.1 Autopsy Room. Layout of a clinical autopsy room for a hospital will depend upon the projected average number of autopsies that will be performed per day. In many hospitals this number has decreased significantly in the past decade, to the point that a major medical center may perform only a few per day, because many physicians do not recommend autopsy, out of their concerns about malpractice suits. Family members not encouraged to permit an autopsy are unlikely to request one. If the number of procedures is low, only one room with one autopsy table may be needed.

In a teaching hospital the autopsy room should be of adequate size to allow a group of physicians or medical students to witness the autopsy without getting in the way. One major teaching hospital has fixed metal bleachers on one side of the room at the foot of the autopsy table to allow a large audience to watch quietly, without causing difficulties by moving about. On the other hand, video cameras can record autopsies very effectively for later presentations to students and physicians. Check with the head of pathology and hospital administration on their particular requirements.

Layout of a clinical autopsy room is based on providing excellent-quality light and ventilation to the autopsy table, bone saw, and tissue preparation sink. The autopsy table(s) should be placed in the center of the room to allow the pathologists to move with ease all around the table. There should be sufficient space for at least one gurney to remain in the room without interfering with circulation. The bone saw may be located in a separate small chamber with local exhaust, within the autopsy room in order to contain the spray of bone particles. Preparation sinks (with local exhausts), instrument cabinets, freezers, and refrigerators are normally arrayed conveniently along the room walls. There should be sufficient space between floor-mounted equipment and storage units so that the floor can be easily and thoroughly scrubbed.

16.2.2.2 Sample Preparation Room. Size and layout of a sample preparation room will also depend on the volume of tissue samples that must be processed daily and on the number of staff normally occupying the room. The concentration of formaldehyde and other fumes may be very high. When sample preparation rooms are also used for long-term storage of pickled samples, they can become dangerously overcrowded and cluttered. Consideration should be given to providing a sample storage area with excellent ventilation, so that the preparation room can function more efficiently and safely. This is particularly recommended for large facilities that may have special tissue collections that require very-long-term storage. The preparation sink(s) and stainless steel work counters should be separated from storage shelves or racks by an aisle that is a minimum of 5 ft wide to allow ease of movement for staff actively handling samples without fear of bumping into or knocking over specimen containers. The 5-ft width should not diminish from the work area to the room exit because this space also allows carts and hand-trucks to safely deliver large containers of chemicals and supplies.

16.2.2.3 Cold Storage Room. The layout of a cold storage room is simple. There can be a single aisle with space for gurneys on one or both sides, according to the number needed per day. The entry from the autopsy room should be separate from an entry to the family viewing room by doors on both ends of the central aisle. Unlike cold storage for a gross anatomy laboratory, this room is for short-term storage. Should the autopsy room be in a trauma center, a civic or forensic morgue, or likely to function as the primary facility in a major disaster or in a war zone, additional cold storage similar to that provided for a gross anatomy laboratory will usually be required.

16.2.2.4 Pathology Laboratory. The layout recommendations contained in Chapters 1 and 2 are generally applicable to research laboratories that are used for pathology research, and they should be followed closely. In addition, where materials classified as biohazard level 2 or above are expected in the pathology laboratory, all the protective measures cited in Chapter 12 for a biosafety laboratory should be considered for use.

16.2.3 Egress

The egress recommendations contained in Chapters 1 and 2 are generally applicable to pathology laboratories and should be followed closely, in addition to the following recommendation.

16.2.3.1 Cold Storage Room. Exits from cold boxes and walk-in freezers should be designed with the egress recommendations contained in Chapters 1, 2, and 9. All doors should be operable from both sides.

16.3 HEATING, VENTILATING, AND AIR CONDITIONING

16.3.1 Introduction

Environmental control in the pathology laboratory could be critical, depending upon the use of the facility. An important difference between a community hospital and a large teaching hospital is the larger number of autopsies performed in the teaching hospital. During an autopsy, organs and tissue samples will be removed for study and exmination. Unless they are immediately frozen at very low temperature, they will be preserved in a formaldehyde solution. This usually results in very high concentrations of formaldehyde in the area. Patients who die with a serious communicable or infectious disease are usually autopsied. This presents a need to control an infection hazard; in some hospitals, a separate area with strict access control and special facilities designated for this service. HVAC systems must deal with each of these environmental control issues.

16.3.2 System Description

The information contained in Section 15.3.1 also applies to hospital pathology laboratories.

16.3.3 Local Exhaust Systems

Local exhaust ventilation hoods or capture hoods are strongly recommended for hospital pathology suites where large quantities of formaldehyde are used in open trays for transfer of tissues. OSHA 29 CFR 1910.1048 sets the limit for long-term exposure at 0.75 ppm for an 8-h exposure and sets the short-term exposure limit at 2 ppm for a 15-min exposure. Local capture hoods placed at work level are effective for controlling formaldehyde exposure (see Figure 16-1). Care must be taken in positioning such hoods to prevent obstruction of the procedure. A good location for them is near the sink area. Stainless steel ducts are recommended for corrosion control. Fan-operated exhaust facilities for autoclaves are strongly recommended. Canopy-type hoods above the access door are commonly employed, and they should be placed as close to the autoclave as possible.

16.3.4 Special Autopsy Requirements

For autopsy rooms dedicated to infectious cadavers, the use of stainless steel for ducts, air outlets, and all other equipment and facilities is desirable for its ease of cleaning. The room must be at a negative pressure with respect to its surroundings. No recirculation of air is permitted. It is strongly recommended that HEPA filters be placed in the exhaust air system. It is also necessary to provide facilities for paraformaldehyde vapor decontamination of the entire room by adding airtight shutoff dampers in the supply and exhaust ducts, gasketing doors, and sealing utility penetrations. All of the precautions and safety regulations described for biosafety laboratories (Chapter 12) should be observed when handling infectious cadavers.

16.3.5 Supply Air Systems

Supply air requirements will be similar to those for other laboratories. Air conditioning is highly desirable.

16.4 LOSS PREVENTION, INDUSTRIAL HYGIENE, AND PERSONAL SAFETY

The loss prevention, industrial hygiene, and personal safety recommendations contained in Chapters 1 and 2 are generally applicable to pathology laboratories and should be followed closely.

16.5 SPECIAL REQUIREMENTS

16.5.1 Waste Fluids

Disinfection of waste fluids from these facilities should be considered. Consult a safety engineer, industrial hygienist, or infection control authority for assistance.

CHAPTER 17

TEAM RESEARCH LABORATORY

17.1 DESCRIPTION

17.1.1 Introduction

A team research laboratory may occupy a single unpartitioned space larger than the average size of a two- to four-module laboratory. It is not defined by the specific activities conducted within, but rather by its large size. The term "team laboratory," as used here, is not interchangeable with the designation "laboratory unit." A laboratory unit, according to NFPA, is identified as an aggregation of laboratory-use spaces defined by the rating of fire separation between the individual spaces within the unit, as well as between the unit and abutting spaces. A team laboratory may also be a laboratory unit, but it need not be. The distinguishing feature of this laboratory type is the interdisciplinary nature of the activities conducted therein. The large size of a team laboratory often poses unique problems in layout, HVAC, and loss prevention.

17.1.2 Work Activities

Activities conducted in a team research laboratory include any or all of those carried on in general chemistry, analytical chemistry, physics, clean room, controlled environment, and radiation.

Guidelines for Laboratory Design: Health and Safety Considerations, Second Edition
By Louis DiBerardinis, Janet S. Baum, Gari T. Gatwood, Anand K. Seth, Melvin W. First, and Edward F. Groden.
ISBN 0-471-55463-4   Copyright © 1993 by John Wiley & Sons, Inc.

17.1.3 Materials and Equipment Used

Equipment used in the team research laboratory may include any of the items listed as characteristic of the several laboratory types listed here in Part II.

17.1.4 Exclusions

Activities, materials, and equipment specifically associated with high-toxicity and biological safety laboratories are not usually conducted in a team research laboratory, because these laboratories require environmental containment that is difficult to achieve in the more open setting of a team laboratory. Refer to Chapters 5 and 12 for discussions on these types of laboratories.

17.2 LABORATORY LAYOUT

17.2.1 Introduction

The recommendations in Sections 1.2 and 2.2 apply generally to team research laboratories. The management of spills and other types of accidents involving hazardous materials is much more difficult in the large open area of a team laboratory (Figure 17-1) than in a room of average size since there is no simple means of containing hazardous fumes or smoke. An incident in one area of the laboratory may not be noticed in time for safeguards to be taken by persons in other parts of the laboratory. In a large open area there is increased risk of flame spread. For this reason the fire protective rating of walls around a team research laboratory should conform to requirements of NFPA 45.

There are several advantages to team laboratories that also have an indirect bearing on safety. Team labs promote interaction among researchers because boundaries between researchers are not "built-in," and team members are more likely to encounter one another in the open environment and to share ideas as well as equipment. Institutions gain a great deal of flexibility in team labs, because they can modulate and shift facilities with changes in funding, research focus, or personnel. After the initial shock of not having walls around their turf, researchers tend to respond in a more cooperative and, sometimes, more considerate manner in use of the common space. Incompatible tastes in music as well as space temperature seem to generate the most problems in team laboratories. Large team laboratories are also less expensive to construct compared to multiple smaller laboratories.

17.2.1.1 *Entire-Floor Laboratory.* A team laboratory that occupies an entire floor of a building and is not divided by walls or egress corridors requires special planning. The space must be organized with a pattern of circulation that is easily perceived. Space allocation in the laboratory may vary from 60 to 200 nsf per person. It is important that the laboratory layout be sufficiently

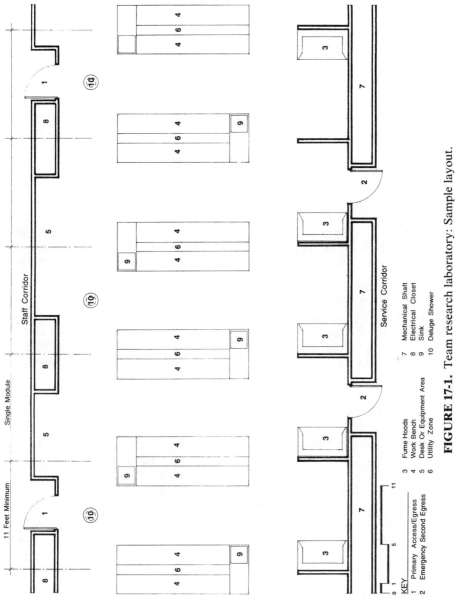

FIGURE 17-1. Team research laboratory: Sample layout.

KEY

1 Primary Access/Egress
2 Emergency Second Egress

3 Fume Hoods
4 Work Bench
5 Desk Or Equipment Area
6 Utility Zone

7 Mechanical Shaft
8 Electrical Closet
9 Sink
10 Deluge Shower

logical and simple that all personnel know where they are within the laboratory at all times and how to get out, even when vision is impaired by darkness, smoke, or chemical irritants. Exotic geometries are inappropriate for laboratories, and odd layouts containing long cul-de-sacs or dead-end aisles can be outright dangerous. They should be avoided in team laboratories. A rectangular grid is recommended for circulation because of its familiarity to all by sight or touch. With no visual clues, people are able literally to feel their way out. Groups can exit safely when the major aisles between benches and equipment are wider than standard aisles. This arrangement is recommended for all team research laboratories. In addition, special patterns in the flooring material (such as a stripe leading in the direction of the exit), along with illuminated directional signs mounted close to the floor, are immensely helpful in evacuating a large team laboratory in a visually obscured or dark environment.

17.2.1.2 *Utility Distribution.* Utility outlets should be aligned with the circulation grid so that all arrangements of fixed and movable benches and equipment will maintain no less than the minimum recommended aisle and egress passageway widths. Sinks should be evenly distributed so that all personnel are within 10 s travel time to an emergency eyewash station and for washing spilled materials off eyes and limited parts of their body extremities.

17.2.1.3 *Zoning.* Because there may be a great variety of activities occurring simultaneously within a team research laboratory, well-defined zones for particular activities should be established early in the planning phase to reduce conflicts and promote safety. As discussed in Chapter 2, zones of activities of increasing hazard should be located farther away from primary egress aisles. Zones can be determined by observing the following characteristics:

1. Areas containing fixed benches will have a range of relatively predictable activities that can be assessed for potential hazards. By contrast, open floor areas containing movable benches and large nonpermanent equipment setups may be less predictable with regard to safety, and consideration should be given to providing a designated open floor area as the more hazardous zone and providing another area as a relatively safe zone.

2. Sterile processes are difficult to protect within a team laboratory unless the total laboratory is operated as a clean area. If this is not possible, sterile processes should be conducted within a full environmental enclosure or some other protective means in an area of the team laboratory zoned and strictly maintained for clean activities. Chapters 8 and 19 contain discussions of clean laboratories. Nonsterile processes generally do not have more than normal cleanliness requirements.

3. Team laboratories may require special exclusion areas for containment of hazardous processes, equipment, or materials. In addition, they may need

special exhaust air requirements to service processes or equipment that produce excessive heat, unpleasant odors or fumes, and acoustic controls to isolate equipment or processes that produce excessive or unpleasant noise.

17.2.2 Egress

The recommendations in Section 2.2.1 apply generally to team research laboratories. When laboratories are larger than four standard modules, the distance to the nearest exit should not exceed 50 ft. Dead-end conditions should not exceed 20 ft.

Egress markings should be in accordance with local and national standards. A special effort should be made to provide supplementary egress signs and visual guides for exit from large team research laboratories. This is particularly important when there are many exits. Exit signs should be visible from all areas and kept clear of pipes, ducts, light fixtures, and similar sight obstructions.

17.3 HEATING, VENTILATING, AND AIR CONDITIONING

The recommendations in Section 2.3 apply generally to team research laboratories.

17.3.1 Exhaust Air Systems

17.3.1.1 Fume Hoods. Large team research laboratories often require many chemical fume hoods. When they are grouped together, a lot of air must be exhausted from a relatively limited area, causing undesirable drafts and turbulence in the airflow patterns. The use of auxiliary air chemical fume hoods, especially when they are placed in alcoves, may be considered. When properly installed and used, they reduce air current velocity and turbulence in the laboratory and also reduce the total amount of conditioned building supply air that must be exhausted to the outside. See Chapter 31 for more information on fume hoods. The alcoves should be considered high hazard zones placed away from the main egress aisles. In addition, the location of the chemical fume hoods must be carefully planned relative to the other local exhaust systems to avoid deflecting or reducing the effective capture velocities of all of these devices.

17.3.1.2 Local Exhaust Air Systems. Recommendations contained in Section 2.3 apply generally to team research laboratories. The distance between a particular benchtop and the nearest fume hood may be significant in a large team laboratory. Ability to exhaust equipment locally is very useful.

17.3.1.3 General Room Air Exhaust Systems. Recommendations contained in Section 2.3.4.1 apply generally to team research laboratories. An even distribution of exhaust grilles is important to reduce the spread of odors and fumes throughout large team research laboratories.

17.3.2   Supply Air Systems

Recommendations contained in Section 2.3.3.1 apply generally to team research laboratories. An adequate volume and even distribution of supply air are especially important in large team research laboratories to provide for many exhaust air systems in a draft-free manner. There are two methods of introducing conditioned air to reduce drafts. One is to use perforated round or oval ducts that can deliver air at far lower velocities over a much larger area than standard diffusers can. Perforated ducts must be hung within the space, not above a suspended ceiling. Tops of these ducts do not collect dust because air sweeps out of perforations on top. Sometimes there is no choice other than to introduce conditioned air above ceilings and let it enter the room via perforated ceiling panels. Care must be taken to assure that the supply plenum is clean and free from dust, duct insulation fibers, or any other contaminents. Standard diffusers installed into lay-in ceiling systems or surface-mounted on ceiling structures should have directional vanes so that high velocity airflows can be better controlled.

17.4   LOSS PREVENTION, INDUSTRIAL HYGIENE, AND PERSONAL SAFETY

The recommendations contained in Sections 1.4 and 2.4 apply generally to team research laboratories.

17.4.1   Emergency Showers

The recommendations contained in Section 2.4.1.4 apply generally to team research laboratories. The emergency showers in team research laboratories that cover an entire floor should be located within the laboratory rather than in the corridors to assure that there will be no more than 25 ft of travel from any point to a shower. The showers should be arranged so that splashing and water runoff will not endanger laboratory equipment, benches, or electrical panels. Appendix I contains additional information on emergency showers.

17.4.2   Emergency Eyewash Facilities

The recommendations contained in Section 2.4.1.5 apply generally to team research laboratories. There should be one emergency eyewash station for every two module equivalents, distributed evenly at sinks throughout the team laboratory. In addition, there should be at least one eyewash facility of the full-face type with water tempered in accordance with recommendations in Appendix II.

17.4.3 Hand-Portable Fire Extinguishers

The recommendations contained in Section 1.4.4.2.2 apply generally to team research laboratories. In addition, large-capacity, multipurpose fire extinguishers may be needed within the laboratory. This should be determined with the aid of fire protection and safety specialists based on the anticipated uses of the laboratory.

17.4.4 Fire Alarm System

It is important that fire alarms be readily heard throughout large team laboratories. Care should be given to selecting fire zones and zones of refuge. In many laboratory facilities with frequent false fire alarms, researchers tend to build up an insensitivity to alarms that could be very dangerous in the situation of a real fire. A good public address system can be very helpful giving building occupants additional critical information under emergency conditions. The BOCA code (BOCA, 1991) requires a public address system in buildings over 70 ft high, and many fire departments also require them. Consideration should be given to including a public address system in team laboratories.

CHAPTER 18

ANIMAL RESEARCH LABORATORY

18.1 DESCRIPTION

18.1.1 Introduction

The design of large research facilities for housing and caring for major laboratory animal colonies, especially when provisions must be made for accommodating many different species in the same facility, is a complex task that should not be undertaken without the active advice and assistance of veterinarians and scientists experienced in animal research. The use of animals for research purposes and the facilities used for their care at universities and research institutions have come under intense criticism by a number of groups in recent years. As a result, new and more stringent regulations covering all aspects of animal care and animal research practices have proliferated, and it appears unlikely that this trend will be reversed in the foreseeable future. The Animal Welfare Act (Title 9 CFR, Subchapter A, Parts 1, 2, and 3) (CFR, 1984) invests the U.S. Department of Agriculture's Animal and Plant Health Inspection Service with responsibility for issuing and enforcing regulations relating to all aspects of laboratory animal use and welfare. The most widely used document is the ''Guide for the Care and Use of Laboratory Animals,'' published by the National Institutes of Health (NIH, 1985). It covers all aspects of the care and use of laboratory animals, including institutional policies for monitoring animals and providing profes-

Guidelines for Laboratory Design: Health and Safety Considerations, Second Edition
By Louis DiBerardinis, Janet S. Baum, Gari T. Gatwood, Anand K. Seth, Melvin W. First, and Edward F. Groden.
ISBN 0-471-55463-4 Copyright © 1993 by John Wiley & Sons, Inc.

sional care. For research calling for the use of animals to qualify for funding by U.S. agencies, the animal facility must conform to the NIH standards and be accredited by the American Association for Accreditation of Laboratory Animal Care (AAALAC). It is extremely important, therefore, that research facilities intended for animal housing and animal research be initially designed to meet the highest foreseeable standards. Anything less is likely to result in rapid obsolescence and to risk serious interference with future research programs. Furthermore, it is difficult to predict which animal species will be needed for future research studies. Therefore, as wide a range of uses as the budget will permit should be built into the facility even though only a specific use is currently contemplated.

Although the construction of large-animal housing and research facilities is beyond the scope of this book, there is frequently a need for the provision of animal facilities of very modest size that are restricted to the housing of small numbers of one or a few similar species of standard animals (e.g., only rodents). The animal laboratory described in this chapter is confined to this small-scale purpose, and the information that is contained here should not be extrapolated to cover the design of larger-sized facilities, housing for pathogen-free and germ-free animals, or multispecies housing. The animal laboratory described here is suitable for the housing and care of the most-used small animals, principally rodents (mice, rats, and guinea pigs). It is not designed for primates, dogs, cats, rabbits, or fowl or for large animals such as sheep, donkeys, and pigs.

The minimum facilities required for a small unit research and teaching laboratory that utilizes small numbers of small animals are the following:

1. New animal reception and quarantine area, usually a closed room isolated from the main animal quarters.
2. Animal holding rooms, where animals undergoing or awaiting experimentation are housed, fed, and cleaned.
3. Sanitation facilities, often designated as a "dirty area," used to (a) wash and sterilize cages, water bottles, feed troughs, and so on, and (b) collect and dispose of animal wastes and soiled bedding. An in-house incinerator or steam sterilizer is needed when dealing with highly infectious human pathogens or potent carcinogens, but otherwise, commercial disposal facilities can be utilized.
4. Storage room with vermin-proof bins for animal feed, clean bedding, and other animal-care supplies.
5. Experimental treatment–surgery–autopsy room. When the size of the facility permits, the autopsy room should be isolated from the treatment and study laboratory. Tissue preparation and pathology facilities can be provided external to the animal laboratory when space is limited.

6. Freezer for holding dead animals prior to disposal and excised tissues prior to examination.

In addition to purely humane considerations, good animal research cannot be conducted with sick, dirty, infested, and poorly tended animals.

Good animal laboratory design includes an efficient layout of facilities for ease of operations, plus selection of materials of construction that make it possible to maintain excellent sanitary conditions. Whatever else may have to be omitted from the animal research laboratory for reasons of inadequate space and insufficient budget, no compromise should be made in provisions for maintaining excellent sanitation throughout the facility. The important factors include the following:

1. Impervious monolithic floors, walls, and ceilings, constructed of materials and finishes resistant to cracking and damage from washing detergents and disinfectants, including chlorine-containing compounds. Walls and floor structure as well as finish must withstand very rough treatment because of the large racks and carts that will be rolled through corridors and holding rooms.

2. Metal window frames that are resistant to damage from moisture and mildew and that are permanently closed to lock out pests. Many animal facilities do not have windows to the exterior to enhance security. Animal rooms should have automatically controlled lighting cycles and therefore do not need windows at all.

3. Doors that are animal-proof when closed, to prevent entry of wild species (attracted by the easy availability of food) and loss of loose laboratory animals.

4. Careful sealing of all penetrations for utilities (electricity, water, drains, fire sprinklers, heating and ventilating ducts, etc.) to close off harboring places for vermin. It should be kept in mind that application of insecticides in active animal colonies is often prohibited because of the unknown influence these poisons may have on the outcome of experiments then underway. Therefore, it should be accepted from the start that the only useful vermin control program is strict prevention of infestation. This is one of the reasons why an isolated quarantine room is essential to prevent the introduction into the main animal colony of infected and infested animals.

5. Some experts in pest control recommend applications of boric acid in wall spaces during construction or renovation in order to reduce cockroach problems.

Each animal species has an ideal air temperature, air humidity, and air movement rate that promotes health and longevity. Drastic average deviations and substantial random variations from these ideal conditions will reflect adversely on the longevity of the animals. The seriousness of uncontrollable environmental conditions within the animal laboratory will vary with the normal animal holding period. For acute toxicity experiments, as for determining LD_{50} data, longevity greater than a week is seldom important. However, for low-level chronic toxicity exposure experiments, extreme longevity of the experimental animals is critical to success, and every possible effort must be expended to promote the health and safety of the animal colony for the duration of long experimental periods. Breeding animals are also extremely sensitive to environmental change. They may lose or destroy their young when under stress. Health, safety, and reduction of stress have been promoted in ingenious ways by the design of individually ventilated cages mounted on self-contained multicage rack systems that continuously bring filtered particle-free air to each cage. These arrangements were developed to maintain sterile conditions for holding transgenic mice, but their use for less valuable animals is advantageous. (See Section 18.2.1.1.)

Optimal spacing and HVAC conditions have been published for a number of commonly used animal species by the U.S. National Institutes of Health (NIH, 1985), as well as by counterpart agencies in other countries. Generally, for best results, it is necessary to maintain the interior laboratory climate with as little variation as possible; therefore, a windowless area is highly recommended and may be a necessity to maintain a correct light cycle for certain animal species. Animal handlers often become dissatisfied with working conditions when they cannot see the outdoors, even if it is only through sealed window glass.

Animal rooms should be capable of an adjustable temperature range between 65°F and 84°F and a relative humidity range between 30% and 50%. In animal rooms containing many closely spaced cage racks, uniform ventilation rates and temperatures from top to bottom and side to center are difficult to maintain. Special HVAC arrangements and diffusers may be needed to assure that every animal will be maintained continuously under the preselected environmental conditions. The temperature and relative humidity maintained in the animal room many not be indicative of conditions inside the cages. Therefore, the arrangement of cage racks and cages must be considered carefully to assure a high degree of air circulation around individual cages at all times.

Because of animal odors and the danger of contamination, it is not advisable to recirculate air from animal rooms unless special air cleaning facilities are provided, as discussed in Sections 18.2.1.1 and 18.3.3. Therefore, all of the ventilation air should normally be discharged to the atmosphere after a single pass through the animal quarters. It may be necessary to filter the air prior to discharge to remove animal hair and dander, bedding fragments, and feces.

When no toxic substances are involved, medium efficiency filters are usually adequate. Certain procedures with animals should be conducted in biological safety cabinets or in a biological safety laboratory (Chapter 12).

18.1.2   Work Activities

The activities performed in a small-animal laboratory facility include: ordinary good animal care of a maintenance and preventive nature; animal breeding for genetics studies; animal experimentation involving administration of drugs, chemicals, and biological agents by inhalation, ingestion, injection, skin application, and surgical procedures; routine pathology preparations and examinations; and record keeping of a highly detailed nature. Much of the daily laboratory routine is occupied with animal care duties such as food preparation, changing animal bedding, washing cages and room surfaces, inspecting animals for illness, registering deaths, and disposing of animal carcasses and other wastes. Direct experimentation with animals may take place inside the animal laboratory when adequate facilities for such work have been provided. Otherwise, the animals may be moved to adjacent laboratories for scientific and surgical procedures, and returned to the animal laboratory daily for housing and routine care. When animals are going to be exposed to hazardous substances or subjected to dangerous procedures, facilities for conducting such work should be made an integral part of the animal laboratory to avoid unnecessary spread of toxic substances and exposure of personnel who are not directly involved.

Only small amounts of drugs, laboratory chemicals, and bottled gases are needed in most modest small-animal laboratories, but substantial quantities of sanitizing and disinfecting materials may be stored for routine use. The nature of the latter materials should be checked carefully to avoid inadvertently introducing toxic or dangerous substances into the animal laboratory.

18.1.3   Equipment and Materials Used

As a minimum, animal laboratories contain cages and racks for housing the animals, a cage washing and sterilizing machine, a steam sterilizer for surgical supplies, food, and other equipment, food preparation machines such as scales, dry feed mixers and vegetable slicers, refrigerators, microscopes, surgical and autopsy tables, sinks, a deep-freeze cabinet for storage of dead animals, and work benches for holding a variety of scientific instruments used to measure animal responses and examine animal tissues. There may also be items usually found in biological safety laboratories such as high-speed blenders, sonicators, and lyophilizers. For handling these devices, reference should be made to Chapter 12. Biosafety cabinets are found in animal facilities where stringent microbiological procedures can be performed. Additional special ventilated work stations are available for changing cage bedding. Correctly designed and used, this equipment can reduce occupational exposure to dust and animal allergens.

18.1.4 Exclusions

Modest small-animal laboratories are not suitable for work with highly infectious biological agents or with more than very small quantities of toxic chemicals, or for housing dangerous animals. Each of these conditions calls for a highly specialized facility that is not covered in this chapter.

For protection of the animals housed in the laboratory and for safeguarding the integrity of the experiments underway, the introduction into and storage of volatile, flammable, and explosive materials in the animal laboratory should be strictly prohibited.

18.2 LABORATORY LAYOUT

18.2.1 Introduction

Animal facilities can potentially be in any part of a laboratory building. Many verterinarians prefer the top floor in order to ventilate the facility with greater ease. Others prefer a basement or other low-level location relatively near a loading dock to make trash removal and material handling much easier. There are advantages and disadvantages to each location. When animal facilities are above grade level, a dedicated service elevator is needed. When animal facilities are on or below grade, dedicated supply and exhaust risers are needed to ventilate the space. There is another major consideration in locating animal facilities. Many rodents and other species are very sensitive to the noise and vibrations produced by laboratory mechanical and fan rooms as well as by elevator machine rooms. There is also concern about the effect of high-voltage transformer vaults and building electrical rooms on animal health. These issues should be considered in locating the facility.

The minimum-sized small-animal laboratory is a suite of several rooms isolated from the rest of the building by doors, preferably pairs that provide a pressurized airlock and out-of-traffic paths. The animal laboratory should be further isolated by maintaining it under negative air pressure relative to connecting corridors and adjacent rooms to prevent the spread of animal-generated odors outside the animal quarters. The recommendations contained in Sections 1.2 and 2.2 are generally applicable to animal research laboratories except as supplemented or modified in the following paragraphs.

A special feature of an animal laboratory is provision of designated "clean" and "dirty" areas and development of a circulation pattern that discourages passage of personnel and equipment from the dirty to the clean side without first passing through a sanitation station. The objective is to avoid introducing infection to the animal colony. A similar purpose is served by providing isolated quarantine and sick bay areas.

Because of the need to bring cage racks periodically to a central cage-cleaning facility, most corridors in the animal laboratory must be unusually wide (7 ft minimum) to accommodate such traffic. A typical layout for a minimum small-animal laboratory using a one-corridor system is shown in Figure 18-1. It may be noted from Figure 18-1 that the receiving and quarantine rooms can be reached from the outside, without entering any other part of the laboratory. The central corridor serves as both a clean and dirty corridor. A better arrangement, the two-corridor system, which utilizes distinct clean and dirty corridors, is also shown schematically in Figure 18-1. The important features of a two-corridor system are (1) corridors confined to one-way movement of personnel, animals, and equipment—always from "clean" to "dirty"—and (2) two doors to each animal room—one for entry, one for egress.

In small-animal laboratories, the operational rigidity imposed by the two-corridor system can become an obstacle to efficiency and speed. Nevertheless, breeding colonies and long-term experiments (such as bioassays for carcinogenicity) that call for the animals to live out a time approaching a normal lifespan require the ultimate in animal protection, and the two-corridor system is designed for such purposes. Because of the design rigidities imposed by the two-corridor system of animal laboratory construction, identification of a need for such a facility must be made early in the design process.

18.2.1.1 *Self-Contained Animal Isolation Housings.*

Many factors in addition to new laws and regulations have focused greater attention on protective animal housings; they include (a) more research interest in long-term chronic studies of toxic substances and aging factors that require maintaining animals to old age in a sterile environment, (b) the greatly increased cost of specially produced animals (e.g., transgenic mice are valued at several thousand dollars each), and (c) a need to protect animal care personnel from infections and allergies. The general methods employed for these purposes include (1) housing one or only a few animals in small transparent plastic cage boxes that may be sealed against cross-contamination by physical barriers or high-efficiency particulate filters, (2) providing positive ventilation to each individual cage at carefully controlled rates from a source of filtered supply air that has passed through a HEPA filter having a minimum efficiency of 99.99% for the most penetrable aerodynamic particle size (i.e., 0.1 μm), (3) self-contained cage racks that provide tempered, humidified, and decontaminated supply air to individual cages and provide positive means to exhaust the contaminated air from each cage, and (4) the use of a Class II biological safety cabinet or similar device for cage changing. Mobile cage-changing stations are commercially available that can be brought close to

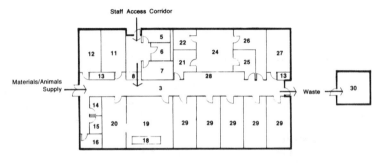

SINGLE CORRIDOR ANIMAL LABORATORY

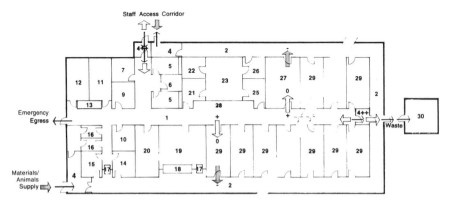

DOUBLE CORRIDOR ANIMAL LABORATORY

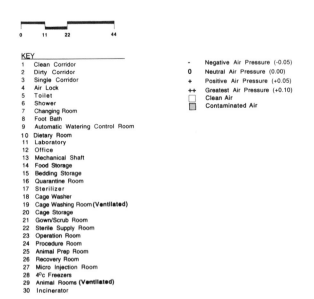

KEY

1	Clean Corridor
2	Dirty Corridor
3	Single Corridor
4	Air Lock
5	Toilet
6	Shower
7	Changing Room
8	Foot Bath
9	Automatic Watering Control Room
10	Dietary Room
11	Laboratory
12	Office
13	Mechanical Shaft
14	Food Storage
15	Bedding Storage
16	Quarantine Room
17	Sterilizer
18	Cage Washer
19	Cage Washing Room (Ventilated)
20	Cage Storage
21	Gown/Scrub Room
22	Sterile Supply Room
23	Operation Room
24	Procedure Room
25	Animal Prep Room
26	Recovery Room
27	Micro Injection Room
28	4°c Freezers
29	Animal Rooms (Ventilated)
30	Incinerator

−	Negative Air Pressure (-0.05)
0	Neutral Air Pressure (0.00)
+	Positive Air Pressure (+0.05)
++	Greatest Air Pressure (+0.10)
☐	Clean Air
▨	Contaminated Air

FIGURE 18-1. Single-corridor and double-corridor animal laboratory: Sample plans.

cage racks to minimize exposure of animals during transfers. In addition to providing effective isolation within each cage unit, use of positive cage ventilation assures steady removal of the ammonia produced in large amounts by rodents. This is said to reduce cage-changing frequency, which is labor saving and protective of the animals.

Figure 18-2 (top) illustrates a self-contained cage rack with individual plastic cages, each connected through internal piping to a filtered air supply unit and a filtered air exhaust unit. Figure 18-2 (bottom) shows the airflow direction through the individual cages. Figure 18-3 is a mobile change station that maintains the animals in a filtered air envelope during cage changing and provides some protection to the animal handler. Use of a Class II biological safety cabinet for cage changing provides the same degree of protection for the animals but provides more effective protection for the animal handler.

The use of self-contained isolator cages and racks with a mobile change station can greatly reduce the size and complexity of animal care facilities for users of small numbers of rodents in a laboratory setting. In addition to providing absolute particulate filtration of all the air withdrawn from the cage, it is possible to add a gas-adsorbent stage and remove ammonia and animal-associated odors as well. It has been stated that the availability of a self-contained cage isolator exhaust air purification unit makes it possible to locate cages as free-standing items in any laboratory or auxiliary room even without the provision of a dedicated air exhaust facility, but animals should remain in a dedicated animal care facility where they can receive adequate care (e.g., watering, feeding, cage changing) from a trained animal care staff. Other needs associated with animal laboratories (e.g., quarantine, operating and autopsy areas, cage-washing facilities, food storage and preparation, provisions for facility and worker sanitation, waste management) should also be provided by a trained animal care staff.

For a facility that has a need for maintaining a limited number of small animals for prolonged periods, the self-contained cage isolator system of animal housing can be adapted to less-than-ideal existing conditions and still comply with animal care regulations. For safeguarding large numbers of clean animals for long periods under ideal conditions, isolator housings represent an advancement in good animal care and preservation even in well-designed animal laboratories.

18.2.2 Floors, Walls, Ceilings

Animal laboratories require impervious surfaces and structural joints that are vermin-proof and easily cleaned and decontaminated. Walls and floors should be monolithic and made of washable and chemical-resistant plastic, baked enamel, epoxy, or polyester coatings. The monolithic floor covering should be carried up 8 in. of the wall to prevent accumulations of dirt and wastes in the corners. Integral cove bases allow more thorough cleaning and use of machines. Corridors subject to heavy traffic from transportation of

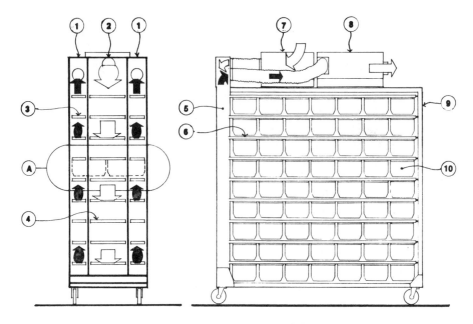

SECTION THROUGH VERTICAL PLENUM   ELEVATION

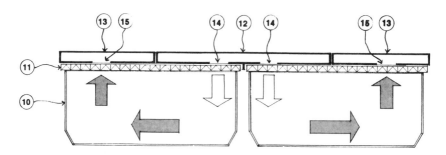

SECTION A, THROUGH A SHELF PLENUM AND CAGES

FIGURE 18-2. Self-contained ventilated cages showing airflow patterns: Sections and elevation. *Source:* Thoren Caging Systems, Inc., Hazelton, Pennsylvania.

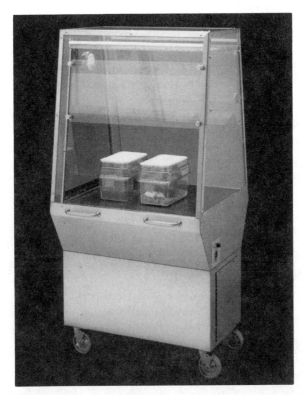

FIGURE 18-3. Mobile cage changing station. *Source:* Thoren Caging Systems, Inc., Hazelton, Pennsylvania.

cage racks and hand trucks handling feed and wastes must be constructed of materials resistant to wear and frequent washing with detergents and disinfectants. Smooth, hard-surfaced concrete and neoprene terrazzo are often recommended for floors, but no ideal floor construction method or material can be identified as totally trouble-free; frequent maintenance and repairs are required.

Bumper guards (or rails) on walls in corridors, animal holding rooms, and so on, that will prevent cage racks and hand carts from colliding with the walls, and thereby gouging the surface and rupturing the monolithic coatings, will go a long way toward maintaining a sanitary and vermin-free animal laboratory.

Floor drains are not essential in animal rooms housing rodents. Some veterinarians who are experienced in rodent care believe that moisture associated with the use of floor drains for cleaning purposes is detrimental to animal health.

Suspended ceilings hide from view the many services to the laboratory and produce a pleasing, finished appearance but they represent a serious impediment to pest control. Solid ceilings are preferred, with exposed heating and ventilating ducts, water and electrical services, and so on, easily accessible for inspection, cleaning, and disinfection (should it be necessary to run these services through the animal laboratory at all). Successful designs arrange utilities and ducts in corridor ceilings where access panels allow servicing. Animal holding room ceilings are then free of these encumbrances.

18.2.3 Access Restrictions

Access should be limited to essential personnel to avoid unnecessary exposure of the animals to infections and contamination. Illegal activities of certain so-called animal welfare groups make it prudent to maintain the animal laboratory behind locked doors.

Care must be taken to prevent the entry of wild rodents that will be attracted by the availability of food, because they bring diseases usually absent from carefully managed laboratory colonies. This is done by keeping laboratory access doors closed when not being used for passage of personnel, supplies, or equipment, and making certain that the crack under the door will not permit passage of even a small mouse. Door accessories that address these problems are automatic closers, weather stripping, and bottom gaskets.

18.2.4 Support Equipment

Cage washers should always be located in the dirty side of the animal facility. When sterilized animal feed is required, a double-door sterilizer should be used, with unsterilized feed being loaded at the dirty side and sterilized feed being removed at the clean side.

18.3 HEATING, VENTILATING, AND AIR CONDITIONING

18.3.1 Introduction

Animal laboratories require rigid control of temperature, humidity, and air movement in animal rooms at all times to provide optimal conditions for the health and growth of the species housed therein. In addition to a need for better-than-usual HVAC control systems, alarm systems are essential to alert responsible personnel to a system failure long before conditions deteriorate to a level that affects the animals adversely. The importance of alarm systems to signal HVAC malfunction will be in direct proportion to the duration of the animal experiments that will be conducted. When animal stress from unfavorable environmental conditions must be avoided at all costs, standby HVAC equipment and emergency power will be needed.

Negative pressure differentials between animal rooms and the remainder of the building housing the laboratory must be maintained to avoid spreading unpleasant animal odors and allergy-producing dander to areas outside the animal laboratory. Within the animal laboratory, pressure differentials should be maintained that always direct airflow from clean areas to dirty areas, never the reverse.

As a rule, air should not be recirculated from animal rooms; it should be discharged to the outdoors, usually after filtration.

18.3.2 Criteria

The HVAC requirements for animal rooms are contained in publications by the National Academy of Sciences and the National Institutes of Health, as well as by similar agencies in other countries. For rodents, the conditions shown in Table 18-1 are recommended (NAS, 1976). The number of air changes per hour for animal rooms is determined in part by the total animal sensible heat contribution to the environment. Heat gain values for commonly used rodents are shown in Table 18-2. Generally, 10–15 air changes per hour will be needed for animal care rooms containing rodents. If the animal rooms are densely populated, as high a ventilation rate as 20–25 air changes per hour may be required. Such high ventilation rates are undesirable because of the difficulty of providing a draft-free air supply. New types of large-area perforated diffusers are designed to solve this problem. Room ventilation requirements for commonly used rodents are shown in Table 18-3.

Air conditioning of animal quarters is a complex consideration involving temperature, relative humidity, and air changes, each of which is known to influence the physiological well-being of the animals irrespective of whether these effects are exerted independently or in combination. Control of environmental conditions in animal facilities may affect not only the usefulness of the animals as a research subject but also the quality of the data obtained from the research efforts (NAS, 1976). It is not necessary to provide extremely high air change rates (120–240 air changes per hour) and clean room filtration in holding rooms for standard animals. Such systems are sometimes appropriate for transgenic and other special animals, but are too expensive to

TABLE 18-1. Recommended Temperature and Relative Humidity for Common Rodents

| Rodent | Temperature | | Relative Humidity (%) |
	°C	°F	
Mouse	20–24	68–75	50–60
Hamster	20–24	68–75	40–55
Rat	18–24	65–75	45–55
Guinea pig	18–24	65–75	45–55

Source: ILAR News, Vol. XIX, No. 4, Summer 1976, National Academy of Sciences, Long-Term Holding of Laboratory Rodents, Institute of Laboratory Resources.

TABLE 18-2. Heat Gain Values for Common Rodents

Rodent	Weight (g)	Heat Gain (kcal/h)	Heat Gain[a] (Btu/h)
Mouse	21	0.403	1.599
Hamster	118	1.470	5.833
Rat	250	2.581	10.242
Guinea pig	350	3.322	13.182

[a] Calculated.

Source: ILAR News, Vol. XIX, No. 4, Summer 1976, National Academy of Sciences, Long-Term Holding of Laboratory Rodents, Institute of Laboratory Resources.

install, operate, and maintain to be cost-effective for standard animals. Microisolator cages (mentioned in Section 18.3.3) usually provide adequate filtered ventilation for most purposes.

18.3.2.1 Cage Washers and Sterilizers. This equipment should be ventilated through stainless steel or aluminum ducts to avoid corrosive attack. Capture hoods should be provided for washers and sterilizers to extract steam and vapors from the animal quarters. ACGIH provides guidelines for hood design (ACGIH, 1992).

18.3.2.2 Survival Surgery Rooms. These facilities, illustrated in Figure 18-4, should be treated like human operating rooms. Ventilation rates of 10–15 air changes per hour and HEPA-filtered supply air are needed. The surgery room should have positive pressure to avoid contamination from the animal holding and servicing areas.

TABLE 18-3. Room Ventilation Recommendations for Common Rodents

Rodent	Weight (g)	Ventilation (m^3/h/animal)	(ft^3/min/animal)
Mouse	21	0.25	0.147
Hamster	118	0.69	0.406
Rat	250	1.38	0.815
Guinea pig	350	1.97	1.15

Source: ILAR News, Vol. XIX, No. 4, Summer 1976, National Academy of Sciences, Long-Term Holding of Laboratory Rodents, Institute of Laboratory Resources.

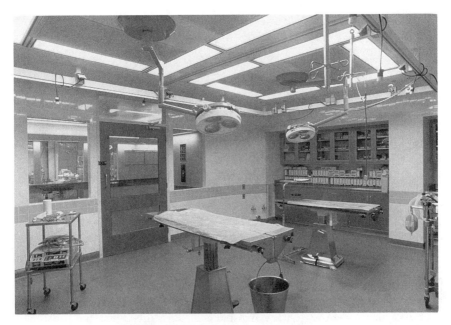

FIGURE 18-4. Animal surgery room.

18.3.3   Filtration

Exhaust air from animal housing and treatment rooms should be filtered before discharge to remove hair, bedding, feces, and so forth. An 85% efficiency filter will be adequate for this service unless the animals are harboring human pathogens. In that event, secondary filtration through HEPA filters will be required in addition (see Chapter 12). If the animals are to be maintained in a germ-free environment, HEPA filtration of all supply air will be required and the use of micro-isolator cages is advised. When using trace quantities of especially toxic volatile chemicals in conjunction with biological experiments, it may be prudent to add an efficient adsorber (generally activated carbon) as an additional effluent air cleaning stage.

18.3.3.1 Exhaust Grille Filters. In animal rooms, disposable filters at all exhaust air grilles provide an excellent way to prevent animal dirt, food particles, hair, etc., from entering exhaust ducts. See Chapter 30. When exhaust grille filters are not provided, the exhaust ducts tend to become plugged with debris. This reduces exhaust efficiency and alters the prearranged pressure relationships between areas. With inadequate air exchange, the whole animal colony tends to become malodorous. Should exhaust ducts become plugged, their cleaning is expensive, difficult, and time-consuming. Generally, animals will have to be moved to some other location while the ducts are being cleaned. Therefore, it is cost-effective to install exhaust grille filters at the outset and to institute a rigorous program of filter maintenance and replacement.

18.3.4 Controls

Pressure control can be maintained within an animal laboratory by providing a constant ratio of supply to return-plus-exhaust air with the aid of differential pressure controllers, modulating dampers, fan inlet vanes, or a combination of all of these. Airflow variations should be minimized as an aid in providing control over room pressure. Continuous operation of hoods, cabinets, and local exhaust systems is an important aid in maintaining reliable pressure control at all times. For odor control, positive-pressure air locks to adjacent areas are recommended. The specifications for air locks should be the same as those required by the General Services Administration (GSA, 1979). Reference should be made to Section 8.2.2.3, a discussion of safety aspects of interlocking doors.

Temperature and humidity control on the other hand may require a constant volume reheat system. All aspects of controls must be considered before deciding on a final control strategy.

Built-in safety aspects should be considered during the design stage. For example, use of N.O. (normally open) control valves on reheat coils should be avoided; otherwise, loss of control signal may overheat the animal space.

18.3.5 HVAC System

In general, a constant-volume terminal reheat (TRM) system is preferred. Zone control for individual areas is provided by the dedicated reheat unit serving the zone. Humidification controls can also be zoned, but care must be exercised to place humidifiers with enough duct length to provide adequate mixing. When adequate duct lengths are not available, a central humidification system should be considered. Packaged room humidifiers and VAV systems should not be considered for this application.

18.3.6 Alarms

Alarms should be provided for the automatic watering systems and HVAC systems as outlined in Sections 1.4.5.2 and 1.4.5.3. An additional alarm should be provided to notify personnel when filters are becoming dust-loaded to a critical point. The information on air cleaning system monitoring instruments and alarms contained in Section 8.3.2.4 for the clean room laboratory also applies to animal laboratories.

18.4 LOSS PREVENTION, INDUSTRIAL HYGIENE, AND PERSONAL SAFETY

The recommendations for loss prevention, industrial hygiene, and personal safety contained in Sections 1.4 and 2.4 apply generally to animal laboratories.

18.4.1 Personal Protective Equipment

Work around animals sometimes results in sensitization to animal dander and other animal products. Should this occur, disposable dust respirators, approved by NIOSH, will generally relieve the symptoms. They will prevent initiation of allergy if worn as a prophylactic measure. Air-supplied respirators may need to be used in more severe cases of animal allergies. Use of Class II biological safety cabinets during cage-changing protects animal handlers as well as the animals from contamination. Personal cleanliness is essential to avoid infection, and it is recommended that animal laboratory personnel (especially those who care for the animals) wear laboratory coats or coveralls and use gloves. These garments should not be taken out of the laboratory. Therefore, provision for storage of clean garments and safe disposal of those that become soiled must be delineated at the laboratory design stage. Also, convenient face- and handwashing facilities are needed inside the animal laboratory for maintaining personal hygiene. Shower and clean clothes locker rooms for animal care personnel are highly recommended.

18.4.2 Fire Suppression

Sprinkler systems intended for animal holding rooms must be designed for rooms that are likely to contain very tall cage racks. Crowded arrays of tall cage racks tend to shield sprinkler heads from prompt exposure to heat and will interfere with the normal spread of the spray patterns, resulting in an unusually severe fire hazard. Additional sprinkler heads, placed around the periphery of animal holding rooms, where there is likely to be permanent clear floor space, may be required. Chemical extinguishers, fixed or portable, are likely to be an additional hazard to animals and are not recommended in animal quarters. Smoke detectors become dirty very quickly in animal quarters and tend to give false alarms; they are not recommended for these areas.

18.4.3 Decontamination

It is sometimes necessary to decontaminate entire animal laboratories when stubborn animal diseases take over. Depending on the organism of concern, formaldehyde decontamination may be required. It is normally conducted by heating paraformaldehyde to release formaldehyde vapors into a space that has been thoroughly isolated from surrounding areas to prevent escape of formaldehyde vapors to occupied areas. National Sanitation Foundation Standard Number 49 contains a "Reecommended Microbiological Decontamination Procedure" (NSF, 1992). For less critical types of decontamination, use of lower toxicity chemicals, such as Alcide ABQ, may be satisfactory.

18.5   SPECIAL REQUIREMENTS

18.5.1   Illumination

Certain animal species, including mice and rats, follow a diurnal pattern of nighttime activity and daytime rest. For these, provisions must be made for reversing the usual illumination sequence by installation of an automatic time-clock controller.

Illumination levels recommended by the National Institutes of Health for rodents are generally 60–80 ft-c. Because many rodents are albinos, higher illumination levels may cause damage to their pigmentless eyes.

Fluorescent lighting is recommended because it generates less heat than do incandescent lamps of equivalent wattage and hence has less effect on HVAC requirements. However, discussion with veterinarians on the color range of fluorescent lamps is recommended. Standard lamps may not be beneficial to certain animals.

18.5.2   Noise and Vibration Control

Most animals are susceptible to audible and electronic noise. Appropriate shielding is necessary to eliminate them when they are present. Vibration also affects animals adversely and should be minimized by providing vibration isolation on moving machinery such as fans and compressors.

18.5.3   Emergency Electrical Power

When emergency power is available, the supply and exhaust fan system should be connected to it. All alarm and temperature control systems should also be connected into the emergency power.

CHAPTER 19

MICROELECTRONICS LABORATORY

19.1 DESCRIPTION

19.1.1 Introduction

A microelectronics or semiconductor laboratory is a specially constructed and tightly enclosed workspace with modulating environmental control systems to maintain the most strict design standards for low airborne particulate matter and maintenance of a preselected constant temperature and humidity. The release of numerous toxic substances associated with the preparation of materials used in semiconductor research and development must also be controlled. Control of airborne particulates, temperature, and humidity is accomplished in the manner described in Chapter 8. The space is maintained at positive pressure relative to its surroundings to prevent intrusion of unconditioned air. Release of toxic airborne substances must be controlled by local and general exhaust ventilation without upsetting airflow distribution and pressure balance.

The design of buildings where hazardous materials are used in semiconductor research and development is regulated by some state building codes; design, construction, and operational requirements generally follow those discussed in Article 80 of the California Uniform Fire Code (California, 1988).

Guidelines for Laboratory Design: Health and Safety Considerations, Second Edition
By Louis DiBerardinis, Janet S. Baum, Gari T. Gatwood, Anand K. Seth, Melvin W. First, and Edward F. Groden.
ISBN 0-471-55463-4   Copyright © 1993 by John Wiley & Sons, Inc.

19.1.2 Work Activities

The activities performed in a microelectronics laboratory are characterized by a need for (a) extreme cleanliness as defined in Chapter 8 and (b) the exercise of extreme caution while using highly toxic materials in the production and testing of "wafers" and "chips." See "Health Assessment of Electronic Component Manufacturing Industry" (NIOSH, 1985) for a detailed description of these processes. Activities performed here will include deposition of SiO_2 on base silicon wafers, etching, cleaning, and measuring the properties of the wafers. Gallium arsenide technology presents more of a health concern than does silicon (SiO_2)-based technology. The laboratory will have access restrictions and require the donning of special lint-free clothing and shoe covers prior to entrance into the work area as an aid to maintaining the required low particulate levels.

19.1.3 Special Requirements

The special requirements for a microelectronics laboratory are similar to those for a clean room. The specific class clean room will depend on the specific activity. The major difference is the need for more critical control of airflow patterns between the work area, anterooms, and adjoining areas. Differential pressure zones should be maintained to allow air to flow from the work areas to the anteroom and from the space outside the anteroom into the anteroom. In some cases, two anterooms, one clean and one dirty, may need to be provided. The remainder of the work area must be sealed to prevent airflow out of the workspace into adjoining areas. The reason for this pressure relationship is to prevent flow of unconditioned air (i.e., dirty air) into the clean room work area while at the same time preventing air potentially contaminated with toxic materials from escaping into surrounding occupied areas. In some cases, a pressurized clean room using hazardous materials is surrounded by a negative-pressure "corridor" or "area" that is not accessible by the general public. This is sometimes called a *hazardous materials corridor*.

The construction quality control for anterooms is critical. Most contractors will not normally seal the penetrations of piping, conduit, and other building systems. Work surfaces and wall seams are not typically sealed. Without special attention to these details, the desired cleanliness or pressure relationships may not be obtained.

19.1.3.1 Structural Materials. Same as in Section 8.1.3.1. There may be some equipment that requires a vibration-free environment (e.g., individual isolation platforms plus structural system stiffness, mass, and local isolation). Common types of vibration originate from machines and motors, air jets and convection currents, excessive fluid velocity in pipes, motor vehicle traffic, and footfalls of people walking normally in corridors and lab aisles. Walking is the least predictable source of vibration and is often the most

disruptive to vibration-sensitive equipment and procedures. In sensitive areas, corridor floors should be supported independently from the microelectronics lab floors from the foundation structure on up. This is an expensive option and should only be employed if local isolation of the equipment is not sufficient. Due to the quantities and nature of toxic materials used in microelectronics, buildings containing these labs should be structured for seismic loads in locations that could possibly experience earthquakes. Some state and local codes require seismic design (California, 1988; Mass, 1990).

When renovating a building to install a microelectronics laboratory, a structural assessment should be done early. If the building does not already meet structural requirements for a microelectronics laboratory, it would likely be cost-prohibitive to adequately retrofit.

19.1.3.2 *Personal Cleanliness.* Same as in Section 8.1.3.2.

19.1.3.3 *Auxiliary Fans.* Same as in Section 8.1.3.3.

19.2 LABORATORY LAYOUT

The layout of a microelectronics laboratory will meet the same general requirements as for a clean room (Section 8.2). In addition, the access restrictions discussed in Section 5.2.6 should be considered, particularly the viewing window. A sample layout is presented in Figure 19-1. The microelectronics lab layout is essentially a series of "modules" within a clean room. Auxiliary equipment (e.g., vacuum pumps, gas cylinders, and HVAC equipment) is located in the service areas adjoining the lab work area.

The clean and dirty corridor concept discussed in Section 18.2 may also be applied here. Generally there is a service corridor that allows access to the auxiliary equipment for maintenance and repair without having to enter the "clean" area.

To meet the general requirement that the lab area be positive to surrounding areas and that airflow be from low-hazard to high-hazard areas, the use of a double-shell structure is recommended. This arrangement allows air to flow from the lab to the lower-pressure inner space between the double shell and from the adjoining corridors or other spaces into the same inner space that is maintained negative to both. The inner space should be designed for seismic loading.

19.3 HEATING, VENTILATING, AND AIR CONDITIONING

19.3.1 Introduction

The items contained in Sections 1.3, 2.3, and 8.3 apply to microelectronics laboratories and should be implemented when appropriate. In addition, the following provisions should be considered for inclusion in building and laboratory plans.

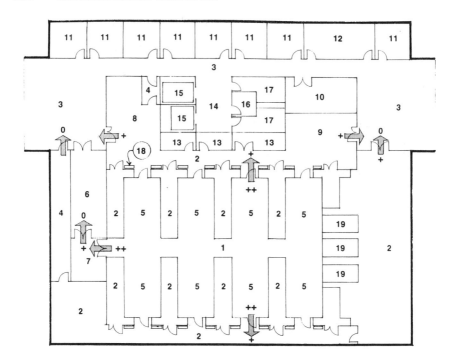

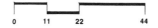

KEY

1	Clean Corridor
2	Semi-Clean Service Corridor
3	Dirty Corridor
4	Air Lock
5	Microelectronic Laboratories
6	Gowning Room
7	Glove Room/Air Lock
8	Staging Laboratory
9	Toxic Gas Alarm Room
10	Supply Room
11	Office
12	Secretary/Reception Area
13	Mechanical Space
14	Elevator Lobby
15	Elevator
16	Janitors Closet
17	Toilet
18	Gas Cabinets
19	Furnaces

- Negative Air Pressure (-0.05)
0 Neutral Air Pressure (0.00)
+ Positive Air Pressure (+0.05)
++ Greatest Air Pressure (+0.10)

FIGURE 19-1. Microelectronics laboratory.

19.3.2 Room Pressure Balance

Because of the need for both personnel and product protection, the directional airflow must be carefully balanced so that air flows from the workspace to the anteroom or outer shell and from the corridor into the anteroom or outer shell (see Figure 19-1).

19.3.3 Contaminant Control Ventilation

Several types of contaminant control exhaust ventilation systems have been designed for use in microelectronics laboratories and manufacturing facilities. Their selection and installation should conform to Article 80 of the Uniform Fire Code. The principles of ventilation design provided in Chapter 3 of the ACGIH *Industrial Ventilation Manual* (ACGIH, 1992) should be followed where no specific guidance is provided here.

19.3.3.1 *Gas Cylinder Storage Cabinets.* Gas cylinder storage cabinets provide a physical separation between gas cylinders and operator's environment and provide a fixed path for an exhaust airflow at the desired capture velocity for the gas. The design of these cabinets has been studied (see Burgess, 1985), and recommended ventilation rates are provided in Article 80 of the Uniform Fire Code (see Figure 19-2).

Vented gas cylinder storage cabinets are usually designed to provide an exhaust quantity of 200 feet per minute face velocity through the window openings (California, 1988) or 80 CFM per square foot of cabinet floor area. This exhaust air quantity is designed to handle small leaks in the systems, not high-volume accidental releases. The best solution of cylinder handling and leakage control is to keep cylinders in a gas cylinder farm either external to the building or in a controlled area on an outside wall and pipe gas to the point of use in the laboratory. Gas cabinets and other gas sources must always be kept out of any recirculated air stream. Restrictive flow orifices and/or excess flow check valves should be used (see Section 19.4.2).

19.3.3.2 *Exhausted Ventilated Work Stations.* Exhaust ventilated work stations are used in microelectronic laboratories to provide product and personnel protection simultaneously. In this regard they resemble Class II biological safety cabinets. However, unlike Class II biological safety cabinets, no specific performance standards have been developed to regulate/ assess their ability to provide the required product and personnel protection. They are constructed in a variety of configurations, depending on their intended use (e.g., solvent cleaning, acid cleaning, or other chemical manipulations). Typical work stations are shown in Figures 19-3 and 19-4. The general control principle for many of these hoods is the use of lip exhaust at

FIGURE 19-2. Vented gas cylinder storage cabinet.

the entire perimeter of a recessed "well" in the work surface used for acid or solvent baths. A slot exhaust is usually present along the back of a work station. Some work stations have no recessed wells, and their operation more closely resembles that of Class II biological safety cabinets. Cabinet manufacturers must be consulted to obtain the design air exhaust flow rates for their equipment. This is usually expressed in cfm per foot of bench length. It is desirable to conduct performance tests on these hoods to confirm manufacturer's recommendations. The performance test commonly used is ASHRAE 110-1985, modified to meet specific design needs (ASHRAE, 1985). When design air exhaust flow rates are not available, consult Chapter 3 of the ACGIH *Industrial Ventilation Manual* (ACGIH, 1992). Since the hoods are typically slot exhausts, one can calculate the required slot velocity and cfm needed for control.

When clean air is supplied directly inside the hood, the air supply and exhaust for these benches should be balanced to prevent escape of contami-

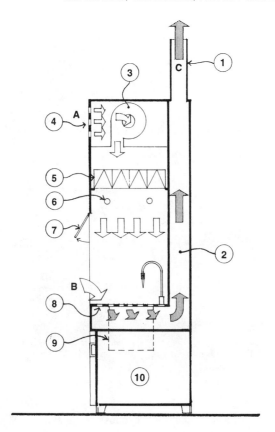

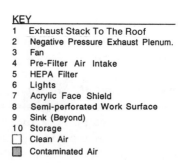

KEY

1   Exhaust Stack To The Roof
2   Negative Pressure Exhaust Plenum.
3   Fan
4   Pre-Filter Air Intake
5   HEPA Filter
6   Lights
7   Acrylic Face Shield
8   Semi-perforated Work Surface
9   Sink (Beyond)
10 Storage
☐   Clean Air
▥   Contaminated Air

AIR FORMULAS
A + B = C

FIGURE 19-3A. Vertical laminar flow exhaust work station: Section.

FIGURE 19-3B. Vertical laminar flow exhaust work station. *Source:* Laminaire Corporation, Rahway, New Jersey.

nants. The exhaust flow quantity must be large enough to exhaust the air supplied directly into the hood and still maintain an adequate flow of air from the room into the face of the hood.

19.3.3.3 Ventilated Equipment. Much of the equipment used in microelectronics research is designed to be operated under vacuum or is provided with a local exhaust outlet to carry away toxic gases and vapors. Therefore, this exhaust capability must be provided. Examples of such equipment includes chemical vapor deposition systems, diffusion furnaces, etchers, ion implanters, vacuum ovens, and spinners. Manufacturers will provide the required exhaust volume flow rates for their equipment. A central exhaust system can be used when effluents from each unit are compatible upon mixing. No recirculation of potentially contaminated air should be allowed. A certified industrial hygienist or certified safety professional can provide assistance in evaluating the compatibility of chemicals in exhaust air manifold systems.

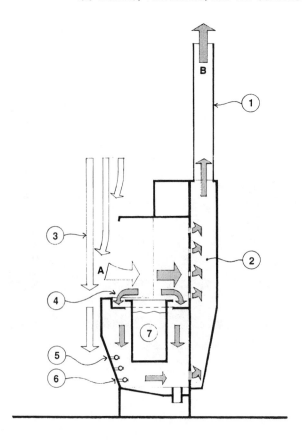

KEY

1   Exhaust Duct
2   Exhaust Slots To Negative
     Pressure Exhaust Plenum.
3   Air Curtain
4   Lip Vent (Around Bath Tank)
5   Water Manifolds
6   Plenum Rinse
7   Heated Bath
☐   Clean Air
▨   Contaminated Air

AIR FORMULAS
A = B

FIGURE 19-4A. Wet processing station: Section.

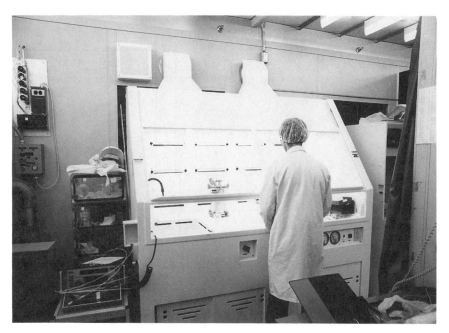

FIGURE 19-4B. Wet processing station.

19.3.4   Treatment of Exhaust Air

Chemical vapor deposition reactors may need to be equipped with effluent gas stream scrubbing systems designed to remove specific particulate and toxic gases. For arsenic and phosphorous trioxide, it is recommended that the air cleaning train be composed of at least one dry trap for condensing solid material, one bubbler for reduction of post-reactions initiated by the presence of oxygen with caustic scrubbing liquid, a nitrogen purge, and a charcoal trap to remove unreacted toxic gases (ACGIH, 1989). Activated charcoal adsorption alone has been used in other applications. Some manufacturers provide air cleaning equipment for the exhaust stream as part of the equipment, and this should be verified prior to purchase. All of this equipment needs to be vented to the atmosphere.

19.3.5   HVAC System Considerations

In general, constant-volume reheat-type systems are most desirable and provide reliable temperature control. Variable volume exhaust systems are not recommended due to the difficulty in maintaining desired pressure relationships, and hense, room cleanliness.

  Makeup or conditioned air distribution is important to prevent cross-drafts and excessive turbulence in these rooms. The recommendations in Section 2.3 should be carefully reviewed.

19.4 LOSS PREVENTION, INDUSTRIAL HYGIENE, AND PERSONNEL SAFETY

The information contained in Sections 1.4, 2.4, and 8.4.1 applies to microelectronics laboratories and should be implemented with the following additions.

19.4.1 Toxic Gas Monitoring

A continuous gas-detection system specific to the gas or gases used should be provided to detect the presence of gas in the room or area in which the gas is stored and used. Generally, this is required for gases with toxicity levels that are below odor thresholds. This may change if the proposed NFPA standard 55 is adopted (NFPA, 1992). The most common gases monitored are the hydrides. Monitors should be capable of responding to one-half the OSHA permissible exposure limit or the ACGIH TLV, whichever is lower. The detection system should shut down gas flow, initiate an audio and visual local alarm, and transmit a signal to a continually staffed remote location that can provide immediate response to the alarm. The alarm should provide warning both inside and immediately outside of the alarmed area or building. Under alarm conditions, the laboratory pressure should become negative with respect to surrounding areas. This is usually accomplished by eliminating or reducing the supply air to the space.

For infrequent uses of small quantities of toxic gases contained in lecture bottles, a portable gas-detection monitor may be used and operated during the use of the gas.

When the gas or gases are used continuously in more than one location, a multipoint detection system can be used. Multiple-sampling-point monitoring systems with 8 to 32 sampling points are commercially available (MDA, 1992). Detection points should be established at every location that has a potential gas release. The key areas for detection points include gas storage cabinets, reactor exhaust enclosures, operator's breathing zone while at specific operational equipment (e.g., OMCVDs), the general room or area, and possibly the exit to the exhaust stack or after the air cleaner.

19.4.2 Toxic Gas Piping Systems

Piping should be designed and fabricated from stainless steel or other materials compatible with the material to be contained and transported (California, 1988). In addition, piping should be of strength and durability sufficient to withstand the pressure, structural and seismic stress, and chemical exposure to which it may be subjected. Piping should be welded. Fittings, if any, should be in an exhaust enclosure.

Double-walled piping should be used throughout for highly toxic gases, such as arsine, stibene, and diborane and for highly flammable gases such as hydrogen as a safety measure. An automatic shutoff valve, of a "fail-safe to close" design, should be provided for each toxic gas cylinder on the main and all branch pipes. Shutoff should be activated by any of the following: gas detection, excessive gas flow, remote location alarm, failure of emergency power, exhaust failure, seismic activity, failure of primary containment, fire, and activation of a manual alarm.

19.4.2.1 Excess Flow Control. Portable tanks and cylinders should be provided with excess flow controls and/or a restricted flow orifice. Valves should be permanently marked to indicate the maximum design flow rate.

19.4.3   Air-Supplied Respirator System

Depending on the toxicity of the gases used and the nature of the work activities performed in these facilities, it may be advantageous to install a breathing air respirator system. During some activities, such as changing gas cylinders and cleaning equipment, it is advisable to wear respiratory protection. Self-contained breathing apparatus could be used, but they are cumbersome. Supplied air respirators, connected to piped in breathing air, are usually better. For such systems, the requirements of OSHA 1910.134 must be followed (OSHA, 1992).

19.4.4   Use of Plastic Ductwork

Where metal cannot be used, several types of plastic hoods and ducts are available. The more common types are polypropylene and PVC. PVC-coated metal is also available. Recent experiences have indicated that fires involving polypropylene and polyethylene have proven disastrous. Use of fire-retardant versions of these plastics is required by some state plumbing codes, including the one in Massachusetts. Therefore, a careful evaluation of the choice of materials should be made based on hood location, length of duct run, materials to be used inside hood, and extent of fire protection system. Generally, we recommend the use of FRP or PVC, but if other plastics must be used, they should be treated for fire resistance and fire spread retardation in conformance with local building codes. Where stainless steel can be used, it is preferred. See Chapter 32.

PVC hoods and duct should be evaluated for the need to provide a sprinkler system in the hood and in the areas where the duct runs.

19.4.5 Signs

Consideration should be given to the installation of lighted signs which can be turned on to warn persons approaching the lab of the present operational hazards.

19.4.6 Standards

The National Fire Protection Association is working on two new standards which will affect the design of microelectronic laboratories. NFPA 318 *Standard for Protection of Cleanrooms* (NFPA, 1991) and NFPA 55 *Standard for the Storage Use and Handling of Compressed and Liquified Gases in Portable Cylinders* NFPA, 1992). Both should be reviewed for their applicability.

19.5 SPECIAL REQUIREMENTS

19.5.1 Safety Storage Cabinets for Cylinders Containing Compressed Arsine, Phosphine, and Diborane

Some of the chemicals used for processing microelectronic chips are acutely toxic at low concentration. Among them are arsine, diborane, and phosphine, which are gases at ambient temperature and pressure and are supplied in pressurized cylinders. Fully charged large storage cylinders contain the following gas volumes when released at normal temperature and pressure:

Arsine (2P cylinder of 100% gas)	75 ft^3
Phosphine (2P cylinder of 100% gas)	75 ft^3
Diborane (1L cylinder of 1% diborane in argon)	275 ft^3

For safety, these compressed gases should be stored in ventilated storage cabinets, and the air from the cabinets should be discharged to the atmosphere from a point well above the roof and surrounding tall buildings. A very small volume cylinder leak from a faulty connection or valve (fraction of a milliliter per minute) can be handled satisfactorily by this method, inasmuch as (a) the negative-pressure cabinet protects the worker in the area and (b) the dilution produced by the cabinet purge air volume, along with the elevated discharge point, prevents significant environmental health exposures. Figure 19-5 shows a pair of ventilated cyclinder storage cabinets connected at the top to a common discharge manifold. Each cabinet is ventilated at the rate of 25 cfm of room air that enters small openings in the bottom of the cabinets. The unenclosed cylinder shown in Figure 19-5 contains instrument air.

A very dangerous situation arises in the event of a major uncontrolled gas emission resulting from a dropped cylinder or broken valve. The gas release rate may reach several liters per minute, and the purge rate would be inade-

FIGURE 19-5. Vented cylinder storage cabinet with emergency vented to scrubber.

quate to dilute the escaping gases to safe levels. Under these conditions, it would be unsafe to discharge untreated cabinet purge air to the atmosphere. Therefore, the discharged air from the cylinder storage cabinets should be connected to a standby purification system that can be activated instantly in the event of a serious gas leak. The wall-mounted switchboard to the left of the cylinder storage cabinets shown in Figure 19-5 can be used to activate the emergency air-purifying system and redirect the air from the cabinets through it before reaching the stack.

It would be ideal to use a passive adsorption bed to collect leaking gases, but it has not been possible to identify a suitable adsorbent up to the present. Although not as convenient for this application as adsorption, gas absorption in an oxidizing solution is an effective air-cleaning method.

Diborane is hydrolyzed rapidly by water to form boric acid and hydrogen. Arsine and phosphine have limited solubility in water (the former is 20 ml/ 100 ml water and the latter is 26 cc/100 ml), but both can be converted to nonvolatile reaction products in an aqueous medium that contains an oxidizing agent. Bromine water is recommended for arsine and hypochlorite for phosphine, but potassium permanganate, a nonvolatile oxidizing agent, is a satisfactory substitute for both. When used in basic solution, neutralization of the acid reaction products can be accomplished, as well. As the concentration of contaminant gas increases, so, too, does the concentration difference driving force between air and scrubber water, and this favors higher contami-

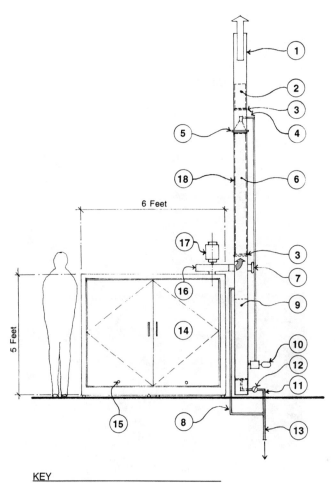

KEY

1   Exhaust Stack To The Roof
2   Droplet Eliminator Packing (1 Foot)
3   Perforated Support Plate
4   Spray Nozzle
5   Flange And Gasket
6   Absorber Packing (5 Feet)
7   Liquid Fill Hatch
8   Safety Overflow Drain Line
9   Liquid Reservoir (6 Gallons $KMNO_4$
    10% Solution)
10 Liquid Pump
11 Bottom Drain
12 Valve
13 Drain To Sewer
14 Cylinder Storage Cabinet
15 1 1/2 Inch Diameter Vent Holes (4)
16 Blower
17 Motor With Thrust Bearings
18 Absorption Tower
☐   Clean Air
▨   Contaminated Air

FIGURE 19-6. Emergency purge air-cleaning system: Elevation.

FIGURE 19-7. Scrubber for leaking gas cylinders in vented cabinets.

nant retention. To an important degree, the increased concentration driving force will offset the reduced retention time in the air cleaner during passage of a high-volume rate cylinder gas release.

A typical emergency purge air-cleaning system is shown diagrammatically in Figure 19-6. It consists of a 6-in. I.D. polyester absorption tower filled with $\frac{1}{2}$-in. polyester saddles to a depth of 5 ft and topped with 1 ft of dry saddles for droplet elimination. Caustic potassium permanganate solution is stored in a reservoir below the absorption tower and spread on top of the wetted saddles with a spray nozzle operated by a chemical pump located below the reservoir liquid level to be self-priming. A spray rate of 2 gpm at 10 psi is satisfactory to achieve at least 95% reduction in gas emissions.

Figure 19-7 is a photograph of the emergency air-cleaning system associated with the ventilated gas-cylinder cabinets shown in Figure 19-5. Because of ceiling height limitations, the single absorption tower shown in Figure 19-6 has been converted into two shorter towers in series. The towers are shown mounted on top of the liquid reservoir. The liquid pump is on the floor beside the reservoir. A supplementary strong caustic containing tank is shown to the right of the reservoir to freshen the absorbing solution whenever it is necessary to operate the system for long periods. Strong caustic solutions absorb CO_2 from the air passing through the unit and become less effective unless freshened. Because there is no commercially available absorption equipment for this service, it will be necessary to have it custom-made.

CHAPTER 20

PRINTMAKING STUDIO

20.1 DESCRIPTION OF PRINTMAKING STUDIOS

20.1.1 Introduction

A studio workroom is the artist's and craftsman's equivalent of the scientist's laboratory. In many types of work, artists and artisans deal with toxic and highly flammable substances and often in quantities that have a potential to cause injury, illness, fire, and explosion. Professional artists and crafts workers tend to work alone or with small groups in private studios that are often combined with their living quarters. Larger numbers customarily work together in art school settings.

Practically all serious artists these days serve an apprenticeship at one or more art schools. Therefore, when professional artists work in "kitchen studios" and have dangerous work habits (as has been widely noted), one must look to the art schools to see what sort of safety training student artists receive while in their formative period. The information contained in this chapter is intended for studios devoted to large group teaching.

Although all artists and artisans work with or generate some amount of toxic materials, printmakers seem to work with the greatest variety and largest amounts of toxic chemicals. Printmaking utilizes a few basic techniques with many variations to prepare a master plate, as in woodblock cutting (perhaps the oldest printmaking technique), engraving and etching of

Guidelines for Laboratory Design: Health and Safety Considerations, Second Edition
By Louis DiBerardinis, Janet S. Baum, Gari T. Gatwood, Anand K. Seth, Melvin W. First, and Edward F. Groden.
ISBN 0-471-55463-4 Copyright © 1993 by John Wiley & Sons, Inc.

metal, lithography, and screen printing (the newest technique). A brief description and a list of materials used in each of four generic processes follow so that the nature of the hazards involved and the protective measures required will be clear to the reader. The basic printmaking processes, as they have been described, are rather simple and straightforward. However, modern materials and artistic ingenuity have developed an almost infinite number of variations and combinations that make it impossible to attempt to cover them adequately in a brief review. Detailed treatment must be sought in specialized texts prepared for art students and professional practitioners. Some are noted in a bibliography section at the end of this chapter.

20.1.2   Work Activities

20.1.2.1   Intaglio Process.   An image is scratched (dry point), engraved, or etched into a flat metal plate (usually copper or zinc). In the dry point and engraving processes, the surface of the plate is cut directly with a cutting tool whereas the etching process requires that the polished plate be precoated with a specially prepared material called a *ground,* and that the image be cut into the ground with an etching needle. The prepared plate is then immersed in an acid bath that chemically cuts the metal surfaces that have been exposed. Finally, the ground is washed off in a suitable solvent and the plate is ready for printing.

The printing process starts by forcing ink into the grooves in the plate and then wiping clean the uncut surface of the plate. A print is made by laying a slightly damp paper on top of the inked surface and passing them together at great pressure through a press that forces the paper into the inked grooves and transfers the inked image to the paper. The plate must be reinked and the surface recleaned for each print. At the conclusion of printmaking, the plate, press, and inking implements must be cleaned by wiping and washing with ink solvents. A list of frequently used materials is shown in Table 20-1; those solely associated with etching are asterisked.

20.1.2.2   Relief Process.   Wood and linoleum are the most frequently used media for preparing relief-cut transfer masters. In this process, the material between the lines is cut away with knives and gouges, leaving the drawing in high relief. The relief surfaces can be inked with a roller and prints prepared by rubbing the back of a piece of paper laid face down on the inked surface, but prints are usually made in a printing press. Inks and cleaning solvents characterize most of the materials used in this process; frequently used materials are listed in Table 20-1.

20.1.2.3   Lithography.   Flat, fine-grained stones are prepared for this use by grinding and etching a horizontal surface until it holds a thin film of water when wetted. When sections of the surface are greasy, water will not wet them; this is the basis of the lithographic process. A master plate is prepared

TABLE 20-1. Materials Commonly Used in the Silk-Screen, Intaglio, Lithography, and Relief Processes

Silk-Screen Supplies

Oil-base inks
Transparent base
Mineral spirits
Polyurethane
Ulano screen degreaser #3
Korn's liquid tusche
Lepage's glue
Ulano Sta-Sharp adhering liquid
Lacquer washup thinner
#60 Ulano water-soluble blockout
Naz Dar Watermask 2000
Ulano direct emulsion 569 with diazo
Serascreen .174WW transparent base
Ulano stencil remover paste #5
Naz Dar binder varnish 5549
Naz Dar retarder thinner 5550
Naz Dar screen wash 2555
Gum turpentine
Sprint black and white film developer
Sprint quick silver print developer
Sprint stop
Sprint fixer
Kodak KPR photo resist Cat. 189
   2074
Kodak KPR photo resist developer
   Cat. 176 3572
Kodak KPR photo resist dye (blue)
   Cat. 147 3362

Intaglio and Relief Supplies

Interior/exterior enamel spray paint
* Oil-base inks (etching inks)
* Transparent base
Mineral spirits
Gum turpentine
Anti-skin spray
* Nitric acid
Easy-wipe compound
* Burnt plate oil
* Hard universal etching ground
* Soft universal etching ground
* Hard graphic ball ground

Intaglio Supplies (Continued)

* Hydrochloric acid
* Asphaltum (powdered)
* Putz pomade
Kodak KPR resist Cat. 189 2074
Kodak KPR developer Cat. 176 3572
Kodak KPR dye (blue) Cat. 147 3362
Whiting 3106

Lithography Supplies

Oil-base inks (lithography inks)
Lithotine
Cellulose gum
Western A.G.E.
Western plate cleaner
Western Diaz-a-Kote
Western PN-RED developer
Hanco tint base
Anti-skin spray
Hanco plate lacquer, deep etch V
Senefelder's asphaltum
Mineral spirits
Lacquer thinner
Turpentine
Denatured alcohol
Nitric acid
Phosphoric acid
Hydrochloric acid
Acetic acid
Tannic acid
Pro Sol fountain solution

by drawing or painting the image directly on the stone surface with a greasy ink, a grease pencil, or a crayon. Now, when the stone is wetted and an inked roller is passed over the wet surface, ink will adhere only to the greasy places. The inked image is transferred to paper in a specially designed traveling bed lithography press that squeezes the paper onto the surface of the stone below. The stone surface is reinked for each print. When the print run is completed, the stone is prepared for reuse by grinding or etching a new surface. A drawing made on paper with a lithographic crayon can be transferred to a stone surface by placing the image face down on the stone and running them together through the lithography press. Materials frequently used in the lithography process are shown in Table 20-1; they include (a) strong mineral acids for stone etching, (b) inks, and (c) organic solvents for ink cleaning and stone degreasing.

20.1.2.4 Screen Printing. Master plates are prepared by placing stencils on the surface of a tightly stretched fine mesh screen (frequently a woven silk fabric) to block out all areas that are not intended to transfer ink. A paper is placed in contact with the under surface of the screen, and ink is placed on top of the screen at one margin. Ink is then squeegeed across the surface to the opposite end, squeezing it through the open screen areas between the stencils and onto the paper directly underneath. Multiple prints can be made by simply adding additional ink to the screen. Because this process may be used with a large variety of colored ink materials, it means that many different chemicals may be present as organic paint solvents and for cleaning up after printing has been completed. In addition, a variety of gums and adhesives are used to stick stencils to the screen. Table 20-1 lists materials commonly used for silk-screen printing.

20.1.3 Equipment and Materials Used

20.1.3.1 Inks and Colors. All of the printmaking processes can use the same variety of monotone inks, usually black, but woodblock and silk-screen printing are especially adaptable to multicolor printing. This is done by pre-paring a master plate for each color and imprinting a paper with each master plate in turn.

20.1.3.2 Photographic Transfer Methods. A modern innovation of wide application to all printmaking processes, with the possible exception of en-graving, is photography. Many methods have been invented for transferring a photographic image to a blank master plate as a first step in preparing etched plates, lithography stones, wood blocks, and silk-screen stencils for printing. This means that a print department at an art school needs to have at least one well-equipped photographic processing laboratory as an adjunct to the printmaking studios. The photographic processing facility will have another list of chemicals that are used there exclusively. Some, such as formaldehyde, chlorine bleach, and methyl cellulose acetate, have well-known toxic properties; others, such as Freon and methyl chloroform, were formerly believed to be of low toxicity but lately have come under suspicion as possible human carcinogens. Contact dermatitis from primary irritants and from substances causing allergic types of hypersensitivity is the largest health concern associated with photographic processing.

Chapter 21, Photographic Darkrooms, should be consulted for a complete list of chemicals and other hazards associated with this activity. A printmaking studio will seldom need as large a facility nor use as wide a range of photographic chemicals as a professional photography laboratory, but the information contained in Chapter 21 provides (a) excellent guidance about potential hazards associated with each photographic process and (b) the protective measures that can be applied, as appropriate, in a print department's photography facility.

20.1.4 Exclusions

Although the information contained in this chapter is intended to apply specifically to institutions handling large student groups in teaching situations, the same principles and protective methods can be adapted to the scale of the small private studio as well as to less structured teaching facilities frequently used for leisure-time classes. Professional assistance (see the section entitled "Information Sources" in the Introduction at the beginning of this book) is advised when questions arise concerning the safe use and storage of art materials and cleaning chemicals that have a potential for health injury or fire and explosion. This is especially advisable when a facility has been in use for a long period and users have become inured to the hazards inherent in their current practices.

20.2 PRINT STUDIO LAYOUT

20.2.1 Introduction

Printmaking studios have a number of common layout features and special requirements for some.

20.2.2 Common Needs

Each studio requires (a) numerous large benches for layout and printing work, (b) storage shelves and cabinets for printing papers, (c) sink facilities for personal washing and for cleaning printmaking materials, (d) approved chemical and solvent storage facilities of limited capacity, (e) safety equipment that includes fire extinguishers and stores of personal protective equipment (goggles, gloves, aprons), (f) an exhaust hood for handling dangerous chemicals, and (g) facilities for receiving and storing safely large quantities of ink-stained and solvent-saturated waste papers and rags.

All studios need access to photography services, but a single facility may be held in common. All studios need a level of general room ventilation with outside air that is adequate to dilute to safe levels all the solvents evaporating directly into the room from inks, paints, and cleaning operations. No part of the air removed from active printmaking studios will be fit for recirculation. In all printmaking studios, electrostatic charge effects can make it difficult to handle large sheets of paper. Therefore, in persistently dry environments, it is necessary to maintain the relative humidity in the supply air near 50% for ease in working.

20.2.2.1 Construction Methods and Materials. The fire-resistive construction requirements of BOCA Article 9 for Class A buildings should be followed. Studios are subject to hard usage by the nature of the activities conducted there, and materials and construction methods should be selected that are resistant to a variety of caustic chemicals and easy to clean when soiled by inks, dyes, and pigments.

20.2.2.2 Floors. Selection of floor materials should also take into consideration a need for (a) resistance to acid and organic solvent spills and (b) ease of removal of inks and paints. Splashes and spills may make flooring slippery. Nonslip finishes should be used.

20.2.2.2 Walls. Studio walls should be totally or partially constructed of materials that allow mounting of unframed artwork produced by students for temporary exhibition. These materials should also have fire-resistive qualities and not fuel a fire.

20.2.2.3 Print-Drying Racks. All printmaking studios need facilities for drying freshly made prints. Generally, print-drying racks are placed around the studio in convenient locations. The need for them can become very substantial when successive classes use the same studio.

20.2.3 Special Needs

In addition to the enumerated common needs, studios have requirements that are unique to the specific nature of each printmaking process.

20.2.3.1 Lithography Studio. Concentrated floor loads for presses are a matter of special concern for lithography, and often the lithography studio will be located at a basement level to take advantage of the strength of a concrete floor laid over compacted fill. Forklifts and carts are used for moving lithography stones around the studio. Their needs must also be considered when evaluating clearances in floor designs.

Additional special facilities for lithography are (a) spacious and very strong shelving to store large, heavy blocks of stone and (b) very large and strong exhaust-ventilated sinks and drainboards for solvent washing and acid etching of stones. Figure 20-1 is a typical layout for a lithography studio.

20.2.3.3 Etching Studio. The etching studio presents special problems with regard to flammable solvent use because grounds such as asphaltum and rosin powder are often applied to metal blanks and heated on an electric hot plate to a temperature that causes melting and fusing. In addition to isolating

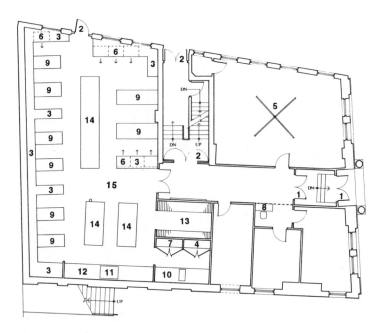

KEY

1	Primary Access/Egress
2	Emergency Second Egress
3	Counter
4	Chemical Storage (Ventilated)
5	Exhibition Room
6	Flat Files (Under Counters)
7	Ink Storage
8	Security Guard
9	Press
10	Plate Processing Ramp (Ventilated)
11	Sink
12	Stone Processing Sink (Ventilated)
13	Stone Storage
14	Table
15	Teaching Studio

FIGURE 20-1. Lithography studio plan: Retrofit into existing structure. (Courtesy of Lerner Associates, Providence, Rhode Island.)

hot plates from areas where solvents are used, it is essential to make certain that the electrical equipment on the hot plate (on/off, thermostatic temperature controller) are spark-proof and acceptable from the standpoint of an explosion proof installation. Figure 20-2 is a typical layout for an etching studio.

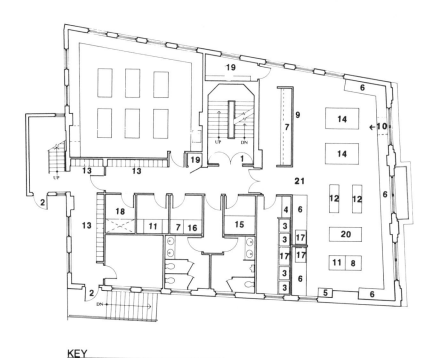

KEY

1	Primary Access/Egress
2	Emergency Second Egress
3	Acid Bin
4	Acid Storage
5	Asphaltum Table **(Ventilated)**
6	Counter
7	Cabinet
8	Cooling Table
9	Display Wall
10	Flat File
11	Hot Table
12	Inking Table
13	Lockers (Wardrobe)
14	Press
15	Plate Cutter
16	Rosin Box **(In Grounded Room With Spark Proof Electric Fixtures)**
17	Sink
18	Spray Booth
19	Storage Room
20	Table
21	Teaching Studio

FIGURE 20-2. Etching studio plan: Retrofit into existing structure. (Courtesy of Lerner Associates, Providence, Rhode Island.)

20.2.3.3 *Screen-Printing Studio.* The large use of volatile organic solvents is a unique characteristic of screen printing compared to the other printmaking processes, and therefore special attention must be paid to pro-

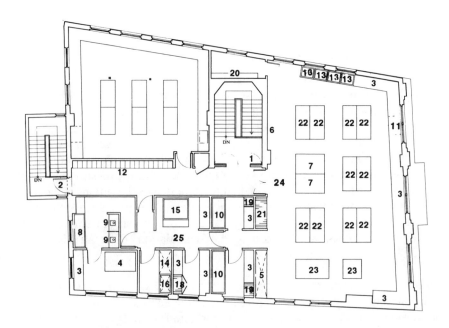

KEY

1	Primary Access/Egress
2	Emergency Second Egress
3	Counter
4	Copy Camera
5	Cleaning Station **(Ventilated)**
6	Display Wall
7	Drying Rack
8	Darkroom Sink **(Ventilated)**
9	Enlarger
10	Enclosed Sink **(Ventilated)**
11	Flat Files
12	Lockers (Wardrobe)
13	Light Table
14	Photo-Etch Counter
15	Plate Maker
16	Photo-Etch Sink **(Ventilated)**
17	Sink
18	Screen Drying Cabinet **(Ventilated)**
19	Screen Drying Rack
20	Shelves
21	Screen Storage
22	Screeding Table
23	Vacuum Table
24	Teaching Studio
25	Photography Suite

FIGURE 20-3. Silk-screen studio plan: Retrofit into existing structure. (Courtesy of Lerner Associates, Providence, Rhode Island.)

viding a level of ventilation that is adequate to maintaining safe solvent concentrations in the screen-printing studio. Figure 20-3 shows a typical layout for a screen-printing studio.

20.3 HVAC

20.3.1 General Requirements

20.3.1.1 Ventilation Rates. A high level of dilution ventilation is needed in large, heavily used printmaking studios that are characteristic of teaching institutions, and none of the air exhausted from studios should be recirculated. Fifteen air changes per hour are recommended as a minimum for screen-printing studios, and ten air changes per hour are recommended as a minimum for lithography and etching studios when the studios are in use. Three air changes per hour are considered adequate for dissipating vapors from drying prints and solvent-soaked trash in disposal containers when the studios are unoccupied. The air change rates cited are based on average student space allotment and chemical usage in a classroom studio setting. More precise values may be estimated by calculations illustrated in Section 20.3.1.3.

20.3.1.2 Coordination of Supply and Exhaust Air Volumes. The local exhaust ventilation systems that are needed in each studio to control point source emissions (e.g., the acid etching baths and grounds heating facilities in an etching studio and the printing benches in a screen-printing studio) are likely to provide adequate volumes of unrecirculated exhaust air to satisfy the air change requirements for occupied studios. Therefore, a low level of general room dilution ventilation need only be provided for nonused periods. This means that the volume of air supplied to each studio must be coordinated with studio use and nonuse periods. In practice, all the studio local exhaust systems and a high studio supply air volume must go on and off together to provide the design air exchange rates and maintain the air balance arrangements that keep the studios at a somewhat lower pressure than surrounding corridors, offices, classrooms, and lunch room to prevent contaminated studio air from flowing into these clean areas. When studio heating, cooling, and humidity are also provided by the ventilation air, the studio control systems tend to become complex. Nevertheless, because these are one-pass systems, energy costs for heating and cooling will be unusually high, and carefully designed control systems that minimize energy utilization are needed.

20.3.1.3 Print Drying. The very essence of printmaking is the production of multiple copies in a short time. In spite of this, the use of instant-drying inks is inimical to orderly printmaking; as a consequence, large numbers of prints must be hung on conveniently located racks while they dry, often for several hours. In busy printmaking studios that have facilities for hanging hundreds of freshly made prints, the amount of solvents evaporated into the studio air from this source can become a matter of health and fire concern unless provision is made for adequate studio ventilation. Not only must the air exchange rate be great enough to dilute the vapors to a low concentration, but the location of the drying racks in relation to the exhaust air vents must draw the vapors away from the breathing zone of the printmakers. Often, prints will be allowed to dry in the racks overnight, making it essential to keep some minimum amount of studio ventilation operational at all times when the studios are in daily use.

The minimum air exchange rate required for handling solvent vapors can be calculated on the basis of annual solvent and ink consumption. Average daily consumption can be calculated by dividing annual consumption by the number of days the studio is used each year. Experience indicates that maximum solvent evaporation will be approximately four times the daily average when the studio is used intensively, and the studio air exchange rate must be based on this number. The minimum air exchange rate for unoccupied studios must be based on the maximum number of drying prints that can be accommodated on the drying racks.

Most of the solvent will be consumed in cleaning operations that take place during, as well as at the conclusion of, printmaking; silk-screen printing probably requires the greatest amount of the several printmaking processes. Therefore, solvent consumption figures must be calculated studio-by-studio, and the ventilation system requirements for each studio is based on the results.

20.3.2 Exhaust Air Systems

20.3.2.1 Fans and Motors. All exhaust air systems from studios using flammable volatile solvents should be equipped with spark-resistant fans and appropriately rated fan motors. Type B construction, as defined in the Air Movement and Control Association's Standard 99-0401-86 (AMCA, 1986) should be adequate for this service. It calls for the fan to have an entirely nonferrous wheel and a nonferrous ring about the opening through which the shaft passes. Spark-resistant construction should not be confused with explosion-proof construction, an electrical term that calls for the device to be able to successfully withstand the effects of an internal explosion and not propagate such energy outside its containment. This is a considerably more costly item to purchase.

20.3.2.2 Local Exhaust Ventilation

20.3.2.2.1 Screen Printing. In addition to good general studio ventilation, exhaust-ventilated screen-printing benches are needed. Figure 20-4 is a typical slot-type exhaust hood serving two screen-printing benches.

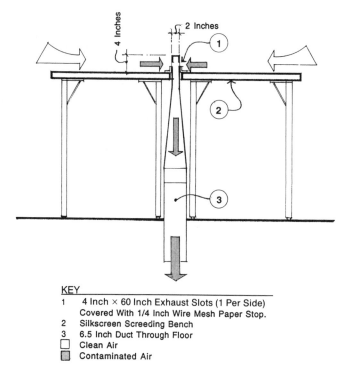

KEY

1   4 Inch × 60 Inch Exhaust Slots (1 Per Side)
     Covered With 1/4 Inch Wire Mesh Paper Stop.
2   Silkscreen Screeding Bench
3   6.5 Inch Duct Through Floor
☐   Clean Air
▨   Contaminated Air

FIGURE 20-4A. Ventilated silkscreen screeding bench: Section. (Sizes will vary according to specific applications.)

20.3.2.2.2   Lithography. An exhaust-ventilated spray booth is needed for spray coating stones with photographic emulsions and for other surface treatments.

20.3.2.2.3   Etching. The special need to heat etching plate grounds on an electric hot plate has already been noted. This operation should be conducted in a laboratory-type fume hood dedicated to this activity. Another facility unique to the etching studio is the acid-etching baths. These may be 4 × 5 ft in cross section and filled with dilute hydrochloric acid. The acid attack on copper and zinc produces bubbles of hydrogen that break the surface carrying fine droplets of acid into the air above the surface. Even when idle, the tanks continuously give off acidic hydrogen chloride gas. Therefore, the acid-etching tanks must be provided with covers and with exhaust-ventilation facilities. Because the need for exhaust ventilation will be greatest during the brief periods that the tank cover is open, a considerable saving in energy requirements can be realized by modulating the exhaust volume rate with the cover movements—that is, maintaining a low air bleed to extract equilibrium acid vapors when the cover is closed and a high-volume exhaust flow for

FIGURE 20-4B. Ventilated silk-screen screeding bench. *Source:* Lerner Associates, Providence, Rhode Island.

effective vapor and droplet control when the cover is open. A simple mechanical arrangement to accomplish this task is shown in Figure 20-5. Figure 20-6 shows a typical ventilated photo-etching bench.

20.4 LOSS PREVENTION AND PERSONAL SAFETY

20.4.1 Introduction

Although it is not practical or wise to attempt to restrict the materials that may be used by professional artists and crafts workers, the situation may be otherwise for students, especially for beginners. There are many art materials available for printmaking that are nonhazardous and that could easily be incorporated into educational programs. Water-based paints and inks come readily to mind; not only are volatile organic solvent vehicles absent from these materials, but none are needed for cleaning purposes. Less drastic

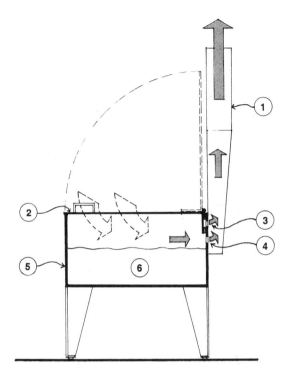

FIGURE 20-5A. Acid-etching tank hood: Section. (Sizes will vary according to specific applications.)

possibilities for reducing solvent exposures reside in the use of more single-service, disposable items (such as screens, scrapers, and brushes) that do not require cleaning. Such material restrictions will probably be considered inappropriate for more advanced students; in any event, advanced students must be taught while at school as learners to handle their materials in safe ways if they are to carry the lessons over to their professional activities. This makes it unlikely that the health and safety safeguards that have been recommended

FIGURE 20-5B. Acid-etching tank hood. *Source:* Lerner Associates, Providence, Rhode Island.

for printmaking studios will become unnecessary in the foreseeable future because of developments in new art materials.

20.4.2 Flammable Solvent Use

20.4.2.1 Permissible Quantities. Printmaking studios used for student instruction come under special rules governing the maximum amounts of flammable and combustible liquids that can be stored in classrooms and the nature of the containers in which they can be stored. Daily-use safety containers must be of small size and correctly labeled. Although local building and fire prevention codes customarily establish minimum safety regulations

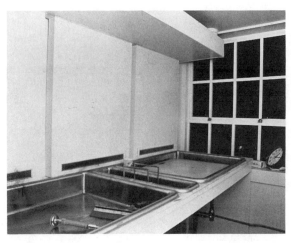

FIGURE 20-6. Ventilated photo-etching bench. *Source:* Lerner Associates, Providence, Rhode Island.

for the storage and use of flammable and combustible liquids, National Fire Protection Association (NFPA) Standard No. 30 (NFPA, 1992) represents a national consensus on this matter. Table 20-2 lists the maximum recommended quantities of five classes of solvents and four classes of containers acceptable in teaching studios. Liquids are classified as *flammable* when the vapor concentration above the liquid at temperatures below 38°C forms a flammable mixture in air. The specific temperature at which this occurs is referred to as the *flashpoint* of the solvent. Solvents with a flash point above 38°C are referred to as *combustible*. The three flammable categories and two combustible categories shown in Table 20-2 are defined in the footnotes of the table. Most solvents used by printmakers are flammable according to these definitions. Excellent nonflammable solvents are commercially available and are greatly preferred from the standpoint of safety, but regrettably they tend to be highly toxic (often carcinogenic), detrimental to the ozone layer, and much more expensive. This means that printmakers must be constantly on guard to prevent fire and explosion.

TABLE 20-2. Maximum Storage Amounts of Flammable and Combustible Liquids Recommended by the National Fire Protection Association for Teaching Studios

	Maximum Container Size				
	Flammable			Combustible	
Type of Container	Class II	Class IB	Class IC	Class II	Class IIIA
Glass or approved plastic	1 pt	1 qt	1 gal	1 gal	1 gal
Metal	1 gal	1 gal	1 gal	1 gal	5 gal
Safety cans	2 gal	2 gal	2 gal	2 gal	5 gal
Metal drums	1 gal	1 gal	1 gal	1 gal	60 gal
	Maximum Storage Quantities: Maximum of 10 gal of Class I and II combined or maximum of 25 gal in safety cans				60 gal

Notes:
1. A *flammable liquid* is defined as having a flashpoint below 38°C and a vapor pressure not exceeding 2070 mm Hg at 38°C.
2. A *combustible liquid* is defined as having a flashpoint at or above 38°C.
3. The flammable and combustible liquid subclasses are the following:
    Class IA—Flashpoint below 22.8°C and boiling point below 37.8°C.
    Class IB—Flashpoint below 22.8°C and boiling point at or above 37.8°C.
    Class IC—Flashpoint at or above 22.8°C and below 37.8°C.
    Class II—Flashpoint at or above 37.8°C and below 60°C.
    Class IIIA—Flashpoint at or above 60°C and below 93.4°C.
4. Metal drums should be approved by Department of Transportation.

20.4.2.2 *Flammable Solvent Protective Measures.* The presence in educational facilities of even the small NFPA code 30 permissible quantities of flammable solvents is certain to become a matter of keen concern for the fire marshal, and this individual should be consulted very early in the facility planning stages to learn what limitations will be imposed and what special requirements will be mandated. Likely requirements will include the following: at least one fire-resistant UL-approved flammable liquid storage cabinet in each studio using and storing flammable solvents and solvent-containing inks and colorants; fire-extinguishing equipment appropriate for flammable liquids and for solvent-saturated waste paper and rags; fire alarms; two or more ways of emergency egress from each studio; and, possibly, installation of sprinkler systems throughout. Appeals of fire marshals' decisions are difficult and usually unrewarding; therefore, advance approvals are essential.

20.4.2.2.1 Flammable Liquid Storage Cabinets. The fire marshal may insist that all flammable liquid storage cabinets in studios be ventilated and maintained under negative pressure as an additional fire protection measure. The cabinets can be connected most conveniently to the studio room exhaust air system inasmuch as none is recirculated, and all exhaust air systems from studios using flammable volatile solvents should be equipped with spark-resistant fans and totally enclosed fan motors.

20.4.2.2.2 Bulk Storage of Flammable Solvents. Although very small amounts of flammable and combustible solvents (scarcely a day's supply for a large and busy studio) should be permitted in the studios, larger quantities must be available for resupply. In addition, there is a significant cost advantage when heavily used solvents can be purchased in 55-gallon drum lots. This means that a bulk solvent storage facility must be constructed outside the areas used for educational purposes but should be reasonably accessible for daily replenishment of small studio safety containers. The preferred location is in a free-standing masonry building constructed specifically for bulk deliveries, safe solvent storage, and small-quantity dispensing facilities. Figure 20-7 shows a bulk chemical solvent, and ink storage facility. Dispensing from a 55-gallon drum may be by transfer pump or by drum faucets installed in a drum mounted horizontally on a wheeled rack. Whichever dispensing method is used, the drum must be thoroughly grounded and bonded to the receiving vessel during filling to avoid static sparks that can ignite flammable vapors. Additional information on flammable liquid storage may be found in Section 2.4.6.3.

20.4.2.3 *Chemical Storage Other than for Flammable Solvents.* The etching studio and the lithography studio use strong mineral acids and acid salts in their operations. Therefore, it is necessary for each of these facilities to provide suitable chemical storage facilities for daily use of small quantities

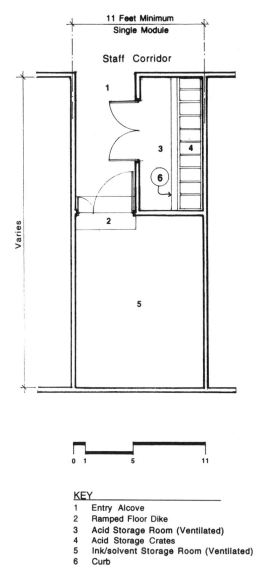

FIGURE 20-7. Chemical solvent and ink storage room: Sample plan.

within the studio and for bulk storage at an easily accessible location. Consult Sections 1.4.7 and 2.4.6 for information on bulk storage of chemicals. An ideal arrangement would be to combine a bulk chemical storage room with a bulk flammable liquid storage room in a free-standing facility.

20.5 SPECIAL REQUIREMENTS

20.5.1 Solvent Degreasers

In view of the large amounts of volatile solvents used for cleaning printing materials (including master plates), and most especially for silk-screen work, the installation of a centrally located industrial-type solvent degreaser needs to be given serious consideration for busy printmaking studios, not only to reduce solvent consumption but to reduce the exposure of those using the studios to a level as low as reasonably achievable. It must be kept in mind that art school students will usually be in their late teens and that the threshold limit values and permissible exposure limits applied to the adult working population may not be entirely appropriate for a younger group. It should also be recognized that the habits of students often include intensive work periods of 12–18 h at a stretch, especially near the end of a term, and these periods must be factored into the interpretation of safe exposure limits for this special population.

20.5.2 Lighting

Satisfactory lighting limits and lighting levels for studios will be similar to those mentioned in Tables 1-10 and 1-11 for drafting rooms (2 W/ft^2) and for cartography, detailed drafting, and designing (100 f-c on the work task).

Bibliography

1. Ivins, W. M., Jr., *How Prints Look: Photographs with a Commentary,* Beacon Press, Boston, MA, 1943, p. 44.
2. Ibid., p. 102.
3. McCann, M., *Health Hazards Manual for Artists,* Nick Lyon Books, New York, 1985.
4. McCann, M., "Printmaking," *Artist Beware,* Watson–Guptill Publications, New York, 1979.
5. National Gallery of Art, "Italian Etchers of the Renaissance and Baroque," Washington, DC, 1989.
6. Rossol, M., *The Artist's Complete Health and Safety Guide,* Allworth Press, New York, 1990.
7. Seeger, N., "A Printmaker's Guide to the Safe Use of Materials," *Alternatives for the Artist Guide at the School of the Art Institute of Chicago,* 108:1, 1981.

PART III

LABORATORY SUPPORT SERVICES

Laboratories are seldom able to function at all, let alone efficiently, as stand-alone units; support services of many types are called for. For example, laboratory wastes can no longer be poured down the laboratory sink or put into the domestic trash bin; they must be handled as hazardous waste and disposed of according to the rules of local, state, and federal agencies (e.g., the Environmental Protection Agency). This calls for the construction and utilization of appropriate facilities, preferably on a centralized basis for the entire organization.

Laboratories have continuing needs for the fabrication and repair of all kinds of items that are not easily or quickly available from commercial sources. Often, they are unique instruments or structures that exist solely in the mind of the research investigator. To translate ideas into hardware requires the services of skilled, research-oriented craftsmen who are experienced in dealing with scientists and engineers. Large laboratory organizations find that in-house shop facilities and personnel to operate them are an essential adjunct for laboratory operations. The health and safety of the personnel must be considered carefully when the shops are designed, constructed, and equipped. The rules of the Occupational Safety and Health Agency will govern in this part of the laboratory facility.

Other support services frequently associated with laboratories are photographic facilities of many types. They are an essential adjunct of many types of scientific investigations and may be operated by laboratory scientists on an ad hoc basis or may have full-time workers when the photographic work load is heavy.

CHAPTER 21

PHOTOGRAPHIC DARKROOMS

21.1 DESCRIPTION OF PHOTOGRAPHIC DARKROOMS

21.1.1 Introduction

Photographic darkrooms are designed to provide a variety of services to support research activities such as filming, developing, printing, enlarging, and cassette loading. Photographic darkrooms are used for processing black and white and color films and print paper of different types. They may be located in the laboratory building or in an ancillary building.

Darkrooms may be one of the following types:
A. A small darkroom for research facilities.
B. Automated photographic systems used for such purposes as medical x-ray processing and some kinds of research.
C. Teaching darkrooms for art schools and photography classes.
D. Commercial facilities.

This chapter is limited to the first three categories, not commercial facilities. In addition, there are numerous packaged photographic development/printing systems that do not require a darkroom. These systems will be discussed also.

21.1.2 Work Activities

Work activities include bulk film handling, mixing of chemicals, developing, washing, rinsing, drying, and printmaking.

Guidelines for Laboratory Design: Health and Safety Considerations, Second Edition
By Louis DiBerardinis, Janet S. Baum, Gari T. Gatwood, Anand K. Seth, Melvin W. First, and Edward F. Groden.
ISBN 0-471-55463-4 Copyright © 1993 by John Wiley & Sons, Inc.

Photographic processes require carefully controlled environmental conditions, including control of light and precise use of chemicals. Film coatings are primarily silver compounds.

21.1.3 Equipment and Materials Used

Equipment may include open tanks, enclosed processing equipment (such as automatic processing units, sometimes referred to as X-O-Mats), and dryers. Where manual operations are performed, eyewash stations and emergency showers are needed. Temperature control of water and chemical solutions is extremely important to the quality of the photographic image. Thermostatic controls are usually set to maintain 75°F ± 1°F.

Many chemicals used in darkrooms and photographic processes are classified as hazardous. They may include solvents, metals, acids, alkalis, aldehydes, and amimes (see Table 21-1 for a partial list). Care should be used in storing these chemicals in cabinets designed for hazardous chemicals. Some have carcinogenic qualities and must be handled in an exhaust-ventilated area.

A more detailed description of specific photograph process and their associated health hazards can be found in *Artist Beware* (McCann, 1979) and *Overexposure: Health Hazards in Photography* (Shaw, 1985).

21.1.4 Packaged Processing Units

There are basically two types of packaged color-processing photographic systems:

1. *Kodachrome.* Chemicals are an integral part of the film; the processors are very expensive.
2. *Echtachrome.* Chemicals are applied to the film; these processors are more widely used.

TABLE 21-1. Typical Chemicals Used in Darkrooms (Partial List)[a]

Silver compounds
Zinc
Cadmium
Hydroquinone
Acetates
Ammonium hydroxide
Thiosulfates
Trisodium phosphate
Hexacynoferrates
Benzyl alcohol

[a] Taken from Kodak Publication J-55, *Disposal and Treatment of Photographic Effluent.*

Packaged processing units are designed for high-volume, assembly-line processing of 35-mm film. They do not require a darkroom facility. Packaged systems are popular because of their compact size and efficient chemical usage. Most machines can process black and white film, develop negatives and slides, and make prints. Packaged processing systems are either table top or floor-mounted; table-top models are smaller. They require a supply of tempered water, cold water, and drainage. Many of the package systems contain an electric heater for solution temperature control. The chemicals are stored in bottles and dispensed as needed. The spent solution is also stored in containers provided with the unit. Waste containers can either be drained for disposal or stored for silver recovery.

The amount of chemical waste produced in these systems is not significant. For example, a typical table-top system that operates all day will generate approximately 2 liters of chemical waste. Some of these chemicals are not considered hazardous waste and can therefore go down the drain. Other chemical wastes, including all silver-bearing wastes, are hazardous and must be disposed of accordingly. Many large and small darkrooms now consolidate waste fixer into 10- to 20-gallon drums for later silver recovery and hazardous waste disposal cost control. An area for collecting and holding waste chemical liquids, perhaps in their original containers or in drums, as with the fixer, should be considered in the design phase. The routine sink disposal of most photo chemicals has stopped. and darkroom design needs to reflect this fact.

An exhaust outlet should be located near the unit for ventilation. Some machines have exhaust connections; otherwise, a canopy-type capture hood may be sufficient.

21.1.5 Exclusions

None.

21.2 PHOTOGRAPHIC DARKROOM LAYOUT

All issues discussed in Sections 1.2 and 2.2 should be reviewed and, if applicable, implemented. Criteria for a good photographic darkroom are capacity for good ventilation, maintenance of light-proof conditions, and ability to store chemicals in a safe manner. Figures 21-1 and 21-2 show typical darkroom layouts.

21.2.1 Personnel Entry and Egress

In large teaching darkrooms, it may be advantageous to divide different types of work into separate rooms—for example, to put color processing, black and white processing, and photo finishing in separate rooms. For rooms of less than 200 net square feet, one entry/egress is sufficient. Larger rooms require a second egress. Rooms should be wheelchair-accessible to disabled persons.

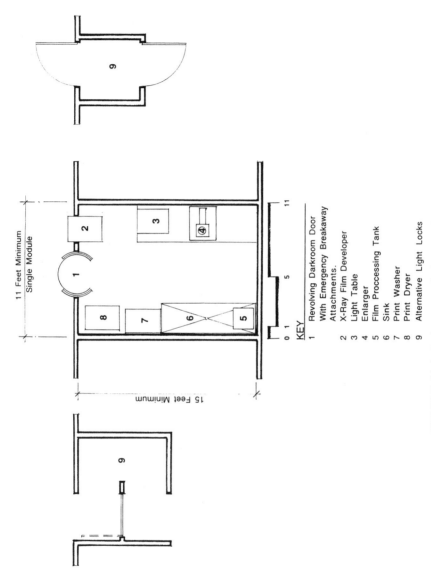

FIGURE 21-1. Laboratory darkroom: Sample layout.

KEY

1 Revolving Darkroom Door
   With Emergency Breakaway
   Attachments.
2 X-Ray Film Developer
3 Light Table
4 Enlarger
5 Film Proccessing Tank
6 Sink
7 Print Washer
8 Print Dryer
9 Alternative Light Locks

11 Feet Minimum
Single Module

15 Feet Minimum

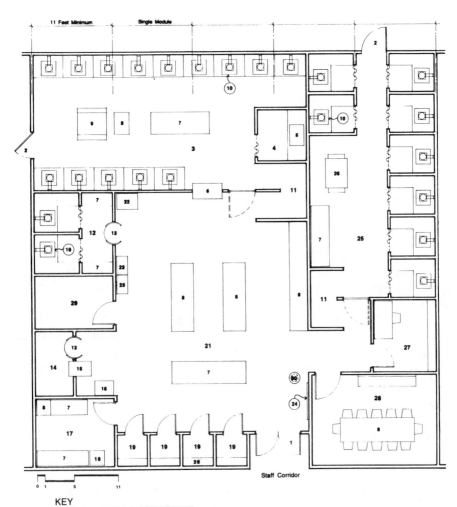

KEY

1 Primary Access/Egress
2 Emergency Second Egress
3 B+W Printing Darkroom
4 Print Inspection Room
5 Viewing Booth
6 Pass-Thru
7 Sink With Eye Washer
8 Table
9 B+W Paper Processor
10 Enlarger Station
11 Light Trap
12 Advanced Student Darkroom
13 2 Way Rotary Light Lock
14 Darkroom
15 B+W Film Processor

16 Film Dryer
17 Film Processing
18 Color Film Processing
19 Film Loading Rooms
20 Counter
21 Finishing Area
22 Wash
23 Dryers
24 Bulletin Board
25 Color Enlarging Darkrooms
26 Color Print Processor
28 Classroom/Conference Room
29 Storage Room
30 Deluge Shower

FIGURE 21-2. Teaching darkroom: Sample layout.

Light control is absolutely necessary. As the name suggests, this room type should be capable of achieving total darkness. In small facilities, a special light-tight revolving door is used for entry and egress, but the revolving doors must have panels that fold back under moderate pressure to provide unimpeded egress in case of an emergency or when a large piece of equipment must be introduced or removed. Revolving doors are available in 1-h fire-rated assemblies for installation in fire-rated egress corridor walls. In large facilities, a zigzag-type light trap maybe used to ensure darkness. This type of entry allows heavier traffic in and out of the darkroom without having to rely on sealed doorways. A sign or signal light outside the darkroom is used to indicate that developing is in process, to assure that light-proof conditions will be maintained.

21.2.2 Laboratory Furniture Locations

Sturdy shelves for storage of chemicals as well as other photographic equipment and supplies are needed. Some photographic supplies may need storage in a light-proof area or cabinet. It is important that sufficient counter-top area be provided for equipment such as enlargers.

Silver recovery systems for the rinse water should be used where volumes are appropriate. Extra space for hazardous waste (5- to 25-gallon containers) may be needed, when *bulking* for waste solutions is required. See Section 23.1.2.1 for definition of *bulking*.

21.2.3 Surface Finish Considerations

Walls in darkrooms are sometimes painted dark colors to reduce chance reflections from an unexpected light source. When safe lights (i.e., special fixtures with a variety of light filters that provide minimum illumination so as not to damage or otherwise affect the photo plates) are switched on, normal working lights cannot accidently be switched on. For floors subject to spills and splashes, surfaces should be slip-resistant, or special mats should be provided with a raised texture or grid provided at sink work areas to reduce risk of falls.

21.3 HEATING, VENTILATING, AND AIR CONDITIONING

21.3.1 Introduction

Heating, ventilating, and air-conditioning systems for processing darkrooms can be very complex due to the high humidity caused by the use of chemical solutions in open tanks or trays and the wash and rinse processes that add additional humidity to the room. Humidity affects the drying time for negatives and film. When humidity is too low, however, it can cause static electricity that will streak processed film. Relative humidity between 40% and 50% is the recommended range. A temperature range of 68–78°F is acceptable for most darkrooms. Conditions in excess of 80°F and 50% RH should be avoided.

Because of the large amount of chemicals used and odors that might be objectionable to others, recirculation of air from the darkroom is not recommended. The pressure in darkrooms should be negative with respect to all other adjacent rooms. Local exhaust should be used to control nuisance level of odorous or irritating chemicals.

Some photographic film processing requires darkrooms with a dust-free environment. For these, use of HEPA filters in a laminar-flow clean room or laminar-flow work-station hood should be considered. Usually, 85% bag-type disposable filters will be adequate (ASHRAE, 1991).

21.3.2 Ventilation Rates

Darkrooms require a minimum of 0.5 cfm of outdoor air per square foot of floor area for ventilation. Alternatively, 10 to 15 air changes per hour are sufficient under most conditions. Supply air outlets should be located so that they will not create drafts or cause short-circuiting of air into the exhaust air registers or systems. Long periods inside a closed, confined darkroom could be unpleasant without adequate ventilation. Therefore a makeup air system should be included to heat and cool the darkroom to provide comfort conditions for those working there.

When calculating ventilation rates it may be enough to match generation rate of contaminants with an adequate flow of dilution air (Crawley, 1984). A rule of thumb is to use 200 cfm per processing machine.

21.3.3 Local Exhaust

A covered tank requires little exhaust; approximately 25–30 cfm per square foot of tank area is adequate. However, an open tank requires an exhaust slot hood at the edge of the tank or in the middle, drawing 150–200 cfm per square foot of tank area

For processes using open trays of chemical solutions, lateral slot exhaust openings or an enclosure hood should be used. The ACGIH *Industrial Ventilation Manual* (ACGIH, 1992) contains design guidance. Local exhaust systems are used because they provide better control of fumes before they reach the breathing zone of workers. Examples are shown in Figure 21-3 and 21-4.

21.4 LOSS PREVENTION, INDUSTRIAL HYGIENE, AND PERSONAL SAFETY

All recommendations contained in Sections 1.4 and 2.4 should be reviewed and implemented.

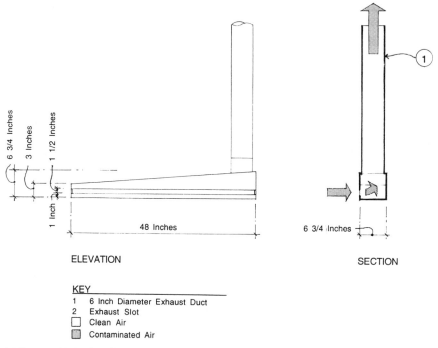

ELEVATION   SECTION

KEY
1   6 Inch Diameter Exhaust Duct
2   Exhaust Slot
☐   Clean Air
▨   Contaminated Air

FIGURE 21-3. Small hood for darkroom: Elevation and section. (Sizes will vary according to specific applications.)

21.4.1   Storage

21.4.1.1 *Storage of Unprocessed Products.* ASHRAE Applications Handbook, chapter 20, pages 20.2 to 20.6 (ASHRAE, 1991) recommends that photographic products not be stored in damp basements or in high-temperature and high-humidity areas. The ideal storage temperature is 60°F with a humidity range of 40–60%, with 40% RH being preferred. In tropical areas, refrigerated storage is recommended. Supplies should be kept in vapor-tight packages or placed in sealed containers.

Black and white printing papers should be stored at 70°F or below, color film or paper at 45–50°F. Long-term storage of color film or papers above 70°F may affect color balance. When products are taken out of long-term storage, warm-up time is necessary (ASHRAE, 1991).

21.4.2   Storage of Processed Film & Paper

Usually, processed film and prints are not stored in the darkroom because storage requirements make it hard for processing and storage to be conducted in the same room. ANSI (1984, 1985) specifies three levels of storage:

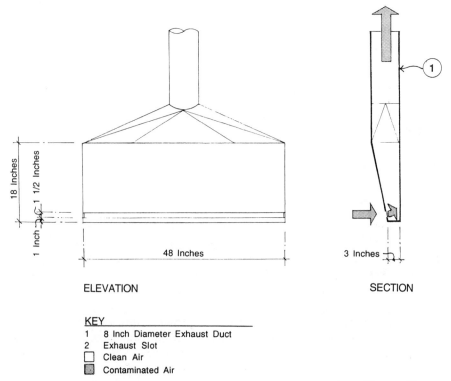

ELEVATION   SECTION

KEY
1   8 Inch Diameter Exhaust Duct
2   Exhaust Slot
☐   Clean Air
▧   Contaminated Air

FIGURE 21-4. Large hood for darkroom: Elevation and section. (Sizes will vary according to specific applications.)

medium-term, long-term, and archival. Specific details for storage environments are available in ANSI (1982, 1984, and 1985) and ASHRAE Applications Handbook, chapter 20 (ASHRAE, 1991).

21.4.3 Safety Considerations

It is important that all electrical receptacles near sinks be provided with GFIC devices to prevent electrical shock hazards. Flooring should be non-skid to reduce the chance of falls due to spills and splashing. A safety shower and emergency eyewash are needed. Emergency eyewash fountains plumbed with tempered potable water should be installed in a large student darkroom.

The water supply to the rinsing bath should have backflow preventers to protect the building water supply from accidental contamination.

21.5 SPECIAL REQUIREMENTS

All items described in Sections 1.4 and 2.4 should be reviewed, and those that are relevant should be implemented.

21.5.1 Personal Protective Equipment

Aprons, face shields, safety glasses, goggles, and gloves should be used, and a storage place provided.

21.5.2 Personal Cleanliness

The greatest health concern that results from continuous exposure to photographic chemicals is contact dermitis or allergic contact dermitis (ACD) (Crawley, 1984). Therefore, a high level of personal hygiene is called for, and washing facilities must be provided in the darkroom.

21.5.3 Plumbing

Wash water discharged from the processes contains chemicals that may not be discharged directly into the sewer system, depending upon the volume. For example, in radiological processors, where a large amount of x-ray films are processed regularly, it will be cost-effective to install a silver recovery system. However, the chemicals from a darkroom used occasionally are not large enough to warrant recovery.

Plumbing fixtures should be selected to provide sufficient width, length, and depth for the number of trays that will normally be used. Thermostatic mixing valves that maintain the correct temperature will be needed. In addition, the water supply may require filtration to remove particulate matter; a 50-mm filter is recommended.

CHAPTER 22

SUPPORT SHOPS

22.1 DESCRIPTION OF SUPPORT SHOP AREAS

22.1.1 Introduction

Support shop areas for laboratory facilities contribute to research activities and instruments. Support services include machine, electronics, glass-blowing, woodworking, plastics, and maintenance shops. At small institutions, several functions may be combined into a single shop. Support shops may be located in a laboratory building or in an ancillary building which may also contain storage or other types of shops for maintenance.

22.1.2 Work Activities

Support shop activities relate to the research or physical plant activities to which the shop is dedicated. Materials in these shops may be milled, cut, sawn, bent, ground, sanded, polished, cleaned, dried, painted, soldered, welded, etched, doped, glued, heated, melted, drilled, screwed, riveted, turned on a lathe, punched, or joined in order to construct or repair a variety of instruments, chambers, stages, or scientific apparatus or facilities used in research. This work will be performed either at a machine, on a work bench, or in the field.

Guidelines for Laboratory Design: Health and Safety Considerations, Second Edition
By Louis DiBerardinis, Janet S. Baum, Gari T. Gatwood, Anand K. Seth, Melvin W. First, and Edward F. Groden.
ISBN 0-471-55463-4 Copyright © 1993 by John Wiley & Sons, Inc.

22.1.3 Equipment and Materials

Equipment used may include any combination of the following: milling machine; metal, glass, or plastic cutter; saws (circular, table, band, jig); grinder; sander; polisher; degreasing or acid dip tank; oven; paint sprayer; furnace; blow torch; soldering iron; welding machine; arc welder; drill press; riveter; punch; anvil; lathe; clamps; joiner; router; crane; hoist; a variety of hand tools; disc grinder; portable hand grinding bench; welding bench; metal cutting band saw; solvent degreasing tanks; spray paint booth; sanders; saws; buffing wheel; and abrasive blasting cabinet. Electronic and machine support shops may contain electrical and electronic analytic instruments, transformers, and oscilloscopes. In addition, support shops may have fume hoods and a variety of local exhaust devices to contain fumes from activities such as welding or acid etching. Materials commonly used in support shops include any combination of the following: sheet metal and metal castings; wire; rods; pipes and extrusions; glass sheets; rods and tubing; plastic sheets; castings; extrusions; pipes; wood planks and plywood sheets; paint; glue; epoxy compounds; solvents; strong acids and other chemicals; volatile and flammable liquids; and compressed gas.

22.1.4 Exclusions

Support shop areas are not production facilities; they are intended only to support laboratory activities. Access to professionally managed support shops areas may be limited to trained shop staff, in order to maintain safety and security of machines, tools, and materials. Often shops and shop areas are specifically designated for use by scientific staff. Support shops open to all scientific staff, faculty, and students should be designed with careful consideration of additional safety features needed for amateur craftspeople.

22.2 LAYOUT

Each item addressed in Section 1.2 should be evaluated for its applicability to the specific needs of those who will use support shops, and items that are relevant should be implemented. Layout considerations in Section 2.2 apply to support shop areas. The shop manager and users should be consulted throughout the design process and should be closely involved in exploring options in shop layout. This interaction will improve safety and function of the space. Because most of the activities involving major shop equipment incur some risk to health and safety, good hazard zoning becomes more difficult to achieve.

22.2.1 Personnel Entry and Egress

Two exits should be provided from support shop areas that use chemicals or that have other potential high hazard conditions such as high voltage or high concentrations of dust and particulate matter. Consideration should be given to access to support shops. Either the support shop areas should have a loading dock or one should be conveniently nearby. Corridor and door widths between shops and loading dock should allow safe transport of machines and materials, and should comply with local building code and NFPA 101 standards.

22.2.2 Aisles

Aisles between benches or machines should have a minimum clearance of 5 ft to allow workers to maneuver materials safely to machines, especially long bar or sheet stock. Adequate clearance also allows workers to safely back away from a machine or process if an unsafe condition develops. Major access/egress aisles in machine and woodworking shops may be much wider than 5 ft, due to the large size of materials and apparatus commonly handled in these types of shops. Aisles that are equipped with cranes may be wider for the same reason. The major aisle width may be determined by the largest machines, to allow servicing, relocation, or replacement without moving other machines aside.

22.2.3 Location of Fume Hoods

As in research laboratories, hazardous operations should be conducted in a chemical fume hood or some other form of local exhaust ventilation, such as a spray paint booth or slot exhaust. Fume hoods must be located away from the primary shop exit and circulation aisle for personnel safety and to reduce adverse effects on hood performance. Similar layout considerations apply to locations of most local exhaust devices at hazardous operations or equipment. Figure 22-1 illustrates a good location of a fume hood within a glass-blowing shop.

22.2.4 Location of Equipment

Layout of support shop equipment, benches, and machines should be guided by several considerations:

1. Sequence of the machine use in typical shop processes
2. Ability to service the machines
3. Size of typical materials and safe materials' approach to the machine
4. Potential break-away or accident modes of the machine itself or of materials on the machine

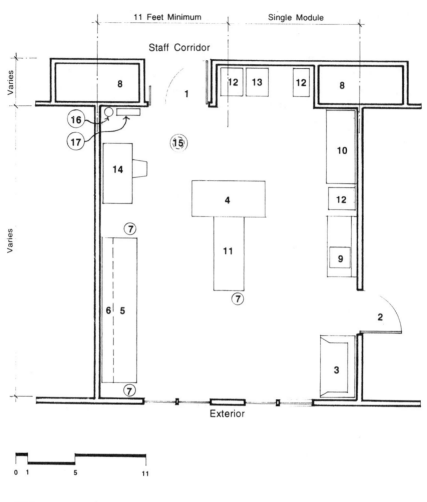

FIGURE 22-1. Laboratory glass-blowing shop: Sample layout.

KEY
___

1	Primary Access/Egress	9	Sink
2	Emergency Second Egress	10	Annealing Oven
3	Fume Hood	11	Lathe
4	Work Bench	12	Grinder
5	High Work Bench	13	Cutter
6	Glass Storage Shelving	14	Desk
7	Gas Tanks	15	Deluge Shower
8	Electrical Closet	16	Fire Extinguisher
		17	Fire Blanket

Equipment manufacturers should be consulted for advice on service access and proper use of the machine may affect location.

22.2.4.1 Sequence of Machine Locations. If machines or pieces of equipment are arranged in the sequence in which they are typically used in a shop, there may be a reduction of cross-traffic in the shop. Persons transporting large pieces of raw materials or working on initial processes are less likely to disturb those executing finer operations on small parts later in the sequence. The goal of this approach is to reduce personnel traffic around machines. This allows shop workers to concentrate on their work and have better freedom of movement around the machine with which they are working. On the other hand, when particular machines are used frequently at many phases of the process, they should be located in a central position, accessible from all parts of the shop. Clearances on all sides of the machines should be ample to allow persons not using the machines to pass around safely without interfering with the operation.

22.2.4.2 Service Clearances. All machines require servicing. The shop layout should allow clearances for safe and convenient servicing of equipment. Refer to equipment installation manuals and consult equipment manufacturers directly for this information to gain sufficient understanding of servicing issues. Many machines require local exhaust ventilation services, and clearances must be allowed for the required hoods, ducts, and piping.

Clearances should also allow easy floor cleaning behind and between machines. Metal particles and shavings can become a hazard when embedded in many flooring materials. Flooring materials such as sealed concrete, heavy-duty vinyl composition tile, or sealed end-grain wood parquet, can be cleaned. Floors must also hold up to grease and solvent spills as well as to destructive scrapes, cuts, and impact loads from falling materials and tools.

22.2.4.3 Materials Handling Clearance. Maneuvering large sheet material of metal, wood, plastic or glass, long bar stock, and pipes can cause safety problems in shops. Many times the first task is to cut the material to a smaller size. These materials are carried to the saw or cutting (or bending) machine, placed and correctly positioned on the work platform, and then processed. Lifting and securing large unwieldy materials onto the machine is difficult and unsafe if there is not sufficient clearance around the machine, particularly if two or more workers must lift and position the material. Machines which are identified for use with large sizes of materials should be positioned so that there are no conflicting machines or activities on the sides of the machine where materials are being maneuvered. Consider putting these machines in single-loaded aisle locations.

Also, machines whose processes produce by-products that may cause harm to people or other machines nearby should be located in a manner that

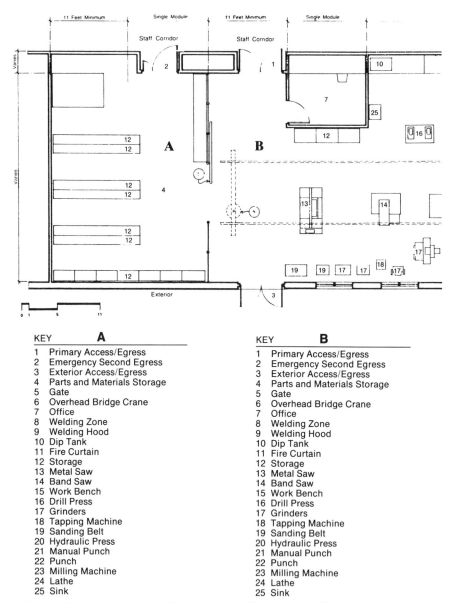

FIGURE 22-2. Laboratory machine shop: **(A)** Storage area; **(B)** Material preparation

KEY **A**

1 Primary Access/Egress
2 Emergency Second Egress
3 Exterior Access/Egress
4 Parts and Materials Storage
5 Gate
6 Overhead Bridge Crane
7 Office
8 Welding Zone
9 Welding Hood
10 Dip Tank
11 Fire Curtain
12 Storage
13 Metal Saw
14 Band Saw
15 Work Bench
16 Drill Press
17 Grinders
18 Tapping Machine
19 Sanding Belt
20 Hydraulic Press
21 Manual Punch
22 Punch
23 Milling Machine
24 Lathe
25 Sink

KEY **B**

1 Primary Access/Egress
2 Emergency Second Egress
3 Exterior Access/Egress
4 Parts and Materials Storage
5 Gate
6 Overhead Bridge Crane
7 Office
8 Welding Zone
9 Welding Hood
10 Dip Tank
11 Fire Curtain
12 Storage
13 Metal Saw
14 Band Saw
15 Work Bench
16 Drill Press
17 Grinders
18 Tapping Machine
19 Sanding Belt
20 Hydraulic Press
21 Manual Punch
22 Punch
23 Milling Machine
24 Lathe
25 Sink

reduces the hazard of the by-product dispersal. Local-exhaust-ventilation devices can be designed and installed to safely remove those by-products. An example of this consideration is grinders, which routinely spin off scrap and particles that could scatter in someone's face or onto an adjacent machine. This type of equipment should be located and protected by a movable screen

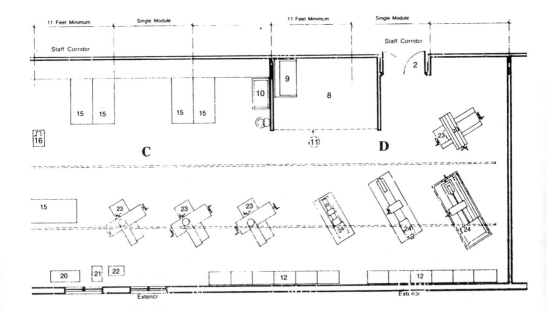

KEY	**C**		KEY	**D**
1	Primary Access/Egress		1	Primary Access/Egress
2	Emergency Second Egress		2	Emergency Second Egress
3	Exterior Access/Egress		3	Exterior Access/Egress
4	Parts and Materials Storage		4	Parts and Materials Storage
5	Gate		5	Gate
6	Overhead Bridge Crane		6	Overhead Bridge Crane
7	Office		7	Office
8	Welding Zone		8	Welding Zone
9	Welding Hood		9	Welding Hood
10	Dip Tank		10	Dip Tank
11	Fire Curtain		11	Fire Curtain
12	Storage		12	Storage
13	Metal Saw		13	Metal Saw
14	Band Saw		14	Band Saw
15	Work Bench		15	Work Bench
16	Drill Press		16	Drill Press
17	Grinders		17	Grinders
18	Tapping Machine		18	Tapping Machine
19	Sanding Belt		19	Sanding Belt
20	Hydraulic Press		20	Hydraulic Press
21	Manual Punch		21	Manual Punch
22	Punch		22	Punch
23	Milling Machine		23	Milling Machine
24	Lathe		24	Lathe
25	Sink		25	Sink

area. **(C)** Milling areas. **(D)** Lathing area.

or curtain so that the surroundings are protected. When a grinder is protected with a shroud enclosure that has a slot exhaust at the bottom to draw down shards and grindings into a collection box, the safety of the operator and those nearby is improved. Refer to the latest edition of the *Industrial Ventilation Manual* for recommended enclosures.

22.2.4.4 Layout for Hazard Reduction. Finally, support shop layout should evaluate potential break-away or accident modes of the machine itself or of materials on the machine. Blades, drill bits, and armatures which are improperly installed, overstressed, broken, or unsecured can be potentially dangerous missiles in a shop. Materials, too, can fly off milling machines and lathes at great speed. Saws, improperly used, are hazardous for this reason also. Locate lathes and saws in positions where ejected materials are not likely to hit a person. It is more difficult to position milling machines and drills to reduce danger from ejected materials because of the potential 360-degree horizontal trajectory. Blow torches for soldering and welding should only be used with appropriate personal protection gear. They should be used on a bench with a slot exhaust hood (see Figure 22-4) or a portable exhaust capture hood. Other workers in the shop should be protected by screens and booths that block harmful light and heat. The goal of this approach in shop layout is to reduce the risk of an accident to others nearby. This allows the shop worker to concentrate on his/her work and not be at as much risk by what may happen at a neighboring machine. This does not absolve shop workers from maintaining vigilance on what is going on around them; it only reduces risk. Clearly, desk work stations should be protected from hazards and located near exits. Figure 22-2 shows a layout for a machine shop in an academic research building that was designed with the concept of hazard zoning in the arrangement of equipment and processes.

22.2.5 Stockroom

Stockrooms should be planned and provided for each type of support shop. Stockrooms provide secure safe storage for materials used in support shops. Various racks, shelves, and cabinets should be structurally sound and adequate to support the materials for which they are intended, and they should be attached to the building structure to prevent tipping. Use industrial, or extra-heavy-duty-grade materials storage components. Avoid standard residential- or commercial-grade storage units; that can fall apart or deform under loading of normal shop materials. Aisles and access doors should be wide enough to move materials in from the loading dock and out to the shop, as illustrated in Figure 22-2.

22.3 HEATING, VENTILATING, AND AIR CONDITIONING

22.3.1 Introduction

All items described in Sections 1.3 and 2.3 should be reviewed, and those that are relevant should be implemented. Machining of some materials, such as metallic sodium or beryllium, might require special controls, atmospheres, or ventilation. Consult the industrial hygienist for directives on special materials.

22.3.2 Additional HVAC Needs

22.3.2.1 Temperature and Humidity Control.

Normal comfort-level temperatures are recommended for laboratory support service areas. Machine and metal shops require, in addition, control of excess humidity to reduce damage to materials and machines from rust. When support shops have direct access to loading docks, use additional heat or curtains to avoid cold drafts in the shop area during periods when shipping doors are open.

Glass-blowing shops often have high heat loads from open flames and high-temperature ovens which run for long periods. Air conditioning is recommended to maintain tolerable working conditions. Airflows must be carefully controlled with reduced exit velocity to avoid turbulence and to maintain safe and consistent blow torch plumes.

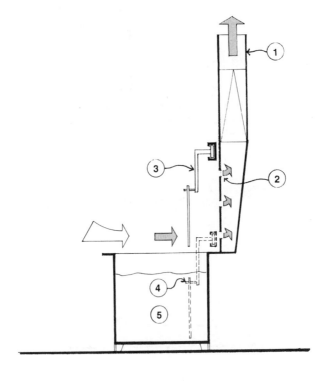

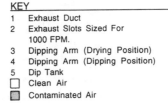

KEY

1   Exhaust Duct
2   Exhaust Slots Sized For
    1000 FPM.
3   Dipping Arm (Drying Position)
4   Dipping Arm (Dipping Position)
5   Dip Tank
☐   Clean Air
▨   Contaminated Air

FIGURE 22-3. Dip tank: Section.

22.3.2.2 *Exhaust Systems.* Manufacturers of specific equipment should be consulted for recommendations on the need for special local exhaust systems. Glass-blowing shops require a fume hood for use with solvents and strong acids. Machine and metal shops may use strong toxic acids (such as hydrogen fluoride) for degreasing metal parts and use flammable solvents for other cleaning. Local exhaust scoops and fume extraction devices are recommended at dip tanks to remove hazardous fumes or particulate matter, as shown in Figure 22-3. Welding and soldering booths require ventilation hoods to remove excessive heat as well as fumes, as shown in Figure 22-4. The *Industrial Ventilation Manual,* 22th edition, has examples of local exhaust systems, machines, and cleaning devices found in shops.

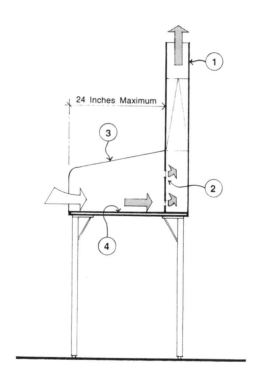

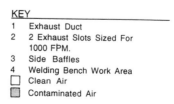

KEY

1   Exhaust Duct
2   2 Exhaust Slots Sized For
    1000 FPM.
3   Side Baffles
4   Welding Bench Work Area
    ☐ Clean Air
    ▨ Contaminated Air

FIGURE 22-4. Welding bench: Section.

22.3.2.3 Dust Collection. Woodworking shop equipment and certain equipment in machine shops will require a dust collection system. Unit dust collectors located beside the machines they serve can help conserve energy when wood and nontoxic metals and plastics are used. Keep in mind that unit dust collectors usually leak dust back into the work room, especially when they are not serviced regularly. Local dust collection at sanders, grinders, and saws are recommended. Dust collectors can be either in a central system or in small local units and are used to reduce amounts of airborne particulate dust at work stations. Under certain conditions, accumulated combustible material may burn or become explosive.

22.4 LOSS PREVENTION AND OCCUPATIONAL SAFETY AND HEALTH PROTECTION

22.4.1 Introduction

Laboratory support shop areas have hazards not typical of most laboratories, such as a wide variety of mechanical hazards and the heavy weight of materials handled. Other hazards are similar to those encountered in labs, such as chemical and electrical hazards, as well as high temperatures and high pressures. All items described in Sections 1.4 and 2.4 should be reviewed, and those that are relevant should be implemented. In addition, refer to the National Safety Council's *Accident Prevention Manual for Industrial Operations* and see the volumes entitled "Engineering and Technology" and "Administration and Programs." The first volume describes the specific devices and considerations for mechanical hazards found at most machines and processes encountered in laboratory support shop areas. It also covers in detail issues concerning equipment used in materials handling. The second volume describes safety programs and procedures that should be implemented in laboratory support service areas.

22.4.2 Safety Station

In each laboratory support shop area, safety stations should be provided with safety equipment and personnel protection gear appropriate to the shop. A typical shop safety station should include a safety information bulletin board, handwash sink, safety glasses dispenser, emergency eyewash fountain, emergency deluge shower, an appropriate type of fire extinguisher, fire blanket, first-aid kit, and chemical-spill kit. Support shop safety stations should be located near the primary shop entry, where they will be visible and not easily blocked by equipment or coats.

CHAPTER 23

HAZARDOUS CHEMICAL, RADIOACTIVE, BIOLOGICAL WASTE HANDLING ROOMS

23.1 DESCRIPTION

Most laboratory chemicals will have to be handled and disposed of as hazardous waste when they are spent, outdated, or no longer needed. Laws no longer permit indiscriminate disposal of such materials into sewers or sanitary landfills, via trash haulers.

Other types of laboratory waste include pathological, radioactive, and normal building or industrial wastes.

Many laboratories generate radioactive wastes which are highly regulated and in need of special attention in the design phase of a laboratory building.

Biological wastes, while not yet as highly regulated, are another form of wastes that need special design attention where infectious or pathological materials are used.

Each of these waste types has its own characteristic hazards but some hazards are common to all, such as combustibility, and for these hazards the controls will be the same. The hazards of radioactive wastes, pathological wastes, and chemical wastes have few similarities and the basic controls for them are likewise not the same.

Chemical waste management and storage must deal with reactivity, toxicity, flammability, chemical compatibility, corrosivity, and explosivity. Radioactive waste may involve some of these same chemical hazards, but with

Guidelines for Laboratory Design: Health and Safety Considerations, Second Edition
By Louis DiBerardinis, Janet S. Baum, Gari T. Gatwood, Anand K. Seth, Melvin W. First, and Edward F. Groden.
ISBN 0-471-55463-4   Copyright © 1993 by John Wiley & Sons, Inc.

the added dimension of radiation exposure. Pathological waste frequently resembles what is known as *municipal waste,* except for its potential to spread harmful organisms. Each of these waste types must be dealt with individually, with emphasis focused on the controls for the unique hazards associated with each. It is seldom useful or safe to combine these wastes into one room or a unified operation for the purpose of disposal management because co-mingling these wastes, without physical separation, subjects all waste types to the hazards of the others.

This chapter presents all three of the waste types: hazardous chemical waste; radioactive waste consisting of isotopes, radiolabeled chemicals, contaminated materials, and such radioactive wastes that may ultimately be declared below levels of concern or NRC regulation; and biological waste which covers infectious as well as physically dangerous medical and biological waste.

23.1.1 Introduction

The task of disposing of hazardous waste is expensive, labor-intensive, and heavily regulated, and it requires adequate work space as well as safe storage space. When adequate facilities, staffing, and programming are provided, a laboratory can save money in two ways: first, through reduced disposal costs achieved by good material management; second, by maintaining strict compliance with regulatory requirements, thereby eliminating fines levied as a consequence of unsatisfactory EPA or state agency inspections. Many states try to inspect every generator of hazardous waste on an annual basis.

23.1.1.1 Hazardous Chemical Waste. Hazardous chemical waste is defined by the U.S. EPA in the Code of Federal Regulations 40 CFR 260 (EPA, 1992) as waste that is hazardous by virtue of its flammability, corrosivity, reactivity, or toxicity (each characteristic has its own definition in the regulation) or that appears on any one of several extensive lists of chemicals and chemical compounds.

23.1.1.2 Radioactive Waste. Many laboratories produce radioactive waste, with medical and research laboratories producing it more than others. Worker safety and, radioactive waste handling are regulated by 10 CFR 20 (NRC, 1991). These regulations are enforced by the U.S. Nuclear Regulatory Commission (NRC). The regulations represent a performance standard insofar as the design of a radioactive waste handling room is concerned, meaning that the NRC leaves the design up to the owner as long as the facility provides adequate protection against accidents, injuries, or unauthorized releases of radioactivity to the environment.

There are four types of radioactive wastes to deal with in a radioactive waste handling facility: scintillation fluids (usually in small plastic or glass vials), absorbed liquids (liquids poured onto absorbent material to meet the criteria of "solid waste"), radioactive dry wastes, and mixed wastes. Mixed waste is defined as a waste controlled by more than one agency or set of regulations. For example, a mixed waste is a highly toxic chemical waste containing an isotope of qualifying radioactivity. It comes under NRC regulations as a radioactive waste and under EPA regulations as a hazardous chemical waste. Mixed wastes are very difficult to dispose of due to the many regulatory requirements imposed on the disposal facility.

23.1.1.3 Biological Waste. Biological waste is defined in state and federal regulations as:

Blood and blood products

Pathogenic waste (human anatomical parts, organs, and fluid)

Cultures of infectious agents

Contaminated animal carcasses and wastes

Sharps (needles, scalpels, broken medical glass)

Biotechnology by-products

Dialysis wastes

Isolation wastes

Many states have, and more are implementing, regulations regarding the handling, sterilizing, and disposal of biological wastes. The federal standard covering handling and disposal of this class of waste is 40 CFR 259 (EPA, 1992).

Disposal techniques for these materials include: incineration; steam sterilization or gas sterilization; ionizing radiation and nonionizing radiation followed by burial; or maceration and discharge to a sewer.

Many hospitals and research laboratories do not have adequate disposal capabilities in-house and must use services provided by commercial haulers, incinerators, and land disposal contractors. Some institutions sterilize the waste and then have it incinerated to assure a safe disposal.

Handling and storage of these wastes even for short periods call for a special facility that needs to be planned carefully.

23.1.2 Work Activities

23.1.2.1 Hazardous Chemical Waste. Activities that take place in a chemical waste management room, suite, or area vary according to the amounts and types of waste chemicals generated. Most chemical wastes are picked up for disposal by licensed waste transporters in what are known as "lab packs." These are drums, made of fiber or steel, into which are placed complete containers, such as bottles of chemicals. Packing material, such as vermiculite, is added to protect against breakage during transportation before the drum is sealed. Due to the limited number of containers which can fit into a drum, this is the most expensive method of disposal available to laboratories, but it is also the only method possible under current regulations in many cases. The drums must be packed according to Department of Transportation (DOT) shipping regulations 40 CFR 100 (DOT, 1992) requiring separate drums for each category of waste: flammable, reactive, corrosive, poisonous.

Preparation of lab packs is most frequently performed by a licensed disposal contractor to assure DOT compliance. Because high packing density is one key to cost containment, adequate space and working accommodations need to be provided to the packers so that they can have several (4 or 5) open drums (30- to 55-gallon capacity) at a time available for sorting and packing.

"Bulking" is the transfer of same or like materials from small or partially filled containers to larger ones, usually 30- to 55-gallon drums. The larger container can usually be disposed of at lower per pound cost (depending on the nature of the material) and can, in some cases, return money or provide chlorofluorocarbon (CFC) credits. Some chemicals such as chlorinated hydrocarbons have a high recycle value, and CFC will eventually be available for purchase only when a company can satisfy a justification requirement imposed by regulation or has adequate recycle credits. Bulking, however, carries the risk of incompatible materials coming together and causing ignition or explosion. Safe bulking of liquids requires maintaining small samples of the total contents of drums and testing for compatibility before adding new materials. Bulking needs unencumbered work space with good ventilation and should be carried out in a room separated from other hazardous waste activities.

Some institutions find it profitable to recycle virgin, unused, or otherwise good chemicals that have become superfluous to a given research group or purchasing laboratory. This process does not require the handling of open containers but it does require a dedicated sorting and storage area.

Diverse activities one may observe in a hazardous chemical waste management area include: filling lab packs; consolidating chemically compatible wastes into larger containers (such as combining two partially filled containers of a single chemical into one full container); consolidation to bulk; transfer of toxic or flammable materials from container to container; sampling; sorting for recycling; limited sink disposal of nonhazardous and nonregulated waste; decontamination; inventory sheet preparation; manifest and shipping document preparation; electronic data entry.

23.1.2.2 Radioactive Wastes.

Scintillation fluids usually enter the waste handling facility in boxes of vials segregated by activity level and nuclide type. They may be packed as is, in steel drums, for shipment to a disposal facility, or processed to reduce their volume by passing them through a vial crusher and a liquid-glass separator. The glass can be washed and disposed of as trash. The bulk scintillation fluids and rinse liquids can then be drummed and shipped off site for disposal as hazardous or radioactive waste.

Absorbed radioactive liquids are bulked in drums and shipped for disposal.

Radioactive dry wastes coming into the waste facility are sorted by radioactivity and type; some can be held in storage for decay of shorter-half-life isotopes and then shipped off as deregulated dry waste (trash). Longer-half-life materials must be disposed of as radioactive waste. Frequently, all of these materials are reduced in volume using a single drum compactor.

To summarize, work activities in a radioactive waste management area for laboratories include: receiving, shipping, drum handling, bulking of liquids, bulking of dry materials, record keeping, labeling, surveying, volume reduction, and document preparation.

23.1.2.3 *Biological Wastes.*

23.1.2.3 Biological Wastes. Lacking on-site disposal facilities, biological waste materials brought into the biological waste handling facility must be segregated by waste class (40 CFR 259) (EPA, 1992) and made ready for shipment. This includes:

Autoclaving

Bagging and boxing

Steam sterilizing

Gas sterilizing

Freezing or refrigerating (to prevent putrefaction during long storage periods on site)

Other activities include record keeping, labeling, spill decontamination, and clean up.

In-house incineration is not considered in this chapter; however the biological waste handling room could be used as a holding area for possible on site incineration or other treatment.

23.1.3 Equipment and Materials Used

23.1.3.1 Hazardous Chemical Waste. Because chemical waste handling is a support type of operation, little laboratory equipment or supplies will be present with the exception of chemicals and compressed gases stored in preparation for disposal. Most of the materials listed below pertain to storing, documenting, handling, and shipping of hazardous waste chemicals needed in a hazardous chemical waste handling area, and space must be provided for them. They are self-explanatory.

Labels, forms, paper supplies

Computer terminal

Desk and chair

Drums—steel, fiber, plastic (5- to 85-gallon sizes)

Packing materials—vermiculite, etc.

Wheeled carts, wheeled drum dollies
Work bench or packaging table
Pneumatic lifts
Scales—up to 500 lb
Plastic and glass containers for repacking
Sink
Gas cylinder rack
Emergency equipment
pH meter
Flashpoint tester
Small wet lab setup

23.1.3.2 Radioactive Waste. Materials needed in a radioactive waste handling area are listed below, and space must be provided for them.

Labels, forms, paper supplies
Desk and chair
Computer terminal
Steel drums (30- and 55-gallon sizes; approximately 50 drums)
Packing materials
Spill cleanup materials
Wheeled drum dolly
Work bench
Scales to 500 lb
Sink
Vial crusher
Radiation monitoring equipment
Trash buckets
Metal detector

23.1.3.3 Biological Waste. Materials needed in a biological waste handling area are listed below, and space must be provided for them.

Steam sterilizer
Desk and chair
Boxes, fiber drums, and other containers
Labels, forms, paper supplies
Spill decontamination equipment
Sink
Storage cabinets
Freezer/refrigerator

23.1.4 Exclusions

23.1.4.1 Hazardous Chemical Waste. Disposal of daily building and laboratory wastes, not otherwise hazardous, are not covered in this chapter. Chemicals not defined by EPA or state agencies as hazardous, either by being "listed" or possessing a hazardous "characteristic," need not be handled as hazardous waste and may be disposed of through normal trash or sink disposal. The local safety or environmental specialist should be consulted to help determine these issues for any specific geographical or political area. A hazardous waste management and storage facility is not designed for treatment or destruction of hazardous waste materials such as distillation, dilution, or reaction. Disposal of nonhazardous materials via the sink is an acceptable procedure.

23.1.4.2 Radioactive Waste. The radioactive waste room is not meant to handle any other types of waste except for the four mentioned in the introduction to radioactive waste, Section 23.1.1.

23.1.4.3 Biological Waste. The biological waste room should not have multipurpose uses. It should be used for biological waste storage and handling only.

23.2 AREA LAYOUT

23.2.1 Guiding Concepts

Generally it is recommended that the hazardous chemical waste, radioactive waste, and biological waste areas be separate rooms.

23.2.1.1 Hazardous Chemical Waste. The health and safety requirements of hazardous chemical waste handling areas are similar to those of the laboratories using the same chemicals and compressed gases. Therefore, the issues discussed in Sections 1.2, 2.2, and 5.2 should be reviewed for their applicability as a supplement to the items discussed here.

A chemical waste handling and storage area or room should be separated from all other building occupancies by not less than 2-h fire-rated construction with 1 1/2-h fire-rated door assemblies. The facility may occupy a separate, detached building removed sufficiently from other buildings or structures if required by code. Waste chemical storage containers should be protected from the weather, excessive heat, and large temperature swings. When the chemical handling and waste storage area is in a multi-use building, at least one wall of the unit should be an exterior wall, above grade level, to provide a pressure relief panel when it is required by local codes.

The best approach to determining room size is through experience. If past records of hazardous waste activity are available, they may be referred to for an accurate determination of waste handling size requirements and room configuration. Where records do not exist, the example following can be used as a guide.

For a science teaching and research complex of five buildings containing 700 or 800 individual laboratory rooms engaged in physics, chemistry, biology, biochemistry, and geology laboratory work, approximately fifty 55-gallon drums of lab packs and bulk wastes will be produced each month. To handle this amount of waste, the bulking room should be separate from other activities and consist of approximately 200 ft^2 of floor space with an outside wall and no in-the-room storage. The remainder of the activities may be carried out in a 800- to 1000 ft^2 room.

Waste handling areas should be located at or above grade for heat venting and loss control in the event of a fire. A sample layout for a hazardous chemical waste room can be seen in Figure 23–1.

23.2.1.2   *Radioactive Waste.*   A radioactive waste handling facility should be separated from all other building occupancies by not less than 2-h fire-rated construction with 1 1/2-h rated door assemblies, or it may occupy a separate detached building removed sufficiently from other buildings or structures if required by code. Waste radioactive storage containers must be protected from the weather, excessive heat, and large temperature swings. If the radioactive waste handling and storage area is located in a multi-use building, at least one wall of the unit should be on an exterior wall, above grade, to provide a pressure relief panel which may be required where flammable liquids are being handled.

The walls may need to be constructed of solid concrete and/or lined with lead sheets to reduce radiation levels in adjacent areas to regulatory requirements that include ALARA (as low as reasonable achievable). Shielding may also be accomplished by locating the room adjacent to nonoccupied areas. Specific site requirements may be obtained from a resident radiation safety officer or a regional office of the Nuclear Regulatory Commission.

For a typical teaching hospital of 500 to 1000 beds, or a research building using radioisotopes, about 200 drums of waste will be shipped per year. To handle this quantity and the activities enumerated in Section 23.1.2, one room with an area of 600 ft^2 is needed for radioactive waste processing and storage. When vial crushing is conducted, a separate small room equipped for highly flammable solvent activities is required. An adjacent room for showers and clothes lockers is desirable.

Waste handling and storage rooms should be located at or above grade for heat, venting, and fire control in the event of a fire. Spill control berms or dikes should be installed at each opening from the room. Floors should be constructed of concrete, be of low porosity, and have a continuous surface, an impervious covering, and no drains. A sample radioactive waste facility is shown in Figure 23–2.

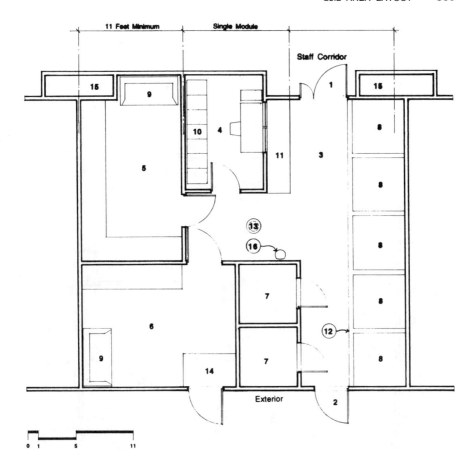

FIGURE 23-1. Hazardous waste storage: Sample layout.

KEY

1	Primary Access/Egress	9	Fume Hood
2	Emergency Second Egress	10	Files
3	Sorting Area	11	Sorting Table
4	Office	12	Curbs
5	Testing Laboratory	13	Deluge Shower
6	Bulking Room	14	Ramped Dike To Loading Dock
7	Primary Drum Storage	15	Electrical Closets
8	Primary Shelf Storage	16	Eyewash Station

23.2.1.3 Biological Waste. The biological waste room has pathological organisms and sharps as primary hazards for waste handlers. There is a minimal fire hazard from the presence of stored combustible shipping boxes and packing materials. This room is a service room to the research which is

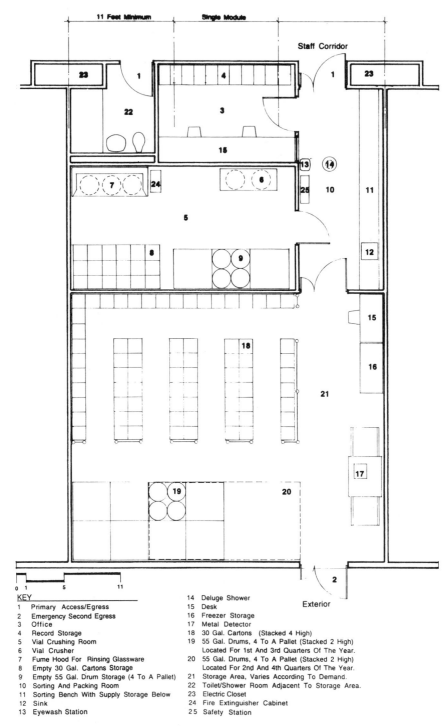

KEY

1	Primary Access/Egress
2	Emergency Second Egress
3	Office
4	Record Storage
5	Vial Crushing Room
6	Vial Crusher
7	Fume Hood For Rinsing Glassware
8	Empty 30 Gal. Cartons Storage
9	Empty 55 Gal. Drum Storage (4 To A Pallet)
10	Sorting And Packing Room
11	Sorting Bench With Supply Storage Below
12	Sink
13	Eyewash Station

14	Deluge Shower
15	Desk
16	Freezer Storage
17	Metal Detector
18	30 Gal. Cartons (Stacked 4 High)
19	55 Gal. Drums, 4 To A Pallet (Stacked 2 High) Located For 1st And 3rd Quarters Of The Year.
20	55 Gal. Drums, 4 To A Pallet (Stacked 2 High) Located For 2nd And 4th Quarters Of The Year.
21	Storage Area, Varies According To Demand.
22	Toilet/Shower Room Adjacent To Storage Area.
23	Electric Closet
24	Fire Extinguisher Cabinet
25	Safety Station

FIGURE 23-2. Radioisotope waste storage: Sample layout.

carried out within the building or group of buildings. Separation from the research areas is recommended. It is desirable to have the biological waste handling room located near a loading dock to facilitate removal of material.

Odor control through ventilation and air discharge away from personnel, the public, and air intakes should be reviewed in the design stage.

The amount of waste generated by a typical biological research facility can be handled and stored temporarily awaiting shipment in a 500-ft^2 room, although the generation of biological waste varies greatly from laboratory to laboratory. Past waste records should be consulted when they exist, or otherwise the records of similar facilities should be used for design purposes. A sample biological waste facility is shown in Figure 23–3.

23.2.2 Personnel Entry and Egress

There should be easy access to the hazardous waste room from waste-generating laboratories and easy transfer of processed wastes from chemical, radiological, and biological waste storage area to a shipping dock with two personnel exits each area. Emergency access should be reviewed with the local fire marshal for preferred entry and access by firefighters.

23.2.3 Materials Handling Access

Receiving should be designed to accommodate the incoming materials from laboratories generating waste. Shipping and receiving access is necessary for the following: materials being received, materials being shipped, and outside contractor movement. A drum access ramp with a maximum slope of 1 in. in 12 should be provided. A separate truck-loading ramp will facilitate the long loading time needed for record-keeping requirements.

23.2.4 Furniture Locations

23.2.4.1 Desks. A desk for paperwork and data entry should be located in or immediately adjacent to the rooms. So much record keeping is required by law that a comfortable work station is necessary. A separate locked office provides improved security for records and data.

23.2.5 Work Areas

The areas described in this section relate to the facility described in Section 23.2.1.1.

23.2.5.1 Hazardous Chemical Waste. All activities except bulking operations can be performed in a 800-ft^2 room. Chemicals for preparing lab packs should be stored on shelves, as detailed below. Flammable liquids should be stored in UL- or FM-approved cabinets. Lab packs prepared on a routine

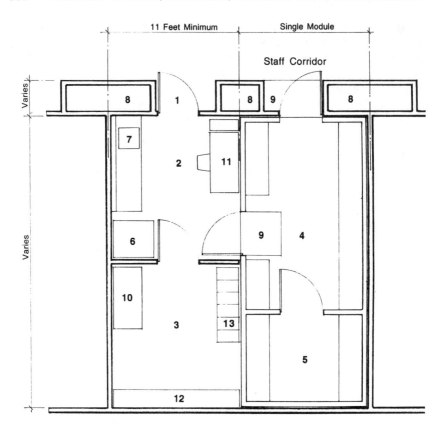

KEY

1	Primary Access/Egress
2	Work Room/Office
3	Dry Waste Storage/Supplies
4	Cold Room (Manufacture Supplied)
5	Freezer Room (Manufacture Supplied)
6	Autoclave
7	Sink
8	Electric Closets
9	Ramp
10	Freezer
11	Desk
12	Shelves
13	Cartons

FIGURE 23-3. Biological waste storage: Sample layout.

basis, such as once per month, should be assembled in proximity to the shelves holding the category of waste being handled.

Shelves should be equipped with bookends and earthquake panels across the front, and they should be constructed of inert materials to preclude reaction with spilled acids or strong oxidizers.

Storage of empty containers may be in or outside of the preparation room. Full lab packs and bulk containers should be stored in the storage room only temporarily until removal from the facility.

Chemical wastes coming into the room from laboratories should be placed on a work bench for identification or sampling, hazard determination, proper containment, and labeling and should then be moved to the bulking room or an appropriate temporary storage shelf.

23.2.5.2 Radioactive Waste. All work except vial crushing can be performed in a 600-ft^2 room. Storage shelves totaling at least 15 ft will be needed to store waste prior to packaging for shipment. A work bench and a writing station in the room would make for efficient operations.

23.2.5.3 Biological Waste All work can be performed in a 500-ft^2 room. A desk for record keeping and refrigerators or freezers should be located in or adjacent to the room. Storage cabinets for supplies and materials will be necessary.

23.2.6 Aisles

Major aisles in waste storage rooms should be a minimum of 5 ft wide to allow clearance for collection carts with laboratory wastes and for dollies carrying heavy 55-gallon drums. Wider aisles of 8 ft will accommodate two-way circulation in busy areas. Storage aisles may be 3 feet wide when containers stored on both sides of the aisle are no larger than 20 liters (5 gallons).

23.2.7 Location of Fume Hoods

A 4-ft or 6-ft chemical fume hood should be located in the work zone for sorting and consolidating chemicals and for chemical testing. Leaking containers could benefit from such an installation. A chemical fume hood may also be helpful in the radioactive waste room for special isolation control. The position of the hood should be protected from movement within the waste storage area to maintain good fume hood containment.

23.2.8 Location of Equipment and Storage Containers

23.2.8.1 Hazardous Chemical Waste. Chemical waste storage areas should be organized into a series of zones in which to store various types and quantities of chemicals. These storage areas need only be minimally separated from work areas to reduce circulation in and around containers. There should be eight separate areas for classified storage of chemicals based on their primary hazard characteristic: toxicity, flammability, corrosivity, reactivity, and combinations of hazards selected with due attention to compatibility concerns. Separation of chemicals that mutually react should be accomplished by assignment of areas to keep incompatible chemicals far apart. Refer to Table 12.1 in "Storage of Laboratory Chemicals," *Improving Safety in the Chemical Laboratory* (Young, 1991).

Each waste handling area must be protected by a floor dike designed to contain a spill from the largest container stored in the area, or 10% of the total stored volume. Each area may be further separated and surrounded by 2-h fire-rated walls with $1\frac{1}{2}$-rated door assemblies. A chemical waste storage area should have full-drum holding areas for each category of chemicals with each protected from other activities. All chemical waste storage areas should be emptied on a regular schedule of not longer than 3 months to comply with EPA regulations. An empty-drum storage area which can handle one cycle of empties and a few larger drums for over-packing of leaking drums in emergencies should be in, or close to, the chemical waste handling room for immediate availability.

There should be an area for storage of materials that can be recycled back to the laboratories or to chemical supply stockrooms. Some storage cabinets within each storage zone may require forced ventilation for highly toxic, not well-contained, or otherwise special chemicals.

23.2.8.2 Radioactive Waste. A storage area for about 20 empty drums (based on the example in Section 23.2) should be established. An equal number of full drums will also need a temporary storage location until they are shipped. Drum tools and daily supplies should also be stored in the area.

Some institutions are utilizing metal detectors, such as those used in airports, to check for lead pigs before shipping low-level solid waste for incineration. The space requirements necessary for this operation should be addressed in the design phase.

23.2.8.3 Biological Waste. The biological waste handling and storage room will need space for a work bench or table, for a desk, and for cold storage, mentioned above, as well as for an area to store dry waste not requiring refrigeration. An autoclave may be located in this room.

23.2.9 Location of Safety Station

Chemical, radioactive, and biological waste storage areas should have safety stations at each exit. At a minimum, the safety station should have a safety bulletin board, an emergency egress plan showing all pathways out of the building (if there is not a direct exit to the exterior from the waste storage area) safety glasses dispenser, fire extinguisher, fire blanket, chemical spill control kits, radioactive and biological decontamination kits, and cleaning materials.

23.3   HEATING, VENTILATING, AND AIR CONDITIONING

23.3.1   Introduction

All the items described in Sections 1.3 and 2.3 should be reviewed, and those that are relevant should be implemented. Additional recommendations are given below. Industrial hygiene personnel and engineers should be consulted for unusual design conditions.

23.3.2   Heating and Cooling

The areas or rooms where hazardous waste chemicals or radioactive materials are handled need to be heated to insure that the material stored does not freeze. Such spaces usually are not air-conditioned. However, if there is a possibility that space temperatures may exceed 100°F for extended periods, air conditioning should be considered. The overall space temperature range should be between 50°F and 100°F, depending on the particular chemicals handled.

Local cooling for personnel who may spend time at the desk may be considered for personal comfort.

Open-flame heating units and electrical resistance heaters should not be used in the chemical and radioactive waste rooms where flammable liquids are handled or where Section 23.4.5 calls for explosion proof installation. For example, direct-fired gas heaters should be avoided.

Due to high ventilation rates described below, the humidity level in the room may be low, especially in colder climates. Experience has shown that humidity levels below 25–30% start to induce static charge problems which may cause explosion or fire in spilled flammables or fugitive vapors. Therefore, this minimum humidity level should be maintained.

23.3.3   Room Ventilation Exhaust

Each waste handling and storage facility should be provided with a minimum ventilation rate of 8–10 air changes per hour (equivalent to 1 cfm per square foot of floor area for rooms 8–10 ft high). The exhaust air requirements may be met by continuously operating laboratory fume hood(s), other local exhaust systems, a general room exhaust system, or combinations of the three.

The cited minimum exhaust rate is required to handle small gas and vapor leaks from containers and spills during material handling. The nature and magnitude of the activities performed often necessitate the use of local exhaust ventilation in addition to the general room exhaust.

To control odors in these rooms, the air pressure should be negative in relation to surrounding areas.

23.3.4   Laboratory Fume Hood

Laboratory fume hood space may be needed in the chemical and radioactive waste handling and storage rooms. The amount of hood space required will depend on the quantity of materials handled and the amount of manipulation of small containers of materials. A 4-ft hood is the minimum recommended.

The hood should meet all the requirements listed in Section 2.3.4.3 and Chapter 31.

23.3.5   Local Exhaust Systems

Other forms of local exhaust ventilation systems, in addition to or in place of chemical fume hoods, may be required, depending on the types of activities performed. It is essential that these systems be separate from the remainder of the building ventilation services and be discharged to the atmosphere with no recirculation.

Generally, they will be low-volume–high-velocity systems or specially designed enclosures. When operations are intermittent, local exhaust system facilities can be designed so that only one local exhaust hood can be used at a time, thereby reducing the size and complexity of the local exhaust installation.

The design criteria contained in Chapter 3 of the *Industrial Ventilation Manual* should be followed (ACGIH, 1992).

23.3.5.1   Barrel Exhausts. When operations require the filling of drums, one or more barrel exhaust systems may be required (Figure 23-4). They should exhaust approximately 150 cfm each. There should be enough flexible duct available to permit their use at various locations, or several exhaust points should be provided.

23.3.5.2   Vial Crushing. This operation should be conducted in a ventilated enclosure for personal protection and fire prevention. The specific design will depend on the size of the crusher and the potential vapor release sources. The procedures outlined in Chapter 3 of the latest edition of the *Industrial Ventilation Manual* should be followed (ACGIH, 1992).

Some vial-crushing equipment may already be designed as an enclosure with an exhaust outlet. In this case the manufacturer should be asked to provide the required exhaust flow rate.

23.3.5.3   Autoclave Exhaust. When an autoclave is used in the biological waste room, local exhaust should be provided. The use of a canopy hood as described in Section 31.9 would be most appropriate.

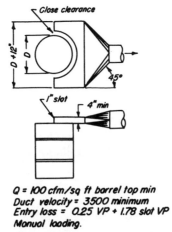

Q = 100 cfm/sq ft barrel top min
Duct velocity = 3500 minimum
Entry loss = 0.25 VP + 1.78 slot VP
Manual loading.

FIGURE 23-4. Barrel filling exhaust. *Source:* American Conference of Governmental Industrial Hygienists, Cincinnatti, Ohio.

23.4 LOSS PREVENTION, INDUSTRIAL HYGIENE, AND PERSONAL SAFETY

23.4.1 Introduction

Careful consideration should be given to the information contained in Sections 1.4.1, 1.4.2, 1.4.4, 1.4.6, 2.4.1, 2.4.2, and 2.4.6 through 2.4.8, and special attention should addressed to recommendations below.

23.4.2 Emergency Equipment Cabinets

An important element of the area design is the establishment and location near the waste handling room of an emergency equipment cabinet containing personal protective equipment, first-aid supplies, materials related to the control and cleanup of chemical spills, fires and explosions, radioactive spills, and biological releases.

23.4.3 Water Supply

The water supply to the waste handling areas should be adequate to handle the sprinkler system. A deluge shower delivering a minimum of 30 gallons per minute, an eyewash fountain requiring 7 gpm, and a handwashing sink are on the potable water system.

23.4.4 Special Fire Suppression

When water-reactive chemicals make up a substantial portion of the waste stream in the hazardous waste room, they should be segregated and confined to an area where sprinklers are replaced by a more appropriate fire suppression system such as dry chemical.

23.4.5   Electrical Service

Electrical services in a waste-bulking room or area and a vial-crushing room should conform to the National Electrical Code for Class I Group C & D Division I installations. The lab pack room and all areas where open containers of flammable liquids are handled should conform to the requirements for Class I Group C & D Division II installations.

23.4.6   Bonding and Grounding System

A static electricity bonding and grounding system should be installed in all areas where flammable liquids are transferred. Bonding systems are used to equalize static electrical charges between objects such as containers (safety cans). Grounding systems are used to bleed such charges to an infinite ground. Information on effective bonding and grounding systems can be found in NFPA 30 Flammable and Combustible Liquids Code (NFPA, 1992).

23.5   SPECIAL REQUIREMENTS

No floor drains should be provided in the waste handling rooms or areas where hazardous waste chemicals or radioactive materials are found. Nevertheless, if a drain is used, it should not be connected directly to the sanitary or storm sewer but be directed to a captured holding sump or tank. Such tanks may fall within the jurisdiction of the Underground Storage Tank Regulations.

PART IV

ADMINISTRATIVE PROCEDURES

Although preparation and evaluation of bidding documents will ordinarily be conducted or supervised by a knowledgeable and experienced professional architect or engineer for all but the most trivial of projects, it is important for the laboratory owners and responsible technical directors to understand the process in order to be sure they have made their needs abundantly clear. When it comes to the matter of acceptance, it is essential that all the technical laboratory directors participate in a detailed examination of every aspect of the construction and furnishings that are about to become their personal work environment and that of their assistants; they will leave this final task solely to a third party at their own peril.

Also included in this part is the subject of energy conservation. Although many aspects of the subject are highly technical, the design decisions that must be made with regard to energy conservation are administrative in nature and readily understood by administrators and laboratory directors inasmuch as they directly affect the safety and comfort of the occupants and the costs of the owners for many years into the future.

It is very important that proposed project scope modifications be carefully examined by all interested parties. In some instances, after initial cost estimates are received from contractors, drastic cutbacks will be recommended. Some may seriously damage the project's integrity if adopted. There is a tendency for ornamental architectural features such as coved ceilings, fancy casework, and an impressive facade to be retained, but compromises tend to be made in the wear resistance of the interior structural elements, the adequacy of the ventilation and temperature control systems, and other essential building services. It is tempting for administrators and owners to allow such compromises, because the architectural features are very visible, whereas the mechanical and electrical systems will be hidden above the ceiling or behind the walls. Such temptations should be resisted.

The project specifications should include good definitions for project commissioning, as well as for documenting all operating and maintenance requirements. Recently proposed guidelines (ASHRAE Committee GPC-4P) can be very helpful in preparing this information.

CHAPTER 24

PROJECT EXECUTION AND BIDDING PROCEDURES

24 GUIDING CONCEPTS

24.1.1.1 Introduction

The intent of this chapter is to provide a brief guideline to the steps to be taken from the inception of a project to completion. The methodology described is by no means mandatory. It only provides a framework, of the principles that can be applied in other methodologies for successful project execution.

24.1.2 Building Committee

For large projects, several owners attempt to form a high-level building team or committee responsible to a Chief Executive Officer or a Board of Directors, etc. This group is responsible to set up the goals and objectives for the project, secure financing, etc. In smaller projects, such a group may not be needed.

24.1.3 Project Director

This should be one person directly responsible to the administration and/or building committee for the success of the project. Any attempts to dilute the

Guidelines for Laboratory Design: Health and Safety Considerations, Second Edition
By Louis DiBerardinis, Janet S. Baum, Gari T. Gatwood, Anand K. Seth, Melvin W. First, and Edward F. Groden.
ISBN 0-471-55463-4   Copyright © 1993 by John Wiley & Sons, Inc.

authority of the project director generally end up costing time and money in the project. Designation of such a person commits the owner to a dedicated implementation of the project.

24.1.4 Project Team

The project director should assemble the project team consisting of qualified architects, engineers, health and safety professionals, and construction experts to design and build the project.

24.1.5 Construction Manager

This role is different than project director. This role (service) is provided by experts who have extensive construction and estimating experience. In essence, the role provides guidance to the owners' project director in evaluating various options and providing cost and construction schedule information.

In some cases, the construction manager accepts a financial stake in the project by assuming the role of general contractor. This causes a conflict of interest and is not a recommended practice.

A project may or may not have a construction manager.

24.2 IMPLEMENTATION

The following options can be considered:

a. Turnkey
b. Fast-track
c. Conventional

24.2.1 Turnkey

Under this approach, a vendor who typically is a commercial developer or general contractor is held responsible for doing all design as well as construction of the project, and at completion "turns over the keys" to the users. There are several advantages to this approach. Most importantly, the one source responsibility for the overall project, means completion can actually be faster. The disadvantages are that the owners do not necessarily get the very best price for the endeavor, the eventual users and owners are less involved in the decision-making process, and sometimes the end product is not exactly what is desired. If the vendor does not have significant and creditable experience in construction of laboratory buildings, the end product may be defective. A similar approach is called *design–build,* where legal ownership of the project remains with the developer.

24.2.2   Fast-Track

This is a fairly common technique. Portions of design and construction of the project are done simultaneously. Construction commences before construction documents are complete. For example, as soon as the building shape and size are decided, foundation work can be designed and commenced while the superstructure is being designed. The superstructure can be constructed while the final mechanical/electrical systems are being designed, and installation occurs later. The research apparatus can be selected last.

There are obvious advantages in reducing the time from overall concept to finish. It means that the building can be commissioned earlier and that productive research work can start accordingly. Problems are as follows:

1. Even with an experienced design team, serious errors can occur in haste, which become very expensive to rectify in the future.
2. No adjustments can be made to early design decisions without financial penalties, and coordination between design phases is critical.
3. The researchers or users are forced to make decisions very early and in a very quick manner. Some of them are not used to making these decisions and change their mind later on. This could result in significant changes in construction with substantial cost additions. There is also a significant potential to compromise safety and health design features as a result of these changes. For example, an increase in scope needing specialized ventilation may be made, but the building layout may not provide adequate space for the additional mechanical systems needed. Some change is inevitable; however, attempts should be made to reduce changes as much as possible.

24.2.3   Conventional

Under this approach, the owners hire an architect who assembles a project team made up of the building users, the research community, engineers, and safety professionals. The following steps usually take place:

1. *Program.* This is a detailed description of the requirements of the research community. This component will decide the kind of laboratory required, type of systems needed, and level of utilities in the building. Refer to Section 1.2.2 for an extensive discussion of the programming process.
2. *Schematic and Design Development Phases.* Schematic drawings indicate the overall concept of the project and that the program fits into the confines of the building or space as designed and could be serviced with mechanical/electrical systems to provide a safe and workable environment. Design development is the phase in which the concept is fully articulated prior to defining all the construction details. It is strongly advisable to estimate the project at completion of both phases

to see if the project estimated construction cost still falls within the projected budget. It could very well be that budget constraints or cost overruns might require program modification or curtailment. Additional funds may need to be acquired for the project. It is our recommendation that the design development sets be "signed-off" with the users and owner to ensure that they fully understand and agree with the program.

24.3 CONSTRUCTION REVIEW SET

After the design development drawings are completed, construction drawings and specifications can be formulated. It is customary to issue these drawings for review by the project team on a regular basis to ensure that the intent of the design concepts, the needs of the investigators, and health and safety concerns are being met. It may be necessary for the user to engage in the use of an independent health and safety consultant to ensure the adequacy of the design. The customary review targets are at 25%, 50%, 75%, and 90% of the drawings, and then a final review before the bids are solicited.

After the drawings have been submitted to contractors for bidding, it is customary that a pre-bid conference be held. At this conference, all pertinent requirements or questions the contractors have should be discussed. The questions may range widely from site access to specific detail on the types of components used or shown on the drawings. It is highly advisable that this conference or conferences be documented and this information be shared equally with all bidders. Otherwise all kinds of misinformation or misunderstanding can occur. Bid clarification documents or addendums or deletions can be issued for clarification of the scope of the project.

24.4 BID FORM

The importance of the bid form cannot be overemphasized. A good bid form summarizes the cost of the project in an easy-to-understand format. At the same time, it documents any component cost (or cost of alternates) in an easy format. Some contractors may take exception to certain segments of the construction documents. The bid form allows a place for them to do so.

24.4.1 Bidding Documents

Bidding documents can be divided into three major sections: (1) contract forms, (2) general conditions, and (3) technical specifications. Contract forms and general conditions address the legal and administrative requirements of the project and vary according to the size of the project and its location. Construction Specification Institute list of normal specification categories is shown in Figure 24-1.

**Uniform
Construction
Index**

Cost Analysis
Format

| 0 | CONDITIONS OF THE CONTRACT |

00000.-00099. unassigned

| 1 | GENERAL REQUIREMENTS |

01020. ALLOWANCES
01021.-01099. unassigned
01100. ALTERNATIVES
01101.-01199. unassigned
01200. PROJECT MEETINGS
01201.-01299. unassigned
01300. SUBMITTALS
01301.-01399. unassigned
01400. QUALITY CONTROL
01401.-01499. unassigned
01500. TEMPORARY FACILITIES AND CONTROLS
01501.-01599. unassigned
01600. MATERIAL AND EQUIPMENT
01601.-01699. unassigned
01700. PROJECT CLOSEOUT
01701.-01999. unassigned

| 2 | SITE WORK |

02000. ALTERNATIVES
02001.-02009. unassigned
02010. SUBSURFACE EXPLORATION
02011. Borings
02012. Core Drilling
02013. Standard Penetration Tests
02014. Seismic Exploration
02015.-02099. unassigned
02100. CLEARING
02101. Structure Moving
02102. Clearing and Grubbing
02103. Tree-Pruning
02104. Shrub and Tree Relocation
02105.-02109. unassigned
02110. DEMOLITION
02111.-02199. unassigned
02200. EARTHWORK
02201.-02209. unassigned
02210. Site Grading
02211. Rock Removal
02212. Embankment
02213.-02219 unassigned
02220. Excavating and Backfilling
02221. Trenching
02222. Structure Excavation
02223. Roadway excavation
02224. Pipe Boring and Jacking
02225.-02226. unassigned
02227. Waste Material Disposal
02228.-02229. unassigned
02230. Soil Compaction Control
02231.-02239. unassigned
02240. Soil Stabilization
02241.-02249. unassigned
02250. SOIL TREATMENT
02251. Termite Control
02252. Vegetation Control
02253.-02299. unassigned
02300. PILE FOUNDATIONS
02301.-02349. unassigned
02350. CAISSONS
02351. Drilled Caissons
02352. Excavated Caissons
02353.-02399. unassigned
02400 SHORING
02401.-02419. unassigned
02420. Underpinning
02421.-02499. unassigned
02500. SITE DRAINAGE
02501.-02549. unassigned
02550. SITE UTILITIES
02551.-02599. unassigned
02600. PAVING & SURFACING
02601.-02609. unassigned
02610. Paving
02611.-02619. unassigned
02620. Curbs and Gutters
02621.-02629. unassigned
02630. Walks

02631.-02639. unassigned
02640. Synthetic Surfacing
02641.-02699. unassigned
02700. SITE IMPROVEMENTS
02701.-02709. unassigned
02710. Fences and Gates
02711.-02719. unassigned
02720. Road and Parking Appurtenances
02721.-02729. unassigned
02730. Playing Fields
02731.-02739. unassigned
02740. Fountains
02741.-02749. unassigned
02750. Irrigation System
02751.-02759. unassigned
02760. Site Furnishings
02761.-02799. unassigned
02800. LANDSCAPING
02801.-02809. unassigned
02810. Soil Preparation
02811.-02819. unassigned
02820. Lawns
02821.-02829. unassigned
02830. Trees, Shrubs, and Ground Cover
02831.-02849. unassigned
02850. RAILROAD WORK
02851. Trackwork
02852. Ballasting
02853.-02899. unassigned
02900. MARINE WORK
02901.-02909. unassigned
02910. Docks
02911.-02919. unassigned
02920. Boat Facilities
02921.-02929. unassigned
02930. Protective Marine Structures
02931. Fenders
02932. Seawalls
02933. Groins
02934. Jettys
02935.-02939. unassigned
02940. Dredging
02941.-02949. unassigned
02950. TUNNELING
02951.-02959. unassigned
02960. Tunnel Excavation
02961.-02969. unassigned
02970. Tunnel Grouting
02971.-02979. unassigned
02980. Support Systems
02981.-02999. unassigned

FIGURE 24-1. Construction Specification Institute Uniform Construction Index (*Continued on following pages*).

3 | CONCRETE

03000. ALTERNATIVES
03001.-03099. unassigned
03100. CONCRETE FORMWORK
03101.-03149. unassigned
03150. EXPANSION &
  CONTRACTION JOINTS
03151.-03199. unassigned
03200. CONCRETE
  REINFORCEMENT
03201.-03209. unassigned
03210. Steel Bar and Welded Wire
  Fabric Reinforcing
03211.-03229. unassigned
03230. Stressing Tendons
03231.-03299. unassigned
03300. CAST-IN-PLACE CONCRETE
03301.-03304. unassigned
03305. Concrete Curing
03306.-03309. unassigned
03310. Concrete
03311.-03319. unassigned
03320. Lightweight Concrete
03321. Insulating Concrete
03322. Lightweight Structural
  Concrete
03323.-03329. unassigned
03330. Heavyweight Concrete
03331.-03339. unassigned
03340. Prestressed Concrete
03341.-03349. unassigned
03350. SPECIALLY FINISHED
  CONCRETE
03351. Exposed Aggregate Concrete
03352. Bushhammered Concrete
03353. Blasted Concrete
03354. Heavy-Duty Concrete Floor
  Finishes
03355. Grooved-Surface Concrete
03356.-03359. unassigned
03360. SPECIALLY PLACED
  CONCRETE
03361.-03369. unassigned
03370. Grout
03371.-03399. unassigned
03400. PRECAST CONCRETE
03401.-03409. unassigned
03410. Precast Concrete Panels
03411. Tilt-Up Wall Panels
03412.-03419. unassigned
03420. Precast Structural Concrete
03421.-03429. unassigned
03430. Precast Prestressed Concrete
03431.-03499. unassigned
03500. CEMENTITIOUS DECKS
03501.-03509. unassigned
03510. Gypsum Concrete
03511.-03519. unassigned
03520. Cementitious Wood Fiber
  Deck
03521.-03999. unassigned

4 | MASONRY

04000. ALTERNATIVES
04001.-04099. unassigned
04100. MORTAR
04101.-04149. unassigned
04150. MASONRY ACCESSORIES
04151.-04159. unassigned
04160. Joint Reinforcement
04161.-04169. unassigned
04170. Anchors and Tie Systems
04171.-04179. unassigned
04180. Control Joints
04181.-04199. unassigned
04200. UNIT MASONRY
04201.-04209. unassigned
04210. Brick Masonry
04211.-04219. unassigned
04220. Concrete Unit Masonry
04221.-04229. unassigned
04230. Reinforced Unit Masonry
04231.-04239. unassigned
04240. Clay Backing Tile
04241.-04244. unassigned
04245. Clay Facing Tile
04246.-04249. unassigned
04250. Ceramic Veneer
04251.-04269. unassigned
04270. Glass Unit Masonry
04271.-04279. unassigned
04280. Gypsum Unit Masonry
04281.-04399. unassigned
04400. STONE
04401.-04409. unassigned
04410. Rough Stone
04411.-04419. unassigned
04420. Cut Stone
04421. unassigned
04422. Marble
04423.-04429. unassigned
04430. Simulated Masonry
04431.-04434. unassigned
04435. Cast Stone
04436.-04439. unassigned
04440. Flagstone
04441.-04449. unassigned
04450. Natural Stone Veneer
04451.-04499. unassigned
04500 MASONRY RESTORATION &
  CLEANING
04501.-04509. unassigned
04510. Masonry Cleaning
04511.-04549. unassigned
04550. REFRACTORIES
04551.-04999. unassigned

5 | METALS

05000. ALTERNATIVES
05001.-05099. unassigned
05100. STRUCTURAL METAL
  FRAMING
05101.-05119. unassigned
05120. Structural Steel
05121.-05129. unassigned
05130. Structural Aluminum
05131.-05199. unassigned
05200. METAL JOISTS
05201.-05299. unassigned
05300. METAL DECKING
05301.-05399. unassigned
05400. LIGHTGAGE METAL
  FRAMING
05401.-05499. unassigned
05500. METAL FABRICATIONS
05501.-05509. unassigned
05510. Metal Stairs
05511.-05519. unassigned
05520. Handrails and Railings
05521. Pipe and Tube Railings
05522.-05529. unassigned
05530. Gratings
05531.-05539. unassigned
05540. Castings
05541.-05699. unassigned
05700. ORNAMENTAL METAL
05701.-05709. unassigned
05710. Ornamental Stairs
05711.-05719. unassigned
05720. Ornamental Handrails and
  Railings
05721.-05729. unassigned
05730. Ornamental Sheet Metal
05731.-05799 unassigned
05800 EXPANSION CONTROL
05801.-05999 unassigned

FIGURE 24-1. (*Continued*)

Uniform Construction Index	Cost Analysis Format

6 WOOD AND PLASTICS

06000. ALTERNATIVES
06001.-06099. unassigned
06100. ROUGH CARPENTRY
06101.-06109. unassigned
06110. Framing and Sheathing
06111. Light Wooden Structures Framing
06112. Preassembled Components
06113. Sheathing
06114. Diaphragms
06115.-06129. unassigned
06130. HEAVY TIMBER CONSTRUCTION
06131. Timber Trusses
06132. Mill-Framed Structures
06133. Pole Construction
06134.-06149. unassigned
06150. TRESTLES
06151.-06169. unassigned
06170. PREFABRICATED STRUCTURAL WOOD
06171.-06179. unassigned
06180. Glued Laminated Construction
06181. Glue-Laminated Structural Units
06182. Glue-Laminated Decking
06183.-06189. unassigned
06190. Wood Trusses
06191.-06199. unassigned
06200. FINISH CARPENTRY
06201.-06219. unassigned
06220. Millwork
06221.-06239. unassigned
06240. Laminated Plastic
06241.-06299. unassigned
06300. WOOD TREATMENT
06301.-06399. unassigned
06400. ARCHITECTURAL WOODWORK
06401.-06409. unassigned
06410. Cabinetwork
06411. Wood Cabinets: Unfinished
06412.-06419. unassigned
06420. Paneling
06421. Architectural Hardwood Plywood Paneling
06422. Softwood Plywood Paneling
06423.-06429. unassigned
06430. Stairwork
06431. Wood Stairs and Railings
06432.-06499. unassigned
06500. PREFABRICATED STRUCTURAL PLASTICS
06501.-06599. unassigned
06600. PLASTIC FABRICATIONS
06601.-06999. unassigned

7 THERMAL & MOISTURE PROTECTION

07000. ALTERNATIVES
07001.-07099. unassigned
07100. WATERPROOFING
07101.-07109. unassigned
07110. Membrane Waterproofing
07111.-07119. unassigned
07120. Fluid Applied Waterproofing
07121. Liquid Waterproofing
07122.-07129. unassigned
07130. Bentonite Waterproofing
07131.-07139. unassigned
07140. Metal Oxide Waterproofing
07141.-07149. unassigned
07150. DAMPPROOFING
07151.-07159. unassigned
07160. Bituminous Dampproofing
07161.-07169. unassigned
07170. Silicone Dampproofing
07171.-07174. unassigned
07175. Water Repellent Coating
07176.-07179. unassigned
07180. Cementitious Dampproofing
07181.-07189. unassigned
07190. Vapor Barriers/Retardants
07191.-07199. unassigned
07200. INSULATION
07210. Building Insulation
07211. Loose Fill Insulation
07212. Rigid Insulation
07213. Fibrous and Reflective Insulation
07214. Foamed-in-Place Insulation
07215. Sprayed-On Insulation
07216.-07229. unassigned
07230. High and Low Temperature Insulation
07231.-07239. unassigned
07240. Roof and Deck Insulation
07241.-07249. unassigned
07250. Perimeter and Under-Slab Insulation
07251.-07299. unassigned
07300. SHINGLES & ROOFING TILES
07301.-07309. unassigned
07310. Shingles
07311.-07319. unassigned
07320. Roofing Tiles
07321.-07399. unassigned
07400. PREFORMED ROOFING & SIDING
07401.-07409. unassigned
07410. Preformed Wall and Roof Panels
07411. Preformed Metal Siding
07412.-07419. unassigned
07420. Composite Building Panels
07421.-07439. unassigned
07440. Preformed Plastic Panels

07441.-07459. unassigned
07460. Cladding/Siding
07461. Wood Siding
07462. Composition Siding
07463. Asbestos-Cement Siding
07464. Plastic Siding
07465.-07499. unassigned
07500. MEMBRANE ROOFING
07501.-07509. unassigned
07510. Built-Up Bituminous Roofing
07511.-07519. unassigned
07520. Prepared Roll Roofing
07521.-07529. unassigned
07530. Elastic Sheet Roofing
07531.-07539. unassigned
07540. Fluid Applied Roofing
07541.-07569. unassigned
07570. TRAFFIC TOPPING
07571.-07599. unassigned
07600. FLASHING & SHEET METAL
07601.-07609. unassigned
07610. Sheet Metal Roofing
07611.-07619. unassigned
07620. Flashing and Trim
07621.-07629. unassigned
07630. Roofing Specialties
07631. Gutters and Downspouts
07632.-07659. unassigned
07660. Gravel Stops
07661.-07699. unassigned
07700. Flashing
07701.-07799. unassigned
07800. ROOF ACCESSORIES
07801.-07809. unassigned
07810. Skylights
07811. Plastic Skylights
07812. Metal-Framed Skylights
07813.-07829. unassigned
07830. Hatches
07831.-07839. unassigned
07840. Gravity Ventilators
07841.-07849. unassigned
07850. Prefabricated Curbs
07851.-07859. unassigned
07860. Prefabricated Expansion Joints
07861.-07899. unassigned
07900. SEALANTS
07901.-07949. unassigned
07950. Gaskets
07951.-07999. unassigned

FIGURE 24-1. (*Continued*)

Uniform Construction Index	Cost Analysis Format

8 DOORS & WINDOWS

08000. ALTERNATIVES
08001.-08099. unassigned
08100. METAL DOORS & FRAMES
08101.-08109. unassigned
08110. Hollow Metal Work
08111. Stock Hollow Metal Work
08112. Custom Hollow Metal Work
08113.-08119. unassigned
08120. Aluminum Doors and Frames
08121.-08129. unassigned
08130. Stainless Steel Doors and Frames
08131.-08139. unassigned
08140. Bronze Doors and Frames
08141.-08199. unassigned ·
08200. WOOD & PLASTIC DOORS
08201.-08299. unassigned
08300. SPECIAL DOORS
08301.-08309. unassigned
08310. Sliding Metal Fire Doors
08311.-08319. unassigned
08320. Metal-Clad Doors
08321.-08329. unassigned
08330. Coiling Doors
08331.-08349. unassigned
08350. Folding Doors
08351.-08354. unassigned
08355. Flexible Doors
08356.-08359. unassigned
08360. Overhead Doors
08361.-08369. unassigned
08370. Sliding Glass Doors
08371.-08374. unassigned
08375. Safety Glass Doors
08376.-08379. unassigned
08380. Sound Retardant Doors
08381.-08389. unassigned
08390. Screen and Storm Doors
08391.-08399. unassigned
08400. ENTRANCES & STOREFRONTS
08401.-08449. unassigned
08450. Revolving Doors
08451.-08499. unassigned
08500. METAL WINDOWS
08501.-08509. unassigned
08510. Steel Windows
08511.-08519. unassigned
08520. Aluminum Windows
08521.-08529. unassigned
08530. Stainless Steel Windows
08531.-08539. unassigned
08540. Bronze Windows
08541.-08599. unassigned
08600. WOOD & PLASTIC WINDOWS
08601.-08609. unassigned
08610. Wood Windows
08611.-08619. unassigned
08620. Plastic Windows
08621.-08649. unassigned

08650. SPECIAL WINDOWS
08651-08669 unassigned
08700. HARDWARE & SPECIALTIES
08701.-08709. unassigned
08710. Finish Hardware
08711.-08719. unassigned
08720. Operators
08721. Automatic Door Equipment
08722.-08724. unassigned
08725. Window Operators
08726.-08729. unassigned
08730. Weatherstripping & Seals
08731.-08739. unassigned
08740. Thresholds
08741.-08799. unassigned
08800. GLAZING
08801.-08809. unassigned
08810. Glass
08811. Plate Glass
08812. Sheet Glass
08813. Tempered Glass
08814. Wired Glass
08815. Rough and Figured Glass
08816.-08819. unassigned
08820. Processed Glass
08821. Coated Glass
08822. Laminated Glass
08823. Insulating Glass
08824.-08829. unassigned
08830. Mirror Glass
08831.-08839. unassigned
08840. Glazing Plastics
08841.-08849. unassigned
08850. Glazing Accessories
08851.-08899. unassigned
08900. WINDOW WALLS/CURTAIN WALLS
08901.-08999. unassigned

9 FINISHES

09000. ALTERNATIVES
09001.-09099. unassigned
09100. LATH & PLASTER
09101.-09109. unassigned
09110. Furring and Lathing
09111.-09149. unassigned
09150. Gypsum Plaster
09151.-09179. unassigned
09167. Gypsum Plaster
09168.-09179. unassigned
09180. Cement Plaster
09181.-09189. unassigned
09190. Acoustical Plaster
09191.-09249. unassigned
09250. GYPSUM WALLBOARD
09251.-09259. unassigned
09260. Gypsum Wallboard Systems
09261.-09279. unassigned
09280. Accessories
09281.-09299. unassigned
09300. TILE
09301.-09309. unassigned
09310. Ceramic Tile
09311.-09319. unassigned
09320. Ceramic Mosaics
09321.-09329. unassigned
09330. Quarry Tile
09331.-09339. unassigned
09340. Marble Tile
09341.-09349. unassigned
09350. Glass Mosaics
09351.-09359. unassigned
09360. Plastic Tile
09361.-09369. unassigned
09370. Metal Tile
09371.-09399. unassigned
09400. TERRAZZO
09401.-09409. unassigned
09410. Portland Cement Terrazzo
09411.-09419. unassigned
09420. Precast Terrazzo
09421.-09429. unassigned
09430. Conductive Terrazzo
09431.-09439. unassigned
09440. Plastic Matrix Terrazzo
09441.-09499. unassigned
09500. ACOUSTICAL TREATMENT
09501.-09509. unassigned
09510. Acoustical Ceilings
09511. Acoustical Panels
09512. Acoustical Tiles
09513.-09519. unassigned
09520. Acoustical Wall Treatment
09521.-09529. unassigned
09530. Acoustical Insulation and Barriers
09531.-09539. unassigned
09540. CEILING SUSPENSION SYSTEMS
09541.-09549. unassigned
09550. WOOD FLOORING
09551.-09559. unassigned
09560. Wood Strip Flooring

FIGURE 24-1. (*Continued*)

Uniform Construction Index	Cost Analysis Format

09561.-09579. unassigned
09580. Plywood Block Flooring
09581.-09589. unassigned
09590. Resilient Wood Floor System
09591.-09599. unassigned
09600. Wood Block Industrial Flooring
09601.-09649. unassigned
09650. RESILIENT FLOORING
09651. Cementitious Underlayment
09660. Resilient Tile Flooring
09661.-09664. unassigned
09665. Resilient Sheet Flooring
09666.-09669. unassigned
09670. Fluid Applied Resilient Flooring
09671.-09679. unassigned
09680. CARPETING
09681. Carpet Cushion
09682. Carpet
09683. Bonded Cushion Carpet
09684. Custom Carpet
09685.-09689. unassigned
09690. Carpet Tile
09691.-09699. unassigned
09700. SPECIAL FLOORING
09701.-09709. unassigned
09710. Magnesium Oxychloride Floors
09711.-09719. unassigned
09720. Epoxy-Marble-Chip Flooring
09721.-09729. unassigned
09730. Elastomeric Liquid Flooring
09731. Conductive Elastomeric Liquid Flooring
09732.-09739. unassigned
09740. Heavy-Duty Concrete Toppings
09741. Armored Floors
09742.-09749. unassigned
09750. Brick Flooring
09751.-09759. unassigned
09760. FLOOR TREATMENT
09761.-09799. unassigned
09800. SPECIAL COATINGS
09801.-09809. unassigned
09810. Abrasion Resistant Coatings
09811.-09819. unassigned
09820. Cementitious Coatings
09821.-09829. unassigned
09830. Elastomeric Coatings
09831.-09839. unassigned
09840. Fire-Resistant Coatings
09841. Sprayed Fireproofing
09842.-09849. unassigned
09850. Aggregate Wall Coatings
09851.-09899. unassigned
09900. PAINTING
09901.-09949. unassigned
09950. WALL COVERING
09951. Vinyl-Coated Fabric Wall Covering
09952. Vinyl Wall Covering
09953. Cork Wall Covering
09954. Wallpaper

09955. Wall Fabrics
09956.-09959. unassigned
09960. Flexible Wood Sheets
09961.-09969. unassigned
09970. PREFINISHED PANELS
09971.-09989. unassigned
09990. Adhesives
09991.-09999. unassigned

10 SPECIALTIES

10000. ALTERNATIVES
10001.-10099. unassigned
10100. CHALKBOARDS AND TACKBOARDS
10101.-10149. unassigned
10150. COMPARTMENTS AND CUBICLES
10151. Hospital Cubicles
10152.-10159. unassigned
10160. Toilet and Shower Partitions
10161. Laminated Plastic Toilet Partitions
10162. Metal Toilet Partitions
10163. Stone Partitions
10164.-10169. unassigned
10170. Shower and Dressing Compartments
10171.-10199. unassigned
10200. LOUVERS AND VENTS
10201.-10239. unassigned
10240. GRILLES AND SCREENS
10241.-10259. unassigned
10260. WALL AND CORNER GUARDS
10261.-10269. unassigned
10270. ACCESS FLOORING
10271.-10279. unassigned
10280. SPECIALTY MODULES
10281.-10289. unassigned
10290. PEST CONTROL
10291.-10299. unassigned
10300. FIREPLACES
10301. Prefabricated Fireplaces
10302. Prefabricated Fireplace Forms
10303.-10309. unassigned
10310. Fireplace Accessories
10311.-10349. unassigned
10350. FLAGPOLES
10351.-10399. unassigned
10400. IDENTIFYING DEVICES
10401.-10409. unassigned
10410. Directories and Bulletin Boards
10411. Directories
10412.-10419. unassigned
10420. Plaques
10421.-10439. unassigned
10440. Signs
10441.-10449. unassigned
10450. PEDESTRIAN CONTROL DEVICES
10451.-10499. unassigned
10500. LOCKERS
10501.-10529. unassigned
10530. PROTECTIVE COVERS
10531. Walkway Covers
10532. Car Shelters
10533.-10549. unassigned
10550. POSTAL SPECIALTIES
10551. Mail Chutes
10552. Mail Boxes

FIGURE 24-1. (*Continued*)

Uniform Construction Index	Cost Analysis Format

10553.-10599. unassigned
10600. PARTITIONS
10601. Mesh Partitions
10602.-10609. unassigned
10610. Demountable Partitions
10611.-10615. unassigned
10616. Movable Gypsum Partitions
10617.-10619. unassigned
10620. Folding Partitions
10621.-10622. unassigned
10623. Accordion Folding Partitions
10624.-10649. unassigned
10650. SCALES
10651.-10669. unassigned
10670. STORAGE SHELVING
10671.-10699. unassigned
10700. SUN CONTROL DEVICES (EXTERIOR)
10701.-10749. unassigned
10750. TELEPHONE ENCLOSURES
10751.-10799. unassigned
10800. TOILET & BATH ACCESSORIES
10801.-10899. unassigned
10900. WARDROBE SPECIALTIES
10901.-10999. unassigned

11 EQUIPMENT

11000. ALTERNATIVES
11001.-11049. unassigned
11050. BUILT-IN MAINTENANCE EQUIPMENT
11051. Vacuum Cleaning System
11052. Powered Window Washing
11053.-11099. unassigned
11100. BANK & VAULT EQUIPMENT
11101.-11149. unassigned
11150. COMMERCIAL EQUIPMENT
11151.-11169. unassigned
11170. CHECKROOM EQUIPMENT
11171.-11179. unassigned
11180. DARKROOM EQUIPMENT
11181.-11199. unassigned
11200. ECCLESIASTICAL EQUIPMENT
11201.-11299. unassigned
11300. EDUCATIONAL EQUIPMENT
11301.-11399. unassigned
11400. FOOD SERVICE EQUIPMENT
11401. Food Service Equipment: Custom Fabricated
11402.-11409. unassigned
11410. Bar Units
11411.-11419. unassigned
11420. Cooking Equipment
11421.-11429. unassigned
11430. Dishwashing Equipment
11431.-11434. unassigned
11435. Garbage Disposers
11436.-11439. unassigned
11440. Food Preparation Machines
11441.-11449. unassigned
11450. Food Preparation Tables
11451.-11459. unassigned
11460. Food Serving Units
11461.-11469. unassigned
11470. Refrigerated Cases
11471.-11479. unassigned
11480. VENDING EQUIPMENT
11481.-11499. unassigned
11500. ATHLETIC EQUIPMENT
11501.-11549. unassigned
11550. INDUSTRIAL EQUIPMENT
11551.-11599. unassigned
11600. LABORATORY EQUIPMENT
11601.-11629. unassigned
11630. LAUNDRY EQUIPMENT
11631.-11649. unassigned
11650. LIBRARY EQUIPMENT
11651.-11699. unassigned
11700. MEDICAL EQUIPMENT
11701.-11799. unassigned
11800. MORTUARY EQUIPMENT
11801.-11829. unassigned
11830. MUSICAL EQUIPMENT
11831.-11849. unassigned
11850. PARKING EQUIPMENT
11851.-11859. unassigned

11860. WASTE HANDLING EQUIPMENT
11861. Packaged Incinerators
11862. Waste Compactors
11863. Bins
11864. Pulping Machines & System
11865.-11869. unassigned
11870. LOADING DOCK EQUIPMENT
11871. Dock Levelers
11872. Leveling Platforms
11873. Portable Ramps, Bridges, & Platforms
11874. Seals & Shelters
11875. Dock Bumpers
11876.-11879. unassigned
11880. DETENTION EQUIPMENT
11881.-11899. unassigned
11900. RESIDENTIAL EQUIPMENT
11901.-11969. unassigned
11970. THEATER & STAGE EQUIPMENT
11971.-11989. unassigned
11990. REGISTRATION EQUIPMENT
11991.-11999. unassigned

FIGURE 24-1. (*Continued*)

12	FURNISHINGS

12000. ALTERNATIVES
12001.-12099. unassigned
12100. ARTWORK
12101.-12109. unassigned
12110. Murals
12111.-12119. unassigned
12120. Photo Murals
12121.-12299. unassigned
12300. CABINETS AND STORAGE
12301.-12499. unassigned
12500. WINDOW TREATMENT
12501.-12549. unassigned
12550. FABRICS
12551.-12599. unassigned
12600. FURNITURE
12601.-12669. unassigned
12670. RUGS & MATS
12671.-12699. unassigned
12700. SEATING
12701.-12709. unassigned
12710. Auditorium Seating
12711.-12729. unassigned
12730. Stadium Seating
12731.-12734. unassigned
12735. Telescoping Bleachers
12736.-12799. unassigned
12800. FURNISHING ACCESSORIES
12801.-12999. unassigned

13	SPECIAL CONSTRUCTION

13000. ALTERNATIVES
13001.-13009. unassigned
13010. AIR-SUPPORTED
STRUCTURES
13011.-13049. unassigned
13050. INTEGRATED ASSEMBLIES
13051.-13099. unassigned
13100. AUDIOMETRIC ROOM
13101.-13249. unassigned
13250. CLEAN ROOM
13251.-13349. unassigned
13350. HYPERBARIC ROOM
13351.-13399. unassigned
13400. INCINERATORS
13401.-13439. unassigned
13440. INSTRUMENTATION
13441.-13449. unassigned
13450. INSULATED ROOM
13451.-13499. unassigned
13500. INTEGRATED CEILINGS
13501.-13539. unassigned
13540. NUCLEAR REACTORS
13541.-13549. unassigned
13550. OBSERVATORY
13551.-13599. unassigned
13600. PREFABRICATED BUILDINGS
13601.-13699. unassigned
13700. SPECIAL PURPOSE ROOMS
& BUILDINGS
13701.-13749. unassigned
13750. RADIATION PROTECTION
13751.-13769. unassigned
13770. SOUND & VIBRATION
CONTROL
13771.-13799. unassigned
13800. VAULTS
13801.-13849. unassigned
13850. SWIMMING POOLS
13851.-13999. unassigned

14	CONVEYING SYSTEMS

14000. ALTERNATIVES
14001.-14099. unassigned
14100. DUMBWAITERS
14101.-14199. unassigned
14200. ELEVATORS
14201. Elevator Hoisting Equipment
14202. Elevator Operation
14203. Elevator Cars and Entrances
14204.-14299. unassigned
14300. HOISTS & CRANES
14301.-14399. unassigned
14400. LIFTS
14401.-14429. unassigned
14430. Platform and Stage Lifts
14431.-14499. unassigned
14500. MATERIAL HANDLING
SYSTEMS
14501.-14549. unassigned
14550. CONVEYORS & CHUTES
14551.-14569. unassigned
14570. TURNTABLES
14571.-14599. unassigned
14600. MOVING STAIRS & WALKS
14601.-14609. unassigned
14610. Escalators
14611.-14699. unassigned
14700. PNEUMATIC TUBE SYSTEMS
14701.-14799. unassigned
14800. POWERED SCAFFOLDING
14801.-14999. unassigned

FIGURE 24-1. (*Continued*)

Uniform Construction Index | **Cost Analysis Format**

15 MECHANICAL

15000. ALTERNATIVES
15001.-15009. unassigned
15010. GENERAL PROVISIONS
15011.-15049. unassigned
15050. BASIC MATERIALS AND METHODS
15051.-15059. unassigned
15060. Pipe and Pipe Fittings
15061.-15074. unassigned
15075. Hose
15076.-15079. unassigned
15080. Piping Specialties
15081.-15099. unassigned
15100. Valves and Cocks (Manual)
15101.-15119. unassigned
15120. Control Valves
15121.-15139. unassigned
15140. Pumps
15141.-15159. unassigned
15160. Vibration Isolation & Expansion Compensation
15161.-15169. unassigned
15170. Meters and Gages
15171.-15174. unassigned
15175. Tanks
15176.-15179. unassigned
15180. INSULATION
15181.-15199. unassigned
15200. WATER SUPPLY & TREATMENT
15201.-15219. unassigned
15220. Pumps and Piping
15221.-15239. unassigned
15230 Booster Pumping Equipment
15231.-15239. unassigned
15240. Water Reservoirs and Tanks
15241.-15249. unassigned
15250. Water Treatment
15251.-15269. unassigned
15270. Distribution and Metering Systems
15271.-15299. unassigned
15300. WASTE WATER DISPOSAL & TREATMENT
15301.-15309. unassigned
15310. Sewage Ejectors
15311.-15319. unassigned
15320. Grease Interceptors
15321.-15329. unassigned
15330. Basins and Manholes
15331.-15339. unassigned
15340. Sewerage
15341.-15349. unassigned
15350. Lift Stations
15351.-15359. unassigned
15360. Septic Tank Systems
15361.-15379. unassigned
15380. Sewage Treatment
15381.-15399. unassigned
15400. PLUMBING
15401.-15419. unassigned
15420. Equipment
15421.-15439. unassigned
15440. Special System Accessories
15441.-15449. unassigned
15450. Plumbing Fixtures
15451. Special Fixtures and Trim
15452.-15454. unassigned
15455. Water Coolers
15456. Washfountains
15457.-15469. unassigned
15470. Pool Equipment
15471.-15499. unassigned
15500. FIRE PROTECTION
15501.-15509. unassigned
15510. Sprinkler Equipment
15511.-15519. unassigned
15520. CO_2 Extinguishing Equipment
15521.-15529. unassigned
15530. Standpipe and Fire Hose Equipment
15531.-15539. unassigned
15540. Pressurized Extinguishers and Fire Blankets
15541.-15549. unassigned
15550. Fire Extinguisher Cabinets
15551.-15559. unassigned
15560. Hood and Duct Fire Protection
15561.-15599. unassigned
15600. POWER OR HEAT GENERATION
15601.-15609. unassigned
15610. Fuel Handling Equipment
15611. Oil Storage Tanks, Controls, and Piping
15612. Bottled Gas Tanks, Controls, and Piping
15613. Oil Piping
15614. Gas Piping
15615. Stokers and Conveyors
15616. Ash Removal System
15617. Breechings
15618. Exhaust Equipment
15619. Draft Control Equipment
15620.-15629. unassigned
15630. Boilers
15631.-15639. unassigned
15640. Boiler Feedwater Equipment
15641.-15649. unassigned
15650. REFRIGERATION
15651.-15657. unassigned
15658. Refrigerant Piping System
15659. unassigned
15660. Compressors
15661.-15669. unassigned
15670. Condensing Units
15671.-15679. unassigned
15680. Chillers
15681.-15689. unassigned
15690. Evaporators
15691.-15697. unassigned
15698. Commercial Ice Making Equipment
15699. Refrigeration Accessories
15700. LIQUID HEAT TRANSFER
15701.-15709. unassigned
15710. Hot Water Specialties
15711.-15719. unassigned
15720. Steam Specialties
15721.-15729. unassigned
15730. Heat Exchangers
15731.-15739. unassigned
15740. Terminal Units
15741.-15759. unassigned
15760. Packaged Heat Pump
15761.-15769. unassigned
15770. Packaged Heating and Cooling
15771.-15779. unassigned
15780. Humidity Control
15781.-15799. unassigned
15800. AIR DISTRIBUTION
15801.-15809. unassigned
15810. Furnaces
15811.-15819 unassigned
15820. Fans
15821.-15834. unassigned
15835. Air Curtains
15836.-15839. unassigned
15840. Ductwork
15841.-15849. unassigned
15850. Special Ductwork Systems Equipment
15851.-15859. unassigned
15860. Duct Accessories
15861.-15869. unassigned
15870. Outlets
15871.-15879. unassigned
15880. Air Treatment Equipment
15881.-15889. unassigned
15890. Sound Attenuators
15891.-15899. unassigned
15900. CONTROLS & INSTRUMENTATION
15901.-15909. unassigned
15910. Control Piping, Tubing, and Wiring
15911.-15919. unassigned
15920. Control Panels
15921.-15929. unassigned
15930. Primary Control Devices
15931.-15949. unassigned
15950. Sequential Controls
15951.-15959. unassigned
15960. Recording Devices
15961.-15969. unassigned
15970. Alarm Devices
15971.-15979. unassigned
15980. Special Process Controls
15981.-15999. unassigned

FIGURE 24-1. *(Continued)*

16 ELECTRICAL

16000. ALTERNATIVES
16001.-16009. unassigned
16010. GENERAL PROVISIONS
16011.-16099. unassigned
16100. BASIC MATERIALS AND
   METHODS
16101.-16109. unassigned
16110. RACEWAYS
16111.-16119. unassigned
16120. CONDUCTORS
16121.-16129. unassigned
16130. Outlet Boxes
16131.-16132. unassigned
16133. Cabinets
16134. Panelboards
16135.-16139. unassigned
16140. Switches and Receptacles
16141.-16149. unassigned
16150. Motors
16151.-16159. unassigned
16160. Motor Starters
16161.-16169. unassigned
16170. Disconnects (Motor and
   Circuit)
16171.-16179. unassigned
16180. Overcurrent Protective
   Devices
16181.-16189. unassigned
16190. SUPPORTING DEVICES
16191.-16198. unassigned
16199. Electronic Devices
16200. POWER GENERATION
16201.-16209. unassigned
16210. Generator
16211.-16219. unassigned
16220. Engine
16221.-16229. unassigned
16230. Cooling Equipment
16231.-16239. unassigned
16240. Exhaust Equipment
16241.-16249. unassigned
16250. Starting Equipment
16251.-16259. unassigned
16260. Automatic Transfer
   Equipment
16261.-16299. unassigned
16300. POWER TRANSMISSION
16301.-16309. unassigned
16310. Substation
16311.-16319. unassigned
16320. Switchgear
16321.-16329. unassigned
16330. Transformer
16331.-16339. unassigned
16340. Vaults
16341.-16349. unassigned
16350. Manholes
16351.-16359. unassigned
16360. Rectifiers
16361.-16369. unassigned
16370. Converters
16371.-16379. unassigned

16380. Capacitors
16381.-16399. unassigned
16400. SERVICE & DISTRIBUTION
16401.-16409. unassigned
16410. Electric Service
16411. Underground Service
16412.-16419. unassigned
16420. Service Entrance
16421. Emergency Service
16422.-16429. unassigned
16430. Service Disconnect
16431.-16439. unassigned
16440. Metering
16441.-16449. unassigned
16450. Grounding
16451.-16459. unassigned
16460. Transformers
16461.-16469. unassigned
16470. Distribution Switchboards
16471.-16479. unassigned
16480. Feeder Circuit
16481.-16489. unassigned
16490. Converters
16491. Rectifiers
16492.-16499. unassigned
16500. LIGHTING
16501.-16509. unassigned
16510. Interior Lighting Fixtures
16511.-16514. unassigned
16515. Signal Lighting
16516.-16529. unassigned
16530. Exterior Lighting Fixtures
16531. Stadium Lighting
16532. Roadway Lighting
16533.-16549. unassigned
16550. Accessories
16551. Lamps
16552. Ballasts and Accessories
16553.-16569. unassigned
16570. Poles and Standards
16571.-16599. unassigned
16600. SPECIAL SYSTEMS
16601.-16609. unassigned
16610. Lightning Protection
16611.-16619. unassigned
16620. Emergency Light and Power
16621.-16639. unassigned
16640. Cathodic Protection
16641.-16699. unassigned
16700. COMMUNICATIONS
16701.-16709. unassigned
16710. Radio Transmission
16711.-16719. unassigned
16720. Alarm and Detection
   Equipment
16721.-16739. unassigned
16740. Clock and Program
   Equipment
16741.-16749. unassigned
16750. Telephone & Telegraph
16751.-16759. unassigned
16760. Intercommunication
   Equipment
16761.-16769. unassigned
16770. Public Address Equipment
16771.-16779. unassigned

16780. Television Systems
16781.-16849. unassigned
16850. HEATING & COOLING
16851.-16857. unassigned
16858. Snow Melting Cable and
   Mat
16859. Heating Cable
16860. Electric Heating Coil
16861.-16864. unassigned
16865. Electric Baseboard
16866.-16869. unassigned
16870. Packaged Room Air
   Conditioners
16871.-16879. unassigned
16880. Radiant Heaters
16881.-16889. unassigned
16890. Electric Heaters (Prop Fan
   Type)
16891.-16899. unassigned
16900. CONTROLS &
   INSTRUMENTATION
16901.-16909. unassigned
16910. Recording and Indicating
   Devices
16911.-16919. unassigned
16920. Motor Control Centers
16921.-16929. unassigned
16930. Lighting Control Equipment
16931.-16939. unassigned
16940. Electrical Interlock
16941.-16949. unassigned
16950. Control of Electric Heating
16951.-16959. unassigned
16960. Limit Switches
16961.-16999. unassigned

FIGURE 24-1. (*Continued*)

Requirements for adhering to all applicable federal, state, and city health and safety regulations during construction should be included, and all unusual hazards, including those likely to be present as a result of construction activities, should be identified here. These unusual hazards may include the presence of asbestos-containing materials during renovation or demolition. The technical specifications may include many of the statements contained in this manual along with non-safety-related items. A careful review of all sections should be made by health and safety professionals to identify all health and safety issues.

To ensure that all of the safety and health considerations will be incorporated into the completed laboratory, building, or renovated area, it is of utmost importance that the criteria delineated in this book be incorporated into the final technical specifications put out for bid. It is equally important that correct bidding procedures be established and closely followed to ensure construction of the facilities as designed. To that end, a set of bidding documents must be prepared. The following guidelines can be used for their preparation. They should be carefully reviewed by a qualified health and safety professional for appropriate additions or deletions.

24.4.2 Types of Bidding

Bidding procedures vary from location to location. In specific cases, due to federal and state regulations, strict guidelines exist.

In many countries, contract terms differ from those in the United States. In Great Britain and many other countries, bidding is called a *tender offering*. The procedures described in this chapter are those used in the United States; however, the general concept and principles are similar and could be applied anywhere. In general, the bidding process can be broken down into competitive and negotiated bidding.

24.4.2.1 *Competitive Bid.* Competitive bidding is when more than one contractor or vendor is requested to provide pricing for the complete building/project package. In general, the contract limits itself to the construction and the mechanical/electrical system. It does not include owner-furnished items like movable research equipment. Competitive bidding may be open or closed. In an open bid process the owner places a public announcement, and any contractor who meets minimum requirements on bonding and financial status may bid on the project. In a closed-bid process a limited number of general contractors and subcontractors who have been carefully researched for their qualifications to do a particular project are invited to bid on that project. Most bids for publically funded buildings are done on an open-bid basis, but many private organizations prefer to close their bid to ensure only well-qualified contractors.

24.4.2.2 *Negotiated Bid.* Under this approach, the owners negotiate with one vendor for the construction of the building and the mechanical/electrical systems. The owner-furnished items described above may or may not be included in the contract.

It is recommended that research equipment be specified and purchased by the research organization rather than by an outside construction vendor. Researchers know best what is needed and the quality level required for scientific equipment. It is not cost-effective to have a third party brought into the middle of that negotiation. When selected, the equipment can be competitively priced from more than one vendor. Bidding offers significant opportunity for project cost savings. The researchers should work with the administrators to look for competitive bidding opportunities for purchase of research equipment where possible instead of using a sole source vendor.

24.4.3 Bid Opening

After the bids are received, bid opening is required. In certain cases the submission of the bid and the opening may be highly structured. For example, certain federal and state regulations require the bid opening to be public, where not only the main general contractor is decided, but the various subcontractors are decided. All the requirements should be carefully noted, reviewed, and followed through to ensure fair bidding practice.

24.4.4 Project Cost Estimate

After the final bids are received, a final project cost should be projected. It is important to note that the project cost is entirely different than construction cost and includes other items like the architects' and engineers' fee, legal and other consultants' fees, site preparation cost, document printing, moving expenses, etc. Once the project cost is estimated, it should be reviewed with the funds available. If there is a discrepancy in funds available; either additional funds need to be raised or the project is required to go through redesign. This is a rather delicate period. Many times the carefully designed health and safety features are compromised for cost containment. A very careful review of the project scope and needs is absolutely vital to ensure that once construction is complete, the intent and requirement of the research are being met.

Unfortunately mechanical/electrical systems tend to present one large category brought to the chopping block. Architectural features tend to be retained instead. We suggest that thorough discussions and careful compromise are necessary. We have had situations where extremely fancy casework, as well as personal office features and furnishings, are retained at the expense of vital health and safety items. The temptation is great, and usually health and safety items (which are behind walls and above ceilings and not readily apparent) are easier for users and owners to justify removing from the project.

24.5   CONTRACT

Once a contractor is selected and adequate funds are available, a contract is signed between the building contractor and the owners. This contract could take several forms. Several jurisdictions have contracts of their own. If none is available, AIA (American Institute of Architects, Washington, D.C.) offers several contracts that should be considered. Legal advice and review should be employed on this and all contracts.

One of the most important duties that requires attention before the contract is signed is a comprehensive review of all sections of the specifications and contract drawings with the contractor. This review of all the building systems will ascertain that all parties understand what is expressed or implied in the documents. All too often items are overlooked which result in additional costs to the project.

This is the time to ensure that all health and safety systems are included and that the contractor understands his obligation in carrying out the conditions of the contract. Careful notes should be taken and signed by both parties concerned. A complete understanding of the intent is a part of the contract. This endeavor will save time and expense, especially on projects that are of long duration.

24.6   CHANGE ORDERS

Change orders are a fact of life. No matter how well the project is designed, there will always be changes due to different site conditions, change in scope, or nonavailability of material, or sometimes a better solution is proposed to solve a problem. It is absolutely vital that change orders be monitored on a very careful basis. Many times change orders have been authorized without due regard to the effect on the overall cash flow or availability of funds or safety and health considerations. Architectural and engineering redesign costs should be added to the contractor's estimate. Excessive change orders can result in the owner and builder running out of money in the middle of the project. It is necessary, therefore, that the project manager be given authority to say no to the requestor of changes when needed. This is not an easy matter because nobody likes to be in that role; however, that role is vital for overall success of the endeavor. At the same time, a balance is necessary. There are some decisions that become irreversible and become a limiting factor for the life of the building. Others can be changed or implemented later on. It is much easier to postpone or defer the items in the second category. For example, the amount of vertical chase space in the building is somewhat irreversible. Once a decision is made, the number of chases is designed and provided. It becomes very expensive to provide additional chases for mechanical/electrical systems. On the other hand, a D.I. water system can easily be added. Additional exhaust systems can also be added, but only if an allowance for expansion was made in the design phase.

AIA has some very good documentation available for help with this issue. The latest edition of the Architects Handbook of Professional Practices (AIA, 1987) should be consulted.

24.7 CONSTRUCTION INSPECTIONS

During the construction process it is vital that the architects and engineers provide routine inspection to ensure that the intent of the design is being met by the contractor. Usually, shop drawings of the equipment used or material to be supplied by the contractor are submitted to the owner and design team for review and approval. This is another stage where the level of quality should be carefully maintained. The contractor or subcontractor, in the desire to increase their profit margin, sometimes tends to shop the market and provide the lowest-priced material for many items in the construction project, not realizing the effect these changes may have on the final building. Some of the items may be clearly unsuitable for the intended use. A good design and construction specification will assist the design team at this level to delete or reject lesser products. Otherwise, it may be necessary for the owner to pay extra to get the desired quality.

It is recommended that construction-site visit notes be kept in consecutive order to be deleted as the problem items are corrected. In this day and age of word processing, it is very easy to keep a running document of the problems noted.

24.8 PUNCHLIST

This is the stage in the construction period when the design team is informed by the contractor that the project is substantially complete and they are invited to inspect the construction for quality as well as completion, and to issue a punchlist, itemizing the problem areas noticed. It is important that this punchlist be comprehensive and thorough. This is almost the very last opportunity the design team and owners will have to request contractors to make major corrections. Refer to Chapter 25.

24.9 ADDITIONAL TESTING AND ACCEPTANCE

Various mechanical/electrical systems in the building require additional testing as part of the acceptance process. They may include (a) short-circuit analysis and testing of the electrical switchboard and (b) pressure testing of the piping system and the exhaust duct systems. It is important that the design team provide clear instructions in the design documents so that the contractor is able to implement such tests. The acceptance of the test should be in written form and should be approved by the design team.

24.10 BENEFICIAL OCCUPANCY

At this stage, the project is substantially complete and the owner is able to move into the space and to connect various research equipment. It is usual to find some deviation from the original intent at this time. There is always a missed utility connection, or wrong voltage is provided. Provisions should be made, therefore, in the project schedule for this last-minute fine-tuning. This shake-down period can extend over several months on complex research laboratory or production buildings, particularly those for the pharmaceutical and chemical industries.

24.11 FINAL ACCEPTANCE

More details are described in the next chapter about the project acceptance. In this chapter the acceptance administration will be discussed. Several state, local, and federal agencies may need to inspect the construction of the project for final approval. Fire alarm systems or fire suppression systems may need to be accepted by the local fire department. After all these required code-related acceptance tests are done, a request is made to the local building department for the occupancy permit. The requirements for this permit vary from locality to locality and may be called something different from place to place; however, it should be recognized at this stage that the building is safe and is habitable by the user without any undue health risk or life safety problems. For example, in a high-rise building the smoke evacuation system or the sprinkler system must be operational. This leads to an interesting situation: In a high-rise building, the various floors could be issued an occupancy permit independent of one another as long as certain building code systems are completed. The requirements vary, and a local regulation should be checked to see how feasible it may be.

CHAPTER 25

PERFORMANCE AND FINAL ACCEPTANCE CRITERIA

25.1 GUIDING CONCEPTS

25.1.1 Introduction

The intent of this chapter is not to describe the entire building construction acceptance and testing procedure, but to concentrate on the building systems that pertain to health and safety. It is assumed, for example, that all structural, foundation, and soil testing necessary for a structurally sound building has already been done. It is critical that the health and safety systems in the laboratory building or renovated space be carefully inspected periodically during construction, because it is easier and more economical to correct defects during this stage than to wait until the final acceptance inspection is made. Such items as proper welding techniques, use of correct construction materials, and precise location of safety equipment storage areas should be observed during construction. For certain systems (e.g., fire suppression) it is not possible for a useful inspection to take place during construction because, for these, only a complete system-wide test after completion will be meaningful. Nevertheless, it is appropriate to keep abreast of preparation of the assigned space and the installation of auxiliary services that will be required for correct operation and ease of maintenance. For large projects, an on-site engineer responsible directly to the building owner should be present continuously during the entire construction period to perform construction inspections in a timely manner. It is recommended that this person be appointed in

Guidelines for Laboratory Design: Health and Safety Considerations, Second Edition
By Louis DiBerardinis, Janet S. Baum, Gari T. Gatwood, Anand K. Seth, Melvin W. First, and Edward F. Groden.
ISBN 0-471-55463-4   Copyright © 1993 by John Wiley & Sons, Inc.

addition to the normal *clerk of the works*. Some suggested test procedures follow. They should be included in the bidding documents.

25.1.2 Regular Testing Prior to Occupancy

To ensure that emergency systems will perform satisfactorily when needed, it is essential that frequent testing be conducted even during the interval before the building is occupied. This is commonly carried out once a week and should include the emergency electrical systems, the fire alarm systems, and all other emergency alarm systems when in service.

25.2 DESIGN, CONSTRUCTION, AND PREOCCUPANCY CHECKLISTS

The list presented below can serve as an aid in focusing on major safety and health items that need to be addressed during various stages of laboratory construction. This is by no means intended as a complete list, but it will serve as an initial checklist to be added to depending upon the special needs of the particular project.

25.2.1 Construction Safety Review Checklist

- Provide adequate chase space for currently specified exhaust ductwork, compressed gas piping, and so on, and allow at least 25% additional space for future additions.
- Avoid horizontal runs of ductwork in all exhaust systems susceptible to internal condensation. Perchloric acid hoods are a special hazard when condensate can collect inside the ducts. Do not combine exhaust ducts from perchloric acid hoods with other exhaust air systems.
- Check locations for portable fire extinguishers and make sure the correct type of unit is at each location.
- Select and identify locations for eye protection dispensers at entrance to eye-hazard areas (e.g., in safety stations at laboratory entrances).
- Check that copper piping was not used for acetylene gas lines.
- Document the safety and health practices that contractors and their subcontractors are expected to comply with while working on the site.
- Check areas requiring emergency lighting capabilities. Restrooms and some laboratory areas are frequently overlooked even though they are included in the NFPA codes.
- Investigate fire escape routes and document for post-occupancy training.
- Check locations for fire alarm installation.

- Check height of storage shelves so that sprinkler systems are not compromised.
- Check locations for waste containers that segregate combustibles, noncombustibles, chemicals, hazardous waste, broken glass, and trash.
- Check locations for use and storage of cryogens (e.g., liquid nitrogen) when these products are needed.
- Identify the systems that will require backup emergency services (e.g., electricity, cooling water, compressed gases/air) and make certain these work properly.
- Identify and record where explosion-proof or laboratory-safe refrigerators are required.
- Check locations and installation of safety stations and first-aid facilities.
- Check outlets for potable water and special laboratory water.
- Check identifications for shutoff valves for all piped utilities, including water services.
- Complete and issue a site safety handbook when occupancy begins.
- Check facilities for receipt, storage, handling, and disposal of radioactive materials and initiate paperwork to obtain the required license.
- Plan for purchase and storage of necessary personal protective devices such as self-contained breathing apparatus and protective clothing.
- Check locations and installation of safety stations and first-aid facilities.
- Verify that floor materials are resistant to slipping and resistant to spills of chemicals and materials that are likely to be used in large amounts.
- Determine whether exhaust air from some fume hoods will have to be cleaned before being emitted to the environment. When air cleaning is needed, the type, size, and location of the equipment and the utilities required (water, sewage, electricity, etc.) should be designated in the plan and installed properly.
- Check the location of air-supply intakes away from the influence of exhaust stacks, parking lots, loading docks, and other sources of contaminated emissions, including those from adjacent buildings.
- Check the chemical storage facilities for segregated storage of small amounts of oxidizing and combustible chemicals in laboratories and for major stores of chemicals and gases in segregated and specially constructed gas and chemical storage sheds. Review compressed gas storage installation for compliance with legal, insurance company, and fire protection association standards.
- Check installation of warning signs in conspicuous locations to identify dangerous areas.
- Check rating and locations of emergency exit doors, and verify direction of door swing with respect to direction of egress.

- Check and identify locations for ground fault interrupters wherever electrical shock hazards may exist.
- Make certain that all solvent storage cabinets are UL- or FM-approved as solvent storage cabinets and are electrically grounded by inspection and test.
- Make certain that all electrical outlets in laboratories are labeled and that corresponding labels are provided at each panel box prior to occupancy.
- Make certain that all fans, ducts, air cleaning devices, and the hoods they serve are labeled with coded tags for easy identification.
- Make certain that all specified flow indicators have been installed in the correct location in every local exhaust system.
- Make certain that the bottled compressed air backup system for equipment that would be damaged by a loss of compressed air is installed and connected to all designated delivery points.
- Clean out all compressed gas lines with a nonflammable solvent followed by compressed nitrogen drying before hookup to compressed gas sources. Cleanliness testing is recommended by microscopic examination of white membrane filters through which 1 m^3 of compressed gas has been passed after passage through the longest branch of the piping systems.
- Make certain that all piping systems have been clearly marked in a readily visible area according to an acceptable standard coding system for easy identification.

25.2.3   Preoccupancy Safety Review Checklist

- Test eye wash fountains and safety showers.
- Test functioning and audibility of fire evacuation alarm system.
- Verify direction of door swing with respect to emergency egress routes.
- Check cup sinks for strainers.
- Check that all equipment items, such as sinks, compressors, cabinets, and shelves, are firmly secured.
- Review all ventilation system balancing records and make certain that all systems are certified to be in conformance with all applicable plans and specifications.
- Verify the placement and operation of fire detection and suppression systems.

25.3 HEATING, VENTILATING, AND AIR CONDITIONING

Review Sections 2.3 and 2.4 for more details on HVAC and Loss Prevention and Safety preoccupancy issues. Chapter 34 ASHRAE HVAC Application Handbook provides excellent guidelines to follow (ASHRAE, 1991).

25.3.1   Air Balancing

In certain types of laboratories where pressure relationships are especially critical (e.g., biosafety laboratories), airtightness of the room enclosure is very important. All penetrations into the room (pipes, electrical conduits, ducts) must be well-sealed. A leakage test, using sulfur hexafluoride tracer gas, may be appropriate when total containment is required. One such test has been described by Billings and Greenley (Billings, 1989).

For ventilation systems to perform as designed, it is necessary that the supply and exhaust systems be balanced after installation is completed. Balancing will necessitate fan tests that include measurements of static pressure, fan and motor rpm, air volume rate, temperature rise, current draw (to determine brake horsepower), and so forth. Adjustments of sheaves, dampers, and so on, will also be necessary to distribute air in accordance with the HVAC drawing specifications. All balancing should be conducted in accordance with the standards of SMACNA and a written report submitted for the design engineer's review and approval. A sample test sheet is shown in Table 25-1.

The contract mechanism for obtaining a balancing contractor is a special concern. One approach is to insist that the testing and balancing contractor work for the owner, and thereby have autonomy and the ability to report freely to the owner or his representative any problems noticed with regard to the work of the mechanical contractor. The other approach is to have the testing and balancing contractor work directly for the mechanical contractor on the assumption that this arrangement leads to closer coordination and greater effectiveness. Our recommendation is for the testing and balancing contractor to work directly for the owner and that a sum of money be designated for this service function right from the start. We do not recommend that any contractors be permitted to monitor their own work.

25.3.1.1   *Air Balancing for VAV Systems.*   Variable air volume (VAV) systems can be very complex, with extensive controls which operate to exacting requirements. Therefore, a complete understanding of the operating characteristics of the system and the controls is mandatory. The system should be balanced at maximum airflow and then at minimum airflow. If outside air quantity at the outside dampers is measured at both maximum and minimum airflow, health and safety concerns will be satisfied. The hood exhaust and general exhaust fans will require checking to ensure that the supply fan will track reduced exhaust air requirements. Duct and terminal box static pressure controllers should be checked to determine that they perform as designed with the proper airflow at maximum and minimum positions. It is important that the balancer be a professional technician who can use instruments for airflow, hydronics, and electricity measurements and who can work with automatic controls that could be electric, electronic, or pneumatic.

TABLE 25-1. Air Moving Equipment Test Sheet

AIR MOVING EQUIPMENT TEST SHEET

Date _____

Project _____

Project
Number _____

SYSTEM NO.								
LOCATION								
MANUFACTURER								
MODEL NO.								
SERIAL NO.								
OPERATING CONDITIONS	SPECIFIED	ACTUAL	SPECIFIED	ACTUAL	SPECIFIED	ACTUAL	SPECIFIED	ACTUAL
TOTAL CFM								
RETURN CFM								
OSA, CFM								
EXHAUST CFM								
TOTAL STATIC								
SUCTION STATIC								
DISCHARGE STATIC								
EXTERNAL STATIC								
BMP								
MOTOR MANUFACTURER								
SIZE (HP)								
VOLTAGE								
RPM MOTOR								
SAFETY FACTOR								
	RATED	RUNNING	RATED	RUNNING	RATED	RUNNING	RATED	RUNNING
AMPERAGE								
RPM FAN								
SHEAVE POSITION								

25.3.2 Fume Hoods

VAV system fume hoods require additional testing and balancing to ensure that the proper airflow is maintained at maximum and minimum airflows. Fume hood control systems are of three types: (1) velocity pressure measurement of the airflow in the exhaust duct and (2) velocity pressure measurement of airflow in the annular space between the outside and inside casing of the hood and (3) by a mechanical means relative to sash height. The control signal from a static pressure controller operates a damper in the exhaust duct which,

TABLE 25-2. Laboratory Fume Hood Inspection Form

BUILDING   DEPARTMENT   ROOM NUMBER   DATE

HOOD NUMBER   PERSON IN CHARGE   LOCATION OF HOOD IN ROOM

USE OF HOOD: HOW OPERATED
   RADIOACTIVE MATERIALS   MANUFACTURER _____
   PERCHLORIC ACID   TYPE OF HOOD _____
   GENERAL CHEMISTRY   SASH: Vertical
   HIGH HAZARD CHEMISTRY   Horizontal
   SPECIAL PURPOSE

RECOMMENDED SASH HEIGHT
VELOCITY FPM _____
HEIGHT _____
DATE _____
SMOKE TEST _____   HOOD MEASUREMENTS

         BOTH SASHES OPEN   ADJACENT SASH OPEN

         AV. VEL. ____   AV. VEL. ____   AV. VEL. ____

EXHAUST FOR CABINET: AIRFOIL _____ AV. VEL. _____
              HIGH   LOW   _____ AV. VEL. _____
LEFT   ____   ____   SMOKE TEST _____
RIGHT   ____   ____
SMOKE TEST   ____   ____   COMMENTS:
                                DATE OF LAST SURVEY _____
                                HEIGHT OF SASH _____

MICROSWITCH _____
ALARM _____
MAGNEHELIC   LOW _____ HIGH _____ "H_2O

in turn, signals the fan to maintain constant static pressure. It is important that the balancing contractor understand the operation of the fume hood and that he perform several airflow measurements from maximum to minimum position to certify that the fume hood is safe. The user's fume hood alarm system requires adjustments to stay in balance and eliminate false alarms.

The face velocity of all fume hoods should be checked to ensure that the airflow rate is in conformity with design requirements. Appropriate labeling is required on each fume hood to indicate correct operation before it is released for service. If the fume hood is an auxiliary air type, it is necessary to test for correct operation of the makeup air system by smoke trails. Fume hood field performance tests are discussed in Section 31.10.3. If the fume hood has an integral flammable liquid storage cabinet, selection of the correct fire rating should be verified. Correct installation and operation of all piped-in gases and electrical fixtures associated with the fume hood should be verified. In hoods equipped with HEPA filters, the integrity of filters and filter mounting should be verified by in-place testing using the standardized techniques recommended for biological safety cabinets in National Sanitation Foundation Standard 49 (NSF, 1992) or the techniques recommended for nuclear applications in ANSI/ASME N510 (ANSI, 1980). An example report is included in Table 25-2. When two-speed, variable-speed, or parallel fan types of arrangements are used for exhausting a fume hood, it is necessary that proper operation be verified in each separate mode.

25.3.3 Ductwork Testing

All exhaust ductwork should be tested to ensure that excessive leakage does not occur. This is most important when the exhaust ducts are under positive pressure and may leak contaminants into occupied areas. ANSI/ASME Standard N510 1980 gives approved procedures for testing the leak-tightness of exhaust ducts and plenums.

25.4 LOSS PREVENTION, INDUSTRIAL HYGIENE, AND PERSONAL SAFETY

25.4.1 Fire/Smoke Alarms

The complete fire alarm system in the building should be checked for correct operation. Each device should be checked individually and as a part of the system by simulation of alarm conditions. Procedures for testing fire alarm systems are described in NFPA 72 "Installation, Maintenance and Use of Protective Signaling Systems" (NFPA, 1990).

25.4.2 Other Alarm Systems

The correct operation of all other alarm systems should be verified. Alarms are frequently used to signal unbalanced airflow, improper operation of mechanical equipment, and so on. Each device should be checked individually and as part of the entire system by simulation of alarm conditions to ensure correct operation even at remote station monitors.

25.4.3 Emergency Electrical System

The emergency electrical generator and associated electrical systems should be started and tested under appropriate load conditions and the engine operated for at least 3 h under 100% overload to ensure that the system will operate as specified when called into service. All transfer switches and ancillary devices should be tested individually and as part of the system and not accepted until found satisfactory.

25.4.4 Eyewash Facilities

The water flow rate of tempered and nontempered eyewash stations should be verified and recorded, and the angle and height-of-rise of the streams should be documented. Tempered eyewash stations should be checked to ensure water temperature is 70 ±5° F.

25.4.5 Emergency Showers

Flow rate should be measured and recorded. Minimum acceptable flow rate is 30 gal/min. Temperature of tempered showers should be between 70 and 90°F.

25.5 PROJECT COMMISSIONING

As building and project get more and more technically complex, it also becomes important for the users or the laboratory owners to be certain that a system as installed is going to perform to the original design intent. The best way this goal can be accomplished is by a very detailed and careful assessment of the start-up of various systems. This process is called *commissioning*. ASHRAE is preparing some guidelines regarding the intent of commissioning. The process can be described as follows:

a. Clear documentation of intent of design by owner and design team.
b. Identification of certain utility meters and expected performance.
c. Confirm that these parameters in a and b have been met.
d. Documentation of test results.
e. Preparation of owners and operators manuals.
f. Operators and maintenance personnel trained to ensure that the parameters can be consistently met over a long period of time.

It is noted that unless the project specification calls for the level of commissioning desired by the occupant or the owner, it is very likely that the contracting team would not be able to supply that level. The commissioning process is time-consuming and expensive for the contractor and unless it is clearly understood that it is part of the building construction process, no contractor will be willing to supply it.

There are several forms available to perform the commissioning process as an independent service. All parties to the contract should be made aware that a commissioning contractor's services have been secured for the project. The general and mechanical subcontractor sign an agreement to abide by the commissioning contractor's recommendations. Unless such agreement has been obtained, procuring such services may cause a traditional contractor relationship to become strained. For example, in the United States the traditional contractural framework of building construction is owner/architect, engineer/safety engineer/industrial hygienist, and other construction documents which are then either bid or given to a general contractor who in turn hires subcontractors to do the work involved. Almost no general contractor has all trades in-house. The mechanical contractor provides equipment, piping, insulation, ductwork, controls, and many other services.

If commissioning is done by a separate contractor or professional, it is quite possible that a conflict could arise between the installing contractor and the commissioning contractor. It is therefore necessary that careful thought be given to the contract language and specific responsibilities before a decision can be made. If a separate commissioning contractor is hired, there will be an additional cost to the project. Many are against adding additional cost, saying that they already have paid it to the architect/engineer team for project supervision. It should be noted that commissioning is something different than traditional construction supervision services provided by the American engineering firms. The intent here is not to recommend a separate commissioning vendor or contractor but to suggest that there is a need for the function and that the owner/operator should review that need.

CHAPTER 26

ENERGY CONSERVATION

26.1 INTRODUCTION

In this chapter, methods for achieving maximum energy conservation are discussed. This chapter should be reviewed during the planning stages of a new laboratory building or during renovation planning for an older building. The selection of one or another of the methods that will be presented will depend on a number of factors. For renovation projects it will include age and geographical location of the building, spatial organization of laboratories and building, size of renovation project, available capital and desired payback period, work schedules, and projected building use. The discussion of the major areas of energy conservation will be divided into five categories: (1) exhaust ventilation for in-laboratory contamination control by the use of chemical fume hoods, biological safety cabinets, and local exhaust points; (2) general laboratory ventilation; (3) lighting; (4) thermal insulation; and (5) humidity control.

Laboratory buildings tend to be energy-intensive. Good engineering practice should allow shutdown of laboratory systems when not in use. Careful control of heating and cooling systems in laboratory buildings results in an environment that is compatible and reduces operating costs.

Guidelines for Laboratory Design: Health and Safety Considerations, Second Edition
By Louis DiBerardinis, Janet S. Baum, Gari T. Gatwood, Anand K. Seth, Melvin W. First, and Edward F. Groden.
ISBN 0-471-55463-4   Copyright © 1993 by John Wiley & Sons, Inc.

26.2 ECONOMICS OF EXHAUST VENTILATION FOR CONTAMINATION CONTROL

A major operating expense for most laboratory buildings is associated with the operation of chemical fume hoods. Therefore, major energy savings can be made by the selection of fume hoods that minimize loss of conditioned air. In addition, operational considerations, such as exhaust air quantity and period of operation, play an important part in determining the energy cost of essential laboratory services.

The laboratory chemical fume hood is the major piece of safety equipment available to laboratory personnel who must, from time to time, work with hazardous chemicals and/or hazardous biological agents. It is estimated that at 1985 Boston area energy costs, a laboratory fume hood costs between $2000 and $4000 per year based on 24-h/day, 7-day/week operation. The range is a reflection of the variation in the size and design of the fume hoods that are installed. The average cost is approximately $3.60/cfm based on the January 1981 cost of electricity and fuel in the Boston area (DiBerardinis, 1983). When there are 700 hoods at a facility, using an average cost of $3000 per hood, the operating cost is over $2,000,000 per year. For 1992, a site-specific energy analysis should be made to calculate the current operating cost. It will, therefore, be clear that substantial energy savings can be realized by installing the most energy-efficient hoods, and equivalent savings can be realized by installing new hoods in older laboratory buildings.

26.2.1 Alternative Energy-Saving Methods

Looking at laboratory chemical fume hoods from an energy conservation point of view, there are six basic alternatives to be evaluated:

1. Reduce operating time.
2. Limit the air quantity exhausted from each hood, including installation of VAV systems.
3. Use auxiliary air hoods.
4. Use heat-recovery systems.
5. Limit hood use.
6. Eliminate inappropriate hood use.

We will evaluate the advantages and disadvantages of each option from an economic and safety viewpoint.

26.2.1.1 Reduce Operating Time. Many laboratory chemical fume hoods are presently operated 24 h a day, 7 days a week. The main reasons for this are:

1. Laboratory personnel work in the laboratories all times of the day and night as well as on weekends and holidays.

2. Volatile hazardous chemicals are stored inside hoods when they are not being used. In fact, a need to continue to vent these chemicals is cited as an important reason for operating hoods in a continuous mode.
3. Some reactions and preparations must be continued in the hood, uninterrupted, for 24 h or longer.

The philosophy behind reduced hood operating time is that when there is no need for a hood to operate (that is, when no one is working actively in a hood, or there is no long-term reaction or preparation taking place), there is no need for that hood to be exhausting the amount of air that it normally does. To reduce operating time when the hood is no longer needed, it is essential either to shut off the fume hood entirely or to lower the quantity of exhaust air. To accomplish this, it is necessary to establish routine working hours when hoods and exhaust points will be fully operational and to establish a procedure whereby legitimate needs during off-hours can be accommodated. It is also necessary to arrange for alternative safe storage for volatile toxic and other hazardous chemicals, including compressed gases.

It is estimated that laboratory fume hoods can be shut off at least 50% of the time without serious interference with research and teaching activities. The advantage is that it can provide a large energy savings. There are also disadvantages that need to be evaluated carefully, and the plan needs to be adapted to individual circumstances; for example:

1. Hood use will be restricted. This can be overcome by providing local control—that is, by giving each user the option of turning the hood on whenever it is needed during off periods. This is best handled by a central building service group because individual control at the hood usually results in the hood never being turned off.

Hood operation connected to light switches offers another means of control. The assumption is that when the lights are off, the hood is not in use. This assumption is mostly correct. However, there may be instances where an experiment is in process inside the fume hood for a long continuous period. Researchers may elect to shut lights off when they are not physically present in the room although the hood should continue to operate.

2. The need for volatile chemical storage can be satisfied by providing alternative safe storage space for the hood user. This can be achieved in several ways: One is to use flammable-liquid storage cabinets that meet NFPA, FM, and OSHA requirements. A second is to provide a separate storage area for nonflammable hazardous liquids that can take the form of (a) a separate, exhausted air storage cabinet, (b) storage under a laboratory fume hood provided with a separate exhaust connection that operates continuously, or (c) storage in a specially designed, passive chemical storage box, developed at the Harvard School of Public Health (DiBerardinis, 1983), as shown in Figure 1-13. An initial capital expense will be required to provide these facilities for each laboratory. In some cases, there may be impediments due to space limitations and/or competing HVAC requirements in exhausting

numerous separate cabinets. Although the air volumes needed to exhaust closed cabinets will be modest (that is, a few cubic feet of air per minute), the Harvard storage box requires no ventilation.

3. A troublesome problem associated with reduced hood operating time is maintaining adequate air balance within the laboratory and the building. Many buildings have one supply system for the entire building, or for each large group of laboratories. The difficulty in maintaining correct pressure relationships when hoods and exhaust points are shut down (that is, maintaining laboratories negative with respect to corridors and maintaining hazardous areas more negative than nonhazardous areas) may be difficult to overcome in certain types of buildings and in laboratory wings of multifunctional buildings. This matter has to be evaluated closely for renovations. In most cases, a modulating air supply system that responds to changes in exhaust air demand will be needed. It is possible to resort to a two-step supply air system on the assumption that not more than a few hoods will be operational during off hours and, therefore, the normal building requirement for outside air will provide sufficient supply air for the few exhaust facilities in operation.

The safety problems associated with reduced hood operating time are encompassed in the three points discussed above, the primary one being provision of alternative storage for hazardous materials.

26.2.1.2 Limit the Air Quantity Exhausted from Hoods. The basis for limiting hood air quantity is that under present good practice conditions, the exhaust air requirement is based on the largest possible hood opening. The largest possible opening is usually the length of the work surface of the fume hood times the height of the fume hood opening when the sash is in the fully raised position. But the maximum possible hood opening can be reduced in two ways:

1. Limit the height of the vertical sash opening. If, for example, under normal conditions, the laboratory fume hood sash can be raised 30 in., the exhaust air requirement must be designed for that full opening. If, however, it were arranged that the sash could only be raised 20 in., there would be a one-third savings in the amount of air to be exhausted. This method of reducing hood face opening may present a problem to hood users because it restricts access to the upper part of the hood. To overcome this restriction, it is possible to equip the hood with an alarm system so that when the sash must be raised above 20 in. for short periods of time to allow installation and construction of apparatus within the hood, an alarm, both audible and visual, will be activated to let the hood user know that the hood is not in a safe operating mode. When the sash needs to be above 20 in. for longer periods, the audible alarm may be turned off, but as soon as the sash is lowered to or below 20 in. the alarms will be reset and become available to respond whenever the hood sash is raised above 20 in. Twenty inches is an arbitrary height for the sash alarm; it can be varied depending on the needs of individuals. In

some cases, less height may be needed, whereas in others, the full opening may be required at all times.

2. Limit the air quantity by the use of horizontal sliding sash rather than vertical sliding sash. Usually two panels of horizontal sliding sash are used. At any given time, only one-half of the full width of a chemical fume hood is open and only the open area needs airflow. About 50% energy savings can be realized by this hood design. A frequently encountered problem with this type of hood sash is nonacceptance among hood users used to working with the more conventional vertical rising sash. A distinct advantage of the horizontal sliding sash over restricting the height of vertical sash is that users can still get to every part of the hood, although they have to move the sash horizontally to do so. An added safety benefit of the horizontal sliding sash is that it provides a safety shield that users can work behind by placing their arms around it. From a safety point of view, certain designs produce disruptive air currents at the edges of the horizontal sliding sash, and this has to be corrected. Unsafe conditions can be created by the ease with which the horizontal sliding sash can be removed from the hood. A method for monitoring and ensuring that the sash remain in place is needed. Some manufacturers provide a combined sash, a vertical sash with horizontal sliding safety glass panels within. This offers an additional feature for safe fume hood usage.

Additional methods to reduce air quantity exhausted from hoods are:

1. Substitute local point exhaust systems for conventional hoods. Certain applications do not require a laboratory chemical fume hood. A local point source of exhaust air fits over such devices as gas chromatographs and may be located on laboratory benches where limited quantities of toxic chemicals will be used. The advantages of local exhaust points are that they exhaust far less air than a conventional hood, even when the face opening is restricted, and the open end can be directed to capture contaminants at the source of generation. Some of the disadvantages are that they are usually designed for specific applications. When different applications are called for, they are sometimes difficult to adapt successfully. In addition, they do not provide the protection that a fume hood does in terms of containment of spills or protection from small explosions or fires.

2. Use multispeed fans to limit air quantity. At high speed, a fan will provide enough exhaust air to give the design face velocity across the entire open face of the hood but when the hood is not in use and the sash is lowered, the fan will go to a lower speed, only one-third to one-tenth full speed, and exhaust just enough air to prevent escape of vapors from the materials or equipment left in the work space. This air volume will be very much less than needed when the hood is in full use. However, multispeed motors operating down to one-third to one-tenth speed are special and not readily available. The most common multispeed motors are full speed and half speed. In many cases, the use of half speed may be adequate. Referring to the fan laws, a speed reduction of 50% will result not only in a volume reduction of one-half but also in an electrical power savings of one-half raised to the third power, or

to one-eighth the power required at full speed. Variable-speed motors are becoming more readily available.

3. Use a very small fan in parallel with the main exhaust fan. During periods of nonuse, the large exhaust fan can be shut off, leaving the small exhaust fan to provide a low-volume continuous exhaust airflow to maintain safe conditions. Care must be taken to prevent short-circuiting with this fan arrangement; for this reason this arrangement is becoming less popular.

Dual fan systems have a common drawback—that is, the possibility of active hood use while it remains in a dormant or nonuse mode. Various alarm systems have been designed and are commercially available to inform the hood operator of potential unsafe conditions. A similar concept is to install a device that locks the sash closed when the hood is not in a safe operating mode.

4. Significant energy conservation opportunities are possible with a variable air volume (VAV) system. A full discussion of options plus advantages and disadvantages is contained in Chapter 33.

Note. Duty cycling of laboratory exhaust systems creates very unsafe conditions and should not be employed for energy conservation purposes.

26.2.1.3 *Limit the Volume of Conditioned Air Exhausted from Hoods (Auxiliary Air Hoods).* Auxiliary air hoods are laboratory chemical fume hoods that supply a major part of the total exhaust air volume from a dedicated supply air duct located above the work opening of the hood, with the remainder coming from the general room supply air. Up to 70% of the air that the hood exhausts is provided at the face of the hood, and the remainder is provided by the building supply air system. The economic advantage of this hood is that in wintertime, the auxiliary air coming into the hood above the hood face needs to be tempered only to 60°F and not to the full temperature of the air supplied to the room. The theory behind this is that all the air is going directly into the hood and not mixing with the room air so that it does not need to be fully heated. It does, however, need to be tempered somewhat because some of the auxiliary air flows down across the head of the person standing at the hood. If it were not tempered to the degree noted, the worker would experience a cold draft. The largest energy savings is realized by the user of these hoods during hot weather in the warmest climates, because in the summertime the auxiliary air does not need to be cooled for comfort. In completely air-conditioned buildings, considerable energy can be saved by installing these hoods. Another advantage associated with the use of auxiliary air hoods is that in areas where there are a large number of hoods they eliminate the need to provide large amounts of makeup air that would represent an excessively large number of air changes per hour in the laboratory and result in uncomfortably high air velocities sweeping through the laboratory.

There are several undesirable features of auxiliary air hoods. First, they require an additional mechanical system, which can generate its own maintenance problems. Second, the design of this type of hood is critical. A number of hoods on the market, whose manufacturers claim they are auxiliary air hoods, provide the auxiliary supply air inside the hood rather than outside the hood and above the work opening. Introducing supply air inside the hood can result in pressurization of the internal volume and result in toxic materials coming out of the open face of the hood. In other unsatisfactory hood designs, the auxiliary air is supplied above the work opening but is incompletely captured by the hood. Instead, it mixes with the room air and defeats the energy conservation purpose of the hood. In northern climates, net savings from installation of auxiliary air hoods may not be realized because the period of hot weather is too brief each year to return a saving of energy equivalent to the added installation and operating costs. During the long cold winter, the auxiliary air has to be tempered to a temperature almost as high as the general room air supply and worthwhile savings may not be realized. One hood manufacturer claims the difference between the cost of an auxiliary air hood and a conventional fume hood is paid back in less than a year through energy savings. There may be locations and usage patterns where such energy savings can be attained, but it is by no means a general rule. If an auxiliary air hood is selected, one must be certain it can operate as designed. The performance tests outlined in Chapter 31 should be used as a guide for hood selection.

26.2.1.4 Heat-Recovery Systems.

Heat-recovery systems employ some type of heat exchanger to extract heat from the exhaust air stream in winter and use the recovered heat to partially warm the incoming air. The reverse cycle is applied in warm climates. The application of this type of system has to be evaluated on a case-by-case basis because the climate in the area of contemplated use may not lend itself to a useful amount of energy savings with these kinds of systems. It should be kept in mind that although a worthwhile savings may be calculated for the days of extreme temperature excursion, such days are often too few each year to pay back the cost of the installation plus the energy cost associated with operating the heat-exchanger system. The additional pressure drop induced by all these devices leads to increased horsepower requirements that need to be considered in the economic analysis. In addition, maintenance and repair requirements when recovering heat from laboratory exhaust air tend to be higher than normal.

Because laboratories are thought by some to be energy-wasteful, some administrative codes mandate heat recovery from laboratory exhaust systems even though small temperature differences make heat recovery inefficient thermodynamically.

Methods available for air-to-air recovery are:

1. Run-around loops
2. Heat wheels
3. Heat pipe systems

4. Plate heat exchangers
5. Air-to-air exchangers (not acceptable due to contamination and other problems)
6. Heat pumps
7. Chemical regenerators

Most heat recovery system installations require the exhaust and supply air streams to be near each other in order to be cost effective. In many buildings this proves to be challenging and expensive. For example, the exhaust is always at the roof level. It is desirable to locate the air intake as far away from exhaust as possible to minimize the reentry problems. The exhaust and supply airstreams then have to be separated and terminated as far away as possible.

Figure 26-1, adapted from Carnes (1984), shows how some options have been accomplished. The example shows dedicated exhaust fans from a fume hood discharging into an exhaust plenum. It could easily be adapted to a VAV exhaust fan system.

Carnes (ASHRAE, 1984) provides a good description of some of these systems, giving their advantages and disadvantages.

26.2.1.4.1 Run-Around Loops. Standard finned–tubed water coils installed in the supply and exhaust airstreams are connected via piping. A pump circulates a water, glycol, or a thermal fluid solution. The intermediate circulating solutions transfer energy between exhaust and supply airstreams. The coils must be constructed to suit the environment and operating conditions to which they are exposed. The effects of condensibles and corrosives may require specialized materials and/or coatings. A 3-way valve provides the control. The system is shown in Figure 26-2.

26.2.1.4.2 Heat Wheel. A "heat wheel" is a revolving cylinder filled with an air-permeable medium. Media material may be selected to recover sensible heat only or sensible and latent heat. The medium material should be reviewed with the properties and contaminants of the airstreams involved. A typical unit is shown in Figure 26-2. Some air carryover from the supply side to the exhaust side is possible.

26.2.1.4.3 Heat Pipe. The heat pipe heat exchanger device is also called a *thermosiphon*. The exchanger consists of a tube that is fabricated with a capillary wick, filled with a refrigerant, and sealed on both ends. Thermal energy applied to either end of the sealed pipe causes the refrigerant at that end to vaporize. The refrigerant vapor travels to the other end of the pipe. At this end, removal of thermal energy causes the vapor to condense into liquid, giving up the latent heat of condensation. By action of the capillary wick, the condensed liquid flows back to where it was originally vaporized, completing the cycle.

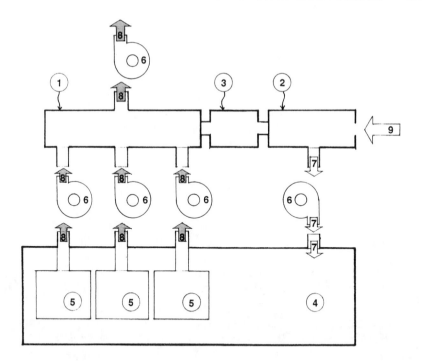

KEY
1	Exhaust Air Plenum
2	Make-up Air Plenum
3	Heat Recovery System
4	Building
5	Fume Hood
6	Fan
7	Make-up Air
8	Exhaust Air
9	Fresh Air Intake
☐	Clean Air
▨	Contaminated Air

FIGURE 26-1. Typical multiple fume hood recovery system: Diagram.

Each heat pipe operates in a closed-loop condensation/evaporation cycle. The heat pipe exchanger appears very similar to the standard HVAC coil, except that each tube is not connected by return bends and headers. Each individual tube of a heat pipe operates independently from the others. The exhaust airstream passes over one end of the heat pipe, while the makeup airstream passes over the other end. A sealed partition normally separates the two airstreams, minimizing any cross-contamination.

26.2.1.4.4 Plate Heat Exchangers. As the name implies, plate heat exchangers consist of plates that provide stationary channels where exhaust

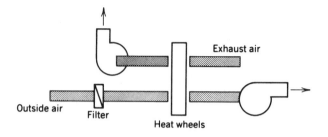

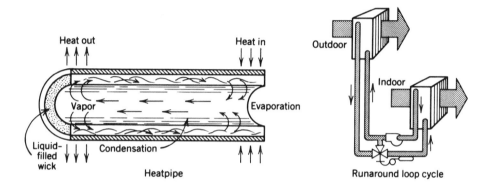

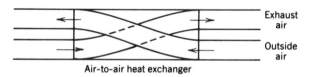

FIGURE 26-2. Heat recovery from exhaust airstreams.

and makeup air (supply air) pass adjacent to each other. Heat transfer takes place through material of construction. No secondary heat transfer media is used.

26.2.1.4.5 Air-To-Air Heat Exchanger. These are similar to plate heat exchangers except the usual design is a honeycomb arrangement. Due to difficult manufacturing quality control in fabricating such systems, these are not recommended for laboratory usage.

Additional concerns that should be noted for air-to-air heat exchangers are as follows:

a. Condensation on a surface will occur when the surface is cooled below the dew point of a warm airstream. In cold conditions, it is possible to freeze the exhaust airstream condensate, and the surface of the coils or heat pipes can frost over. Exhaust flow will be reduced. Design consideration should be made for frost prevention.
b. Backup or emergency heating systems should be installed in the supply airstreams.
c. Heat-recovery systems can be damaged. Methods of replacement should be considered during the original building design.
d. Temperature stratification is possible and should be considered in locating other components of the air-handling system.

Studies have shown that a rotary air-to-air regenerative heat exchanger with either a metallic or desiccant-impregnated matrix is the most effective device available because of its total heat-transfer capability. Although airside cross-contamination can be reduced to less than 1% by volume when the device is designed with a purge chamber, the major disadvantage is that the device is subject to exhaust-side condensation during winter operation when serving chemistry laboratories. Water-soluble hydrocarbons and other chemicals will revaporize and enter the makeup airstream, producing inlet air contamination. Furthermore, because of the nature of the chemicals used, a reaction may take place that could destroy the metallic or desiccant material. Experience has indicated that after some period of operation, a substantial amount of leaking occurs between the exhaust and supply sides of this system. Because of these problems, rotary heat exchangers are not recommended.

26.2.1.4.6 Heat Pumps. In a direct expansion heat pump refrigeration system, the evaporator and condenser section can be reversed. This allows exhaust and supply airstreams to pass through either the condenser or the evaporator. In the wintertime the exhaust air stream passes over the evaporator and releases its heat to the refrigerant. Heat is transferred into the supply air on the condenser side. In the summertime the reverse occurs.

26.2.1.4.7 Chemical Regenerators. Chemical spray headers are installed in both supply and exhaust air streams. The chemical is usually LiBr (lithium bromide). The chemical becomes the heat transfer media.

Exhaust air from biomedical laboratories may have special requirements that must be addressed:
1. In some cases the exhaust air must be incinerated to eliminate the danger of biological contamination.

2. The exhaust air from animal areas must be carefully filtered to eliminate animal hairs, food particles, and so forth, before it enters a heat-recovery device.

The selection process for any type of air-to-air heat exchangers requires a thorough analysis of the costs associated with (1) equipment's purchase and installation, (2) equipment's operating (electrical energy resulting from increased air and any circulated liquid pressure drop) and maintenance costs, (3) steam and chilled water consumption, and (4) capital deferrals. If buildings are served from a central steam and chilled water generating plant, the present worth cost of a future central utility plant expansion should be applied for the capital deferral values.

Only sensible heat-transfer devices are recommended. Stationary plate exchangers provide good separation between makeup and exhaust streams, but they are heavy and bulky and are difficult to build for large installations. In addition, they have only a 40–60% range of efficiency. Coil run-around systems have good potential when makeup and exhaust streams are physically remote. However, the range of effectiveness is also only 40–60%. A nonregenerative heat pipe in a coil configuration can be more effective, ranging from 60% to 70%, and the airstreams are isolated from each other by a center baffle or separator, preventing cross-contamination. Liquid lithium bromide systems are not used for laboratory air heat recovery because they are unlikely to avoid contamination of the incoming air. With the cost of fuel coming down, it is unclear how long these techniques will remain cost-effective.

26.2.1.5 *Limit Hood Use.*

In large laboratory buildings containing numerous fume hoods, it has been found that the number of fume hoods in operation at any one time is fairly small. Studies (Lentz, 1989; Moyer, 1987) indicate that only about 30% of available hoods are in use at any one time during the day. The criteria for assessing hood use were:

a. Someone working at the hood
b. An active experiment being conducted inside the fume hood

The lower percentage of hood use suggests the following opportunities for conservation.

1. Use a two-speed motor to make it possible for a hood to be switched to low speed when not in active use, thereby reducing the volume of air exhausted. As discussed in Chapter 33, it is also necessary that the makeup air be modulated simultaneously to ensure a negative pressure in the laboratory at all times.
2. Reduce the sash opening by lowering the sash to the position that represents the minimum exhaust volume for that hood design.

3. Use a variable-volume-exhaust system that modulates a damper in the exhaust duct or uses a variable-volume fan to adjust exhaust quantity to demand. The systems are discussed in more detail in Section 2.3.4.4.6. The variable-fan-speed system may be a single-hood variable-speed fan or a multihood exhaust system with an infinitely variable-speed fan instantly responsive to changes in hood demand.

Whatever system is used for controlling exhaust air volume, it must be kept in mind that there needs to be a minimum ventilation rate in occupied laboratories to provide safe conditions (See Section 2.3.4). Therefore no matter how low the exhaust air requirements from laboratories become, sufficient supply air must be provided (and removed) to maintain approved air exchange rates and still maintain the desired pressure relationship.

26.2.1.6 Eliminate Inappropriate Hood Use. In many laboratories the fume hoods are rarely used for anything other than storage of chemicals or equipment.

This is a very energy-wasteful practice. Chemical storage cabinets are much less expensive in terms of both first cost and operating cost. Many items of equipment often found in hoods, such as gas chromotographs, could be provided with a bench-mounted local exhaust point at the gas outlet port instead of locating the entire unit in a hood.

The research director as well as the contractor and entire design team should discuss real needs frankly and in detail. Often, immediate needs can be reduced when space is provided for the future installation of additional fume hoods.

26.2.2 Conclusions

It can be seen that there are several alternatives that may be applied to realize energy savings when dealing with laboratory chemical fume hoods. There is no single answer as to which may be the best method. In many cases a combination of methods may be required, particularly in new buildings. In retrofitting or renovation of older buildings, it may be difficult to apply some of the methods that have been reviewed, and therefore each situation has to be evaluated as a unique problem. The issues that need to be evaluated to make the best decision include the following:

1. Cost of retrofit for each of the alternatives.
2. Acceptance by hood users of a variety of restrictions that may be placed on hood use.
3. Effect of each alternative on the functioning of the building HVAC systems, on the comfort of the occupants, and on the type of work that must be conducted in the laboratory fume hoods within each building.

4. Effect of each alternative on safety in the laboratory. This factor must be evaluated carefully, including consideration for (a) providing additional storage space, (b) acceptance by laboratory personnel of restricted operating time, and (c) use of different types of laboratory fume hoods.

Any program that seeks to alter the traditional use of laboratory chemical fume hoods must include a detailed education and training program for laboratory personnel that will include information on how a laboratory chemical fume hood operates, what restrictions, if any, are being placed on its use, and how the restrictions may affect their research activities.

Sizing Equipment for a Manifolded VAV System. Because, according to Lentz and Seth (Lentz, 1989), the number of fume hoods at any given time is significantly lower than the installed number, there is an opportunity to reduce the total installed exhaust and supply air volumes to conserve chase and shaft space as well as air-tempering equipment. Care must be observed, however, not to block a future expansion of activities and people by undersizing equipment significantly.

Lentz and Seth (Lentz, 1989) describe some methods to estimate future capacity needs and keep as many options as possible available.

26.3 LIGHTING

A reduction in the energy required for lighting can be accomplished in laboratories by using task lighting at desks, laboratory benches, and work stations. The use of energy-efficient fluorescent tubes and ballasts, along with multiple switching, within a laboratory building will also conserve electrical energy. Lighting should be maintained at the levels outlined in Section 1.5.

26.4 THERMAL INSULATION

This energy-conservation measure is very effective for residences, but is not as useful for laboratory buildings because the heat transferred through the structure is a small fraction of the energy expended for ventilation air conditioning and contamination control. Nevertheless, the energy savings achievable through the use of good building thermal insulation are worthwhile and should be realized. Solar heat loads through windows are particularly troublesome for laboratories. Thermal windows with low-E glass are desirable in almost all climates for south and west building exposures.

26.5 HUMIDITY CONTROL

Needs for humidity control should be considered carefully. In summer, for example, to maintain excessively close humidity tolerances or exceptionally low humidity means more intensive use of the air-conditioning system because the usual method for reducing humidity in the air is to subcool it to reduce the moisture content and then to reheat it to the required room temperature. Because this process is an enormous energy consumer, it should be avoided wherever possible. To maintain high humidity in winter, moisture must be added to the air, and this is also energy-intensive. Air humidification methods are reviewed in Section 27.5.

PART V

HVAC SYSTEMS

Well-designed and well-operated HVAC systems are essential for laboratory health and safety protection as well as for an environment that promotes comfort and good work practices. In some laboratories, such as clean room and microelectronics laboratories, the HVAC system contributes importantly to the functionality of the facility. This part of the book is concerned with the design of satisfactory HVAC systems for laboratories in general and for a number of specific applications. It is also concerned with the selection and installation of HVAC equipment that is specifically designed for laboratory use; that is, equipment that has been found to be reliable, long-lived, and efficient for the task.

Some general information on the nature of HVAC systems and equipment is included here to assist those not professionally involved with HVAC matters to understand the issues and to be prepared to assist in reaching critical decisions on cost and function.

CHAPTER 27

BACKGROUND ON HVAC

27.1 LABORATORY HVAC SYSTEM DESCRIPTIONS

This chapter is intended to provide a general background for those less familiar with the terminology used in describing HVAC systems. It is by no means an exhaustive discussion of HVAC systems. For more detailed information, see ASHRAE (ASHRAE, 1989–1992). Additional reference sources include McQuiston (1977), Stoeker (1958), and Thuman (1977).

To meet design conditions in all laboratory areas, several independent control variables must be satisfied simultaneously under all conditions of operation. They are (1) temperature, (2) differential space pressures, (3) humidity, and (4) air exchange rate. Five different system approaches are usually considered in the design stage. They differ significantly with respect to system configuration, space requirements, control psychometrics, first cost, operating cost, and flexibility.

The system design options are designated:

a. Constant-volume terminal reheat (TRH).
b. Variable-volume terminal reheat (VVTRH).
c. Fan-coil variable air-volume system (FCVAV).
d. Fan-coil constant-volume system (FCCV).
e. Dual duct (DD).

Guidelines for Laboratory Design: Health and Safety Considerations, Second Edition
By Louis DiBerardinis, Janet S. Baum, Gari T. Gatwood, Anand K. Seth, Melvin W. First, and Edward F. Groden.
ISBN 0-471-55463-4   Copyright © 1993 by John Wiley & Sons, Inc.

27.1.1 Constant-Volume Terminal Reheat (TRH) Systems

A constant volume of air is supplied to the ventilated laboratory space through a high- or low-velocity ductwork system and is introduced into the laboratory through an air-volume-regulating device and a reheat coil. The air-volume-regulating device may be as simple as a manual balancing damper (in low-velocity systems) or as complex as a pressure-independent terminal box (in high-velocity systems). The reheat coil tempers the supply air when cooling requirements and internal heat gains are less than design conditions. Building cooling loads and internal heat gains do not normally peak simultaneously; therefore, reheat is usually needed most of the time. Because a constant volume of air is supplied to a laboratory using the TRH system, a constant volume of air must be exhausted to maintain design pressure relationships. Air may be exhausted through a central system, through individual exhaust hoods, or a combination of both.

The TRH system is relatively simple in concept. When it is used to meet all requirements for ventilation, heating, cooling, and space pressurization, it must be designed and operated to meet the full load climatic design conditions even though it seldom operates under this condition. Therefore, it calls for large systems, high design load conditions, and high operating costs which results in inefficient use of system capacity. In addition, TRH systems tend to be inflexible and to have limited ability to support expansion without significant cost penalties. When the terminal equipment does not employ pressure-independent volume control regulators at each space, a change in flow conditions in one space will normally cause a deviation in another, resulting in a need for frequent costly system balancing. A constant volume terminal reheat system is shown in Figure 27-1.

27.1.2 Variable-Volume Terminal Reheat (VVTRH) Systems

This is an all-air system that uses a central air-handling system to meet all facility needs for ventilation, heating, cooling, and space pressurization. However, VVTRH systems use variable-volume-regulating devices as terminal equipment instead of a constant-volume-regulating device to control the introduction of air into each space. A variable-volume terminal reheat system is similar in configuration to a constant-volume terminal reheat system except for the use of an air volume controller. A typical room terminal system is illustrated in Figure 27-2. The supply air volume may not be reduced below a minimum needed to maintain laboratory temperature, humidity, and pressure differential.

27.1.3 Fan-Coil Variable-Air-Volume (FCVAV) Systems

A fan-coil variable-air-volume system differs from terminal reheat and variable-volume systems in system function and the configuration of the terminal unit. A typical room fan-coil variable air volume system with eco-

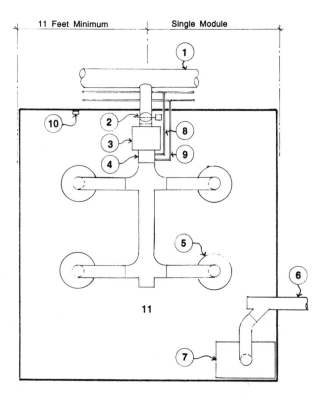

11 Feet Minimum   Single Module

KEY

1 Primary Air Supply
2 Damper
3 Terminal Reheat Box
4 Reheat Coil
   (Typically Located In Reheat Box)
5 Diffusers
6 Central Exhaust System
7 Fume Hood
8 Hot Water Supply
9 Hot Water Return
10 Thermostat to Reheat Coil
11 Typical Laboratory

FIGURE 27-1. Constant volume terminal reheat system.

nomizer exhaust is illustrated in Figure 27-3. FCVAV systems employ a central variable-volume air-handling system that distributes tempered air through supply ductwork and returns it for recirculation through exhaust ductwork. Alternatively, heating and cooling functions may also be provided in each space with fan-powered VAV units serving as fan coils. By separating heating and cooling functions from space pressurization and ventilation func-

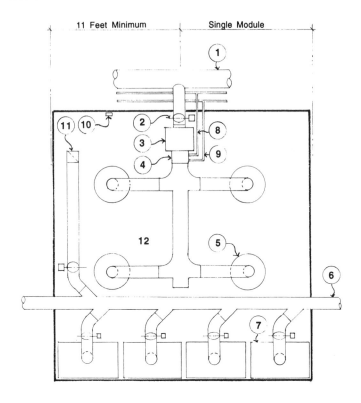

KEY
1 Primary Air Supply
2 Damper
3 VAV Box
4 Reheat Coil
  (Typically Located In VAV Box)
5 Diffusers
6 Central Exhaust System
7 Fume Hoods
8 Hot Water Supply
9 Hot Water Return
10 Thermostat to Reheat Coil
   And Economizer Exhaust Damper
11 Economizer Exhaust Duct
12 Typical Laboratory

FIGURE 27-2. Variable-volume system.

tions, outside air requirements are reduced and this results in a major reduction in operating costs. The FCVAV system design results in the smallest
possible central HVAC units and primary distribution ductwork, but requires
more ceiling space in the served areas. Heating and cooling is accomplished
by the sequential operation of hot and chilled water control valves at the
fan-coil unit. Ventilation and space pressurization functions are provided by
a controller, located at the room boundary that regulates the rate of air

introduction into each space to maintain a preset pressure relationship relative to contiguous spaces. When air is exhausted from a pressure-regulated space, the pressure controller senses an imbalance and compensates for it by increasing the supply airflow rate. Space pressurization control is recommended over other methodologies for reasons of responsiveness and minimum cost.

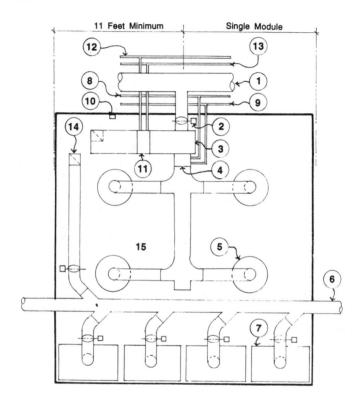

KEY

1	Primary Air Supply
2	Damper
3	Fancoil VAV Box
4	Reheat Coil
	(Typically Located In VAV Box)
5	Diffusers
6	Central Exhaust System
7	Fume Hoods
8	Hot Water Supply
9	Hot Water Return
10	Thermostat to Reheat Coil
	And Economizer Exhaust Damper
11	Recooling Coil
12	Cold Water Supply
13	Cold Water Return
14	Economizer Exhaust Duct
15	Typical Laboratory

FIGURE 27-3. Fan-coil VAV with economizer exhaust system.

27.1.4   Fan-Coil Constant Volume (FCCV) Systems

Fan-coil constant-volume systems differ from terminal reheat and variable-volume systems in terminal unit configuration and system function. A typical FCCV system is illustrated in Figure 27-4. Just as for FCVAV systems, the separation of heating and cooling from space pressurization and ventilation functions results in a major reduction in outside air requirements and operating costs. The result is a small central system and a reduced air distribution network but greater use of ceiling space in the served areas. Heating and cooling is accomplished by the sequential operation of hot and chilled water control valves at the fan-coil unit. The ventilation and space pressurization functions are provided by a controller located at the room boundary.

27.1.5   Dual-Duct Systems (DD)

Two sets of ducts are needed—one carrying cold air, the other hot to terminal boxes that mix hot and cold air to satisfy the preset room thermostat and regulate total volume to satisfy a room differential pressure controller. The two airstreams are used for temperature control and are referred to as the *hot deck* and the *cold deck*. DD systems may be constant-volume or variable-volume. In variable-volume systems, there is a minimum turndown level at the terminal box to maintain temperature and pressure control at all times. A Dual-Duct system is illustrated in Figure 27-5.

27.2   SPACE PRESSURE CONTROL

27.2.1   Introduction

The laboratory HVAC systems must be designed to satisfy space pressure needs under all conditions of operation. There are four pressure control strategies that can be used:

1. Space pressure control
2. Flow totalization
3. Flow synchronization
4. Constant-volume reset control

Each strategy has advantages and limitations, but when correctly applied, each can result in satisfactory facility performance at reasonable installation and operational costs. Any of the strategies can be applied to existing facilities using constant-volume systems to correct operational deficiencies and support a moderate degree of expansion without necessarily increasing physical plant or primary air-handling system requirements. A combination of more than one strategy is possible.

When the maximum exhaust air requirements of all variable volume hoods

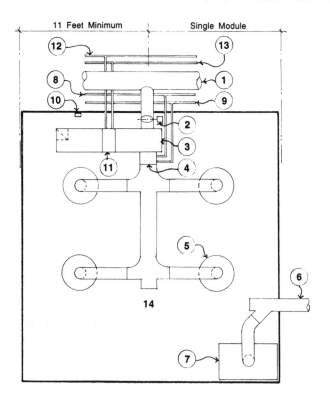

KEY

1 Primary Air Supply
2 Damper
3 Fancoil VAV Box
4 Reheat Coil
   (Typically Located In VAV Box)
5 Diffusers
6 Central Exhaust System
7 Fume Hood
8 Hot Water Supply
9 Hot Water Return
10 Thermostat to Reheat Coil
   And Economizer Exhaust Damper
11 Recooling Coil
12 Cold Water Supply
13 Cold Water Return
14 Typical Laboratory

FIGURE 27-4. Fan-coil CV system.

in a laboratory exceed 100% of the normal room air-supply volume, and the laboratory is designed for controlled air exchange between adjacent spaces, automatic flow control should be provided to modulate the supply air volume.

27.2.2 Pressure Control Systems

Pressure control systems use velocity or differential pressure sensors at the boundary of the room to control the introduction of air into the space. In addition to the hood exhausts that are normally served by the air supply

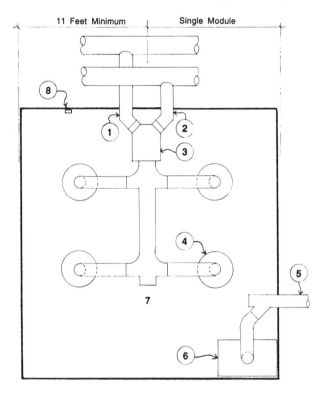

KEY

1	Hot Air Supply Duct
2	Cold Air Supply Duct
3	Dual Duct Mixing Box
4	Diffusers
5	Central Exhaust System
6	Fume Hood
7	Typical Laboratory
8	Thermostat

FIGURE 27-5. Dual-duct system.

system, an economizer exhaust is needed to make sure that space heating and cooling requirements are met whenever hood exhaust air quantities are inadequate to induce sufficient tempered airflow.

27.2.2.1 Space Pressure Control. Pressure control systems have the following advantages:

a. The simplest type of control.
b. Provide positive separation of environments from room to room because they require compartmentalization.
c. Easily support system expansion and contraction of facilities without the addition or modification of controls.
d. Are not limited in turndown except by the leakage rate of the volume regulators used.
e. Eliminate a need to place sensitive control instruments in corrosive environments, resulting in a high degree of reliability.
f. Only one supply system is required per regulated space.
g. A minimum of supply air is used to meet the changing pressure needs of the space, thereby conserving primary air use.
h. They measure differential pressure (the control variable) directly. Therefore, they are self-balancing and accommodate external disturbances such as stack effects and air infiltration from wind.
i. Least costly control method to implement.

Pressure control systems have the following disadvantages:

a. Require maintenance of laboratory compartmentalization.
b. Are highly dependent on (1) the continuous sensitivity of the pressure sensor to small pressure changes that directly affect infiltration and exfiltration between compartments and (2) how the system responds to transient disturbances such as traffic through the pressure controlled boundary.
c. Response to pressure disturbances can call for sudden supply air increases that may cause localized air turbulence that affects fume hood performance adversely.
d. Establishing minimum ventilation rates is difficult.

27.2.2.2 Flow Totalization Control of Space Pressure. Flow totalization pressure control employs a pitot tube or hot-wire anemometer velocity sensor in the airstreams entering and leaving the controlled space. The signals received from these sensors are processed by a square root extractor (pneu-

matic systems), a pressure/electric transducer (pneumatic/digital system), or an analog–digital processor to provide a signal that is used to regulate the total pressure-controlled laboratory supply air and exhaust volumes. The physical configuration of the air distribution system will be the same as for the space pressurization approach. Flow totalization systems with differential pressure control require a high order of complexity with high first cost and high maintenance requirements.

Flow totalization control systems have the following advantages:

a. Do not require compartmentalization of the facility or maintenance of compartmentalization for system control. Infiltration and exfiltration rates are a matter of system balance and are not subject to transient disturbances in maintenance or compartmentalization (i.e., traffic through doors).

b. Only one supply system is required to support a space.

c. Minimum ventilation rates are easily established.

d. Disturbances do not cause localized air turbulence that affects fume hood performance adversely.

Flow totalization systems have the following disadvantages:

a. Sensitive control instruments are placed within corrosive environments, thereby reducing reliability.

b. Maybe limited to a 4 : 1 turndown before the ability of the equipment to measure air velocity is affected adversely.

c. Do not readily support the addition or removal of exhaust systems without the addition or modification of system controls.

d. Technically complex.

e. Expensive.

f. Are not self-balancing because the control method does not directly measure the important control variable.

g. Subject to degradation of balance from control drift and sensor fouling; require regular rebalancing.

h. Air balancing is difficult and expensive. The system must be balanced throughout its entire range of modulation.

i. Insensitive to nonquantifiable disturbances such as infiltration and building stack effect.

27.2.2.3 *Flow Synchronization Control Systems.*

Flow synchronization control systems of laboratory differential pressure control differ from space pressurization and flow totalization control methods in physical configuration. Each system is served by dedicated air ducts that employ pitot tube or hot-wire anemometer velocity sensors in every airstream entering and leaving the controlled space. The signals received from the sensors are fed into a proportional plus integral laboratory differential pressure controller to eliminate output signal offset error. Controllers providing derivative functions are useful in eliminating control system transients.

Flow synchronization control systems have the following advantages:

a. Do not require compartmentalization of the facility.
b. Room infiltration and exfiltration rates are a matter of system balance and are not subject to transient disturbances such as from door traffic in the maintenance of compartmentalization.
c. Minimum ventilation rates are easily established.
d. Pressure transients do not cause localized air turbulence that affects fume hood performance adversely.

Flow synchronization systems have the following disadvantages:

a. Sensitive control instruments are placed within corrosive environments, thereby reducing reliability.
b. Maybe limited to a 4:1 turndown ratio before the ability of the equipment to measure air velocity is affected adversely.
c. Do not permit easy addition or removal of exhaust system components.
d. Complexity.
e. Expensive.
f. The control system does not respond directly to the control variable, namely, pressure differential. Therefore, the system is not self-balancing and is subject to degradation from controller drift and sensor fouling, requiring regular maintenance and repeated air balancing.
g. Air balancing is difficult and expensive because the system must be balanced through its entire range of modulation.
h. More than one air supply system may be required for each regulated space because each exhaust system, such as fume hoods, requires a separate supply system.
i. Insensitive to nonquantifiable disturbances such as infiltration and the building stack effect.

When correctly designed, installed, and calibrated, space pressurization, flow synchronization, and flow totalization systems are as effective as variable-air-volume systems for laboratories and each improves hood efficiency.

27.2.2.4 Constant Volume Reset Control Systems. Laboratory pressure control is maintained by volume control alone. Unlike other methods, volume control does not require the installation of mechanical or electronic devices. Instead, the volume of exhaust air is kept larger than the supply volume by the preset amount, and as long as the preset air balance is maintained, the laboratory will remain at the correct differential pressure. For laboratories that must be kept at a positive pressure relative to adjacent areas, the pressure relations will be reversed.

27.3 AIR-CONDITIONING SYSTEMS

Air conditioning is commonly understood to mean the supply of tempered (heated or cooled) air into a room to offset heat losses or heat gains, but air motion, relative humidity, and air purity controls are also included in a fully implemented system. A conditioned zone is a boundary inside the conditioned space that is controlled by a single control point or thermostat.

To a remarkable degree all warm-blooded animals, including man, are able to maintain a constant internal environment, called *homeostasis,* while living in a changeable external environment characterized by extremes of temperature and humidity. Man alone, however, is unique in having the ability to regulate the external environment in accordance with his own ideas. Warming the interior of buildings during cold weather has been carried on since time immemorial, but cooling is a twentieth-century phenomenon, initiated for factory production of heat- and humidity-sensitive items, adopted on a large scale in the United States for commercial properties during the 1930s, and adopted for residential use after 1950. Although the original objective in cooling commercial and residential properties was to provide a lower indoor temperature during hot weather, it was soon learned that the moisture content of the cooled air and its rate of motion over the body were additional factors in regulating human comfort in air-cooled spaces. This knowledge is expressed in "effective temperature charts" (see Chapter 28).

Heat energy flows from a region of higher temperature to one of lower temperature, meaning that building interiors lose heat when the weather is cold and gain heat when the weather is hot. The rate at which heat flows from the interior to the exterior, or vice versa, is variable and depends partly on the conductivity of the structure and partly on the amount of air exchange between inside and outside. Outside, unconditioned air is brought into the structure in two ways: purposefully for ventilation supply air and inadvertently as infiltration air that enters through open doors, windows, and structural cracks under the influence of an inside pressure that is lower than the pressure outside the structure. Wind action has an important influence on infiltration rates as well as on heat conductivity.

During the past decade, energy conservation measures have been influential in substantially decreasing the heat conductivity of new and old structures by the addition of heat insulation materials, and by reducing building porosity to decrease infiltration. Both measures have been effective in reducing building air-conditioning costs, and it is unlikely that this trend toward energy efficiency will be reversed in the foreseeable future.

Heat gains by insolation are welcomed during cold weather, but they are guarded against during hot weather by window and structural shading, reflective roof and wall treatments, and building orientation relative to the sun. Heat gains by insolation are generally neglected when calculating heating requirements but must be carefully accounted for when calculating cooling loads. A similar comment applies to outside air humidity. Together, heat gains from insolation and the need to reduce the moisture content of outside air entering the air-conditioned spaces through mechanical ventilation and infiltration often represent a major fraction of the total summer cooling load. Heat sources within the building itself, from lighting, equipment, and personnel, decrease the winter heating load but increase the summer cooling load. In some laboratories, the heat generated internally by equipment, people, and lights may be greater than the summertime heat gains from the sun and by conduction through the building's structure. It is critical, therefore, to inquire closely into the nature and amount of laboratory equipment that emits heat when designing the HVAC facilities that will be included in the building design. In extreme cases, a cooling capability may be required year round, even in cold climates, and this requirement must be provided when the heat load from equipment is unusually large.

The usual design procedure is first to establish extreme operating conditions in order to select equipment capable of satisfactory full-load operation. Next, decisions must be made regarding (a) the type of system required (e.g., multiple zones, individual room controls) and (b) the nature of the equipment that will be selected to deliver the preselected design conditions (e.g., central heating, humidifying, and cooling systems, under-window units to provide final temperature control).

In a *central system*, cooling and heating are performed with central chillers and boilers connected to large air-handling equipment that serves the whole or a major part of the building. The heating or cooling sources may be remote from the buildings they serve.

In a *modular system*, heating and cooling are done locally, sometimes room by room. Central and noncentral cooling systems used in modern buildings are described below.

27.3.1 Cooling Systems

The most commonly used cooling system, referred to as *mechanical cooling*, is a thermodynamic process called *direct expansion*. A gaseous heat-transfer medium such as Freon or ammonia is compressed to a state of high temperature and pressure by a compressor and then transformed into a cool liquid in a condenser, which may be air- or water-cooled. The cooled liquid, a refrigerant under high pressure, is allowed to flow to a low-pressure region through an expansion valve where the refrigerant evaporates. The latent heat of evaporation is extracted from the fluid (air or water) being cooled, thereby producing the temperature reduction step. The warm evaporated refrigerant gas is drawn next into the suction side of a compressor, and the cycle is repeated. This most basic of thermodynamic cycles is the heart of all direct expansion, or DX, refrigeration cooling systems. Some of the variations are described below.

27.3.1.1 *Compressors and Prime Movers.*

There are many types of compressors in use for air-cooling systems. Compressors can be hermetic or semihermetic. In hermetic compressors, the electric motor windings are cooled by the refrigerant suction gas. This prevents overheating, and at the same time it increases efficiency by heating up the cool suction gas, evaporating any liquid droplets that may be in the stream. A true hermetic compressor is integrally sealed. In smaller sizes it is called a "tin can." A semi-hermetic compressor is similar to a hermetic one, except that it can be taken apart for maintenance. For larger systems, it is almost mandatory that a semi-hermetic compressor be used. Another classification is an open compressor where the prime mover operating the compressor is separate from it. Although there is more flexibility in the choice of prime movers, the most frequent choice is an electric motor. Other kinds of prime movers, such as gasoline engines and steam or gas turbines, are also employed.

In large systems, the voltage to the electrical motor may be substantially higher than usual to avoid unusually large wire sizes and motor windings.

Reciprocating Compressors. The most common compressor is a reciprocating one in which the gas is compressed in cylinders by pistons operated by an electric motor.

Centrifugal Compressors. This machine compresses the refrigerant gas by centrifugal action rather than by reciprocal piston action but is otherwise similar to a reciprocal pump in its function. It is used primarily in systems having more than 100 tons of refrigeration capacity because it provides a more efficient ratio of energy consumption.

Screw Compressors. In this type, the refrigerant gas is compressed between two turning helical screws. The system is generally used from medium to high refrigeration tonnage because the energy ratio (i.e., kilowatts of electricity expended per ton of refrigeration produced) is attractive and the system offers a very reliable alternative to reciprocating and centrifugal compressors. The choice of prime mover is predominately electrical, but others have been used.

27.3.1.2 Condensers. The system is named according to the heat-transfer medium, either air or water. An air-cooled condenser, as the name implies, cools the hot compressed gas phase by forced air around the refrigerant heat exchanger. In a water-cooled system, the refrigerant condenser is cooled by water that can be sent directly to waste or recycled after itself being cooled in a direct or indirect water cooler. In a water-cooling tower, hot condenser water is sprayed or otherwise distributed, over packings or pans that allow the water to trickle downward in thin films countercurrent to a rising airflow. Some of the heated water evaporates, extracting the heat of vaporization from the water and rejecting it to the ambient air. The water returns to the condenser cooler than when it left. In an evaporative cooler, water from a secondary source is sprayed over a closed loop coil in which the hot condenser water flows. Air is forced over the outside wetted coil surfaces to evaporate the secondary water, which, in turn, cools the primary condenser water inside the coil. One of the advantages of this system is that the primary condenser water is not contaminated, but this system is slightly less efficient than the cooling tower configuration.

27.3.1.3 Expansion Devices. There are two commonly used expansion devices: thermostatically controlled expansion valves and capillary tubes.

27.3.1.4 Evaporators. The liquid refrigerant at low temperature is drawn into a low-pressure region by compressor suction where it evaporates, withdrawing the heat of vaporization from the air being cooled or from water used to cool the ventilation air. There are various kinds of evaporators in use, but the most widely used are refrigerant-to-air or refrigerant-to-water types.

27.3.1.4.1 Refrigerant-to-Air Evaporators. The liquid refrigerant is evaporated directly into a coil located inside an air handler that supplies cool air directly into conditioned spaces. In this case, a thermostatically controlled expansion valve is normally used.

27.3.1.4.2 Refrigerant-to-Water Evaporators. In this type of equipment the liquid refrigerant is evaporated into a coil that cools water surrounding the coil. The cooled water is then pumped through cooling coils located in the space to be cooled. This system is normally referred to as a "chiller." It is commonly used for large, widely distributed areas or buildings and when close control of environmental conditions is required.

27.4 SYSTEM DESCRIPTIONS AND STRATEGIES

After the cooling and heating equipment selection is made, the distribution system must be selected. There are three possible combinations: an all-air system, an all-water system, or a combination of the two.

Some commonly used combinations are as follows:

1. A *reciprocating-type direct expansion system* consists of a reciprocating hermetic or semihermetic compressor connected to an air-cooled condenser followed by an expansion device and directly into a refrigerant-to-air coil in an air handler. A small version of this system is the common window air conditioner. Large systems of this nature are known as "package units," or "rooftop package units," because they are often installed on roofs. A variation of this system is a so-called *split system,* where some of the components of the refrigeration cycle are inside and the heat rejection, or condenser, parts are outside.

2. A *chiller/cooling tower combination* is the most popular of the central refrigeration systems. A complete reciprocating or centrifugal compressor system is used to chill a secondary cooling medium (water) that is then pumped throughout the spaces to be cooled. The chilled water is incorporated into air handlers that consist of large housings containing fans, cooling coils, and filters. They cool the air and move it into the conditioned spaces. Heating coils are usually installed for winter heating and to reheat air for summer air conditioning that has been cooled below the comfort level for the purpose of condensing out excessive water vapor. These air handlers can be either low-pressure or high-pressure systems. A low-pressure system has less than 3 in. w.g. static pressure capability, whereas a high-pressure system exceeds this value.

27.4.1 All-Air Systems

In all-air systems, one or several air handlers are used to condition the entire building. A further classification can be made among the following:

1. In *constant temperature systems,* air is cooled to a constant discharge temperature. Reheat coils to temper the air to any desired temperature are installed downstream to accommodate variations in requirements.

2. In *Dual-duct systems,* a single or a cluster of air handlers have a split air system, consisting of a *hot deck* and a *cold deck,* that simultaneously produces separate hot and cold airstreams. Both are distributed to each conditioned zone where a preset thermostatic controller positions dampers in a mixing box to mix the hot and cold airstreams to produce the desired air temperature in the room.

3. *Multizone Systems* are similar to dual-duct systems in that the air handler has a hot and a cold deck. However, the mixing zone dampers will be located at the air handler.

4. *Variable-volume systems* supply air at a constant temperature into the conditioned space where the zone thermostat controls a volume box that reduces or increases the amount of air supplied to the conditioned space to satisfy the thermostat setting.

27.4.2 All-Water Systems

Chilled water for cooling, and in some cases hot water for heating, is piped throughout the conditioned building. In each conditioned space there are fan units to blow air over the coils to condition the area. The terminal fan-coil units are small, unitized cabinets with filters, a small fan, and coils for cooling and/or heating. All-water systems of this type are classified as:

1. *Two-pipe systems* in which a single set of water-containing pipes is distributed to each of the fan coils. The circulating water may be cool or hot, depending upon the season. Changeovers from hot to cold water and vice versa are conducted as the seasons require.
2. *Four-pipe systems* in which hot and cold water are supplied to each fan-coil unit through a separate set of pipes for instantaneous heating or cooling at each fan-coil unit as needed.

Whenever a terminal fan-coil unit is used, a condensate drain is normally needed to remove water that condenses on the outside of the cooling coil during humid weather.

27.4.3 Combined Systems

Mixed air-and-water systems are sometimes employed when special conditions or unusual requirements dictate their use, but they are less common.

27.5 HUMIDIFICATION

27.5.1 Humidity

Humidity is an ambient air condition that affects human comfort and environmental control in laboratories. Water vapor is usually present in the air. Section 1.5.6 on water sources for humidification should be consulted.

Relative humidity (R.H.) is the amount of water vapor present in the air compared to the maximum amount it could hold at the same temperature expressed as percent. Air at 100% R.H. is saturated. For information on the effect of humidity on human comfort see Chapter 28. In general, indoor relative humidity should not exceed 60% or be less than 30% (Sterling, 1984). Relative humidity in excess of 45% reduces or eliminates static electricity for most materials; wools and some synthetic materials may require higher humidity. Best results are obtained in the control of airborne infections at 50% R.H.

27.5.2 Humidification Systems

Humidification can be accomplished by local or central systems. Central systems are recommended for laboratories. Special humidity control systems

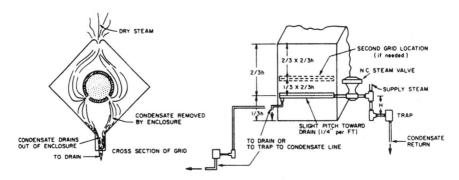

Enclosed Steam Grid Humidifier

(a)

Jacketed Steam Humidifier

(b)

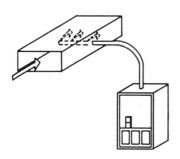

Self-contained Steam Humidifier

(c)

Pneumatic Atomizing Humidifier

(d)

FIGURE 27-6. Humidification systems. *Source:* ASHRAE.

may be needed for very precise humidity regulation. Chapter 20, in ASHRAE Systems Handbook (ASHRAE, 1992), describes various system options. Some are illustrated in Figure 27.6.

27.5.2.1 *Direct Steam Injection.* Steam at relatively low pressure and temperature is introduced directly into the air to be humidified without affecting air temperature significantly. This is the most popular system. The control system can be a modulating or an on-off type. Following are the most frequently recommended options.

a. *Enclosed Steam Grid Humidifier:* A grid containing one or more steam injection tubes is installed in supply ducts. A steam control valve driven by a laboratory humidity sensor provides R.H. control. Prevention of steam condensate splash-over inside the duct requires careful steam trap installation design.
b. *Jacketed Steam Humidifier.* An integral steam control valve with a steam jacketed tube that disperses steam into the supply duct. It contains a separator to prevent water from being introduced in the air stream.
c. *Self Contained Steam Humidifier:* Converts water directly to steam by an electric heating device.

27.5.2.2 *Water Atomization.* Atomizing humidifiers introduce a fine mist of water directly into the air, where the water evaporates. The ability of air to evaporate all the mist depends on air temperature, air velocity, and entering R.H.

27.5.3 Duct-Installed Humidifiers

Care must be taken to avoid water condensation inside ducts because the persistent or intermittent presence of liquid water can initiate and sustain the growth of fungi and bacteria that degrade air quality. When condensation is persistent, water may drip out of air outlets. The location of a humidifier installation is critical for optimum safe performance and manufacturers' recommendations should be followed.

27.5.4 Humidifiers Installed in Laboratories

Devices are available for direct humidification of the laboratory space but care must be taken in their selection because uniform dispersion of water vapor is difficult to achieve by this means.

CHAPTER 28

ASHRAE COMFORT STANDARDS

28.1 COMFORT INDEXES

Although the air-conditioning engineer strives to satisfy the largest possible percentage of occupants, some may complain of being too hot while others complain of being too cold, no matter how well-designed the system may be or how great an effort is made to satisfy everyone. This is because feelings of discomfort can occur even when the external environment is well within limits universally accepted as beneficial for enjoyment of life. Research on indoor comfort conditions has shown that various combinations of temperature, humidity, air movement, and heat radiation tend to produce equal feelings of comfort, and this has led to the development of a number of indexes referred to as "effective temperature," "operative temperature," "equivalent temperature," "resultant temperature," and "equivalent-warmth index," each of which seems to account for the ways that combinations of these factors affect the subjective sensation of comfort.

28.2 EFFECTIVE TEMPERATURE

Effective temperature combines indoor temperature, relative humidity, and air velocity into a single value that has been correlated with the results of

Guidelines for Laboratory Design: Health and Safety Considerations, Second Edition
By Louis DiBerardinis, Janet S. Baum, Gari T. Gatwood, Anand K. Seth, Melvin W. First, and Edward F. Groden.
ISBN 0-471-55463-4   Copyright © 1993 by John Wiley & Sons, Inc.

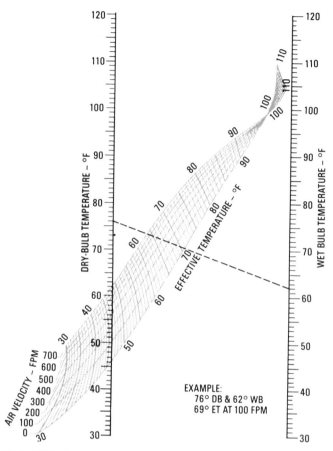

FIGURE 28-1. Effective temperature (at rest). From *Fan Engineering* (Buffalo Forge, 1983). From data of C. P. Yaglou and W. E. Miller, "Effective temperature with clothing" *Trans. ASHVE,* **31,** 1925, pp. 89–99.

empirically determined comfort evaluation polls. Values differ for men and women, for summer and winter, and for different areas of the continental United States and Canada with somewhat higher temperatures being preferred in the north than in the south. As clothing habits and preferences of the affected populations have changed over the decades, the effective temperature charts have been modified accordingly (ASHRAE, 1989). Effective temperature is the most widely used comfort index, perhaps because it is the oldest, having been published in 1925 (Yaglou, 1925a). It has been criticized for neglecting radiant heat effects and for tending to magnify the warming effect of high-humidity conditions within the normal indoor temperature range. In spite of these defects, it remains a useful guide of comfort ventilation conditions for men and women at rest or engaged in light physical activity, the condition for which it is designed. Some of the other comfort

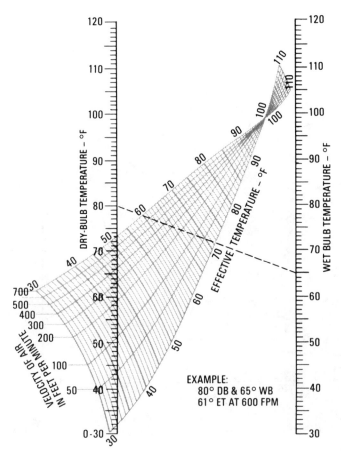

FIGURE 28-2. Effective temperature (light work). From *Fan Engineering* (Buffalo Forge, 1983). From data of C. P. Yaglou, "Comfort Zones for Men at Rest and Stripped Waist," *Trans ASHVE,* **33,** 1927, pp. 165–179.

indices are more appropriate for those engaged in hard physical labor or for work under high-heat-stress situations such as occur in foundries, smelters, and drop forge shops.

Figures 28-1 and 28-2 are typical effective-temperature charts for people at rest and when engaged in light work that have been adapted from the original effective-temperature charts of Yaglou and Miller (Yaglou, 1925b).

28.3 ASHRAE COMFORT STANDARD

Figure 28-3 is a version of the effective-temperature scale that has been adopted as Comfort Standard 55-74 by the American Society of Heating, Refrigerating, and Air-Conditioning Engineers (ASHRAE, 1977). It applies

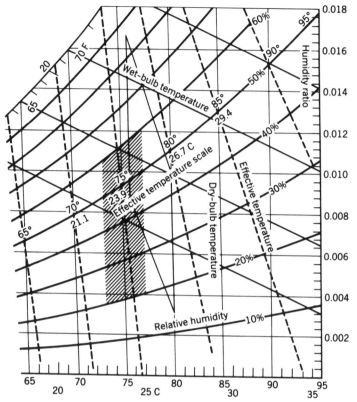

FIGURE 28-3. New effective-temperature scale (ET). With permission of the American Society of Heating, Refrigerating and Air-Conditioning Engineers, Inc., Atlanta, GA.

to lightly clothed, sedentary individuals in areas with low air movement, and avoids involvement with the effects of radiant energy by specifying that the mean radiant temperature (MRT) will be equal to air temperature. It is defined as the temperature at 50% RH that results in the same total heat loss from the skin as in the experienced environment. ASHRAE Comfort Standard 55-74 applies to the hatched area shown in the center of the chart in Figure 28-3.

The comfort standard was revised in 1981 by ASHRAE as standard 55-81 (Thermal Environmental Conditions for Human Occupancy) (ASHRAE, 1981). Figure 28-4 shows the 1981 effective-temperature line and comfort zone. Considerably more research is ongoing to further define, in a quantitative and qualitative manner, how human comfort is perceived.

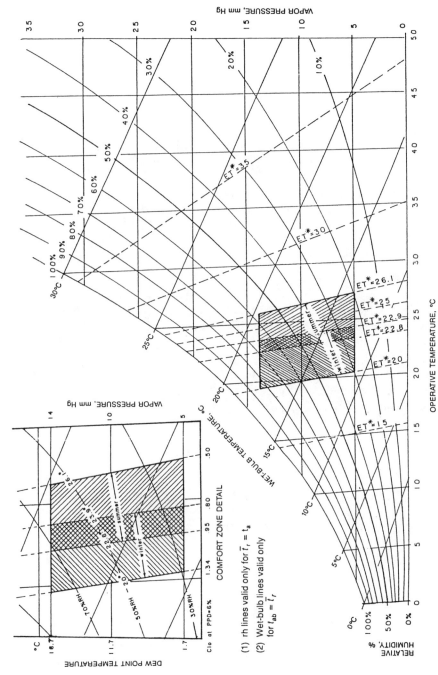

FIGURE 28-4. Standard effective temperature and ASHRAE comfort zones.

415

ASHRAE COMFORT STANDARDS

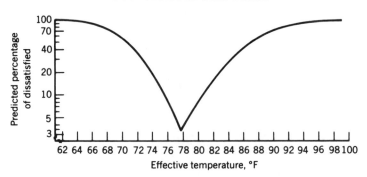

FIGURE 28-5. Standard effective temperature and ASHRAE comfort zones. Adapted from ASHRAE, 1977. With permission of the American Society of Heating, Refrigerating and Air-Conditioning Engineers, Inc., Atlanta, GA.

28.4 PERCEPTION OF COMFORT

Figure 28-4, adapted from ASHRAE, 1977, shows the percentage of individuals who are predicted to experience dissatisfaction over a range of effective temperatures calculated from the chart shown in Figure 28-3. It is clear that no environment is judged to be satisfactory by everyone. The comfort zone Figure 28-5, adapted from ASHRAE, 1977, shows the percentage of individature is greatly influenced by clothing and activity, it is not possible to develop a single, all-purpose scale.

CHAPTER 29

FANS

29.1 FAN TERMINOLOGY

The following definitions have been adapted from the 1983 edition of *Fan Engineering, An Engineer's Handbook on Fans and Their Applications* (Buffalo Forge, 1983).

FAN
: *Generally,* any device that produces a current of air by the movement of a broad surface.
  Specifically, a turbo machine for the movement of air having a rotating impeller at least partially encased in a stationary housing.

VENTILATOR
: A very-low-pressure-rise fan.

EXHAUSTER
: A fan used to remove air or gases from something or some place.

BLOWER
: A fan used to supply air or gases to something or some place.

IMPELLER
: The rotating element of a fan that transfers energy to the air (also called a wheel, rotor, squirrel cage, or propeller).

BLADES
: The principal working surfaces of the impeller (also called a vane, paddle, float, or bucket).

Guidelines for Laboratory Design: Health and Safety Considerations, Second Edition
By Louis DiBerardinis, Janet S. Baum, Gari T. Gatwood, Anand K. Seth, Melvin W. First, and Edward F. Groden.
ISBN 0-471-55463-4   Copyright © 1993 by John Wiley & Sons, Inc.

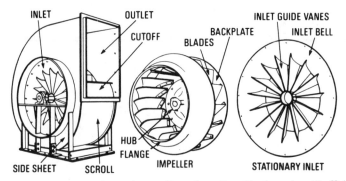

FIGURE 29-1. Exploded view of centrifuge fan. *Fan Engineering* (Buffalo Forge, 1983).

SHROUD	A portion of the impeller used to support the blades (also called a cover, disk, rim, flange, inlet plate, back plate, or center plate).
HUB	The central part of the impeller that attaches to the shaft and supports the blades directly or through a shroud to the shaft.
HOUSING	The stationary element of a fan that guides the air before it enters the impeller and after it leaves the impeller (also called a casing, stator, scroll, scroll casing, ring, or volute). Centrifugal fan housings include side sheets and scroll sheets. Axial-fan housings include the outer cylinder, inner cylinder, belt fairing, guide vanes, and tail piece.
CUTOFF	The point of the housing closest to the impeller (also called the tongue).
INLET	The opening through which air enters the fan (also called the eye or the suction).
OUTLET	The opening through which air leaves the fan (also called the discharge).
DIFFUSER	A device attached to the outlet of a fan to transform kinetic energy to static energy (also called a discharge cone or evasé).
INLET BOX	A device attached to the inlet of a fan to make it possible to use a side entry into a centrifugal fan (also called a suction box).
VANES	Stationary blades used upstream or downstream of a fan to guide airflow. (When used upstream they are also called inlet guide vanes; when used downstream they are also called straightening vanes or discharge guide vanes.)

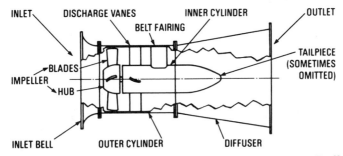

FIGURE 29-2. Cutaway view of a vane-axial fan. *Fan Engineering* (Buffalo Forge, 1983).

AXIAL-FLOW FAN   A fan contained in a cylindrical housing and characterized by flow through an impeller that is parallel to the shaft axis.
Vane axial fans have stator vanes.
Tube axial fans do not have stator vanes.
Propeller fans are pedestal or panel-mounted low-static-pressure axial-flow fans.

CENTRIFUGAL   A fan contained in a scroll-shaped housing character-
   FAN   ized by radially inward and outward flow through the impeller.

Typical centrifugal and axial-flow fans are illustrated in Figures 29-1 and 29-2.

29.2 EXHAUST FAN SPECIFICATIONS

Centrifugal fans, complete with motor and drive, should be used for general exhaust service as well as for exhausting fume hoods. They should be of Class I construction, in accordance with the standards established by the Air Moving and Conditioning Association (AMCA, 1986). The fan impeller should be backwardly inclined and should have a true self-limiting horsepower characteristic. The fan should be single width, single inlet with ball bearings and overhung pulley for V-belt drive. In some cases the fan/motor drive combination should be a nonsparking type. Fan housing should be constructed of steel, and all parts should be bonderized and then coated with baked primer-finisher especially formulated to meet stringent corrosion-resistance standards. The coating should have a thickness of at least 1–2 mils without voids. The fan housing should be weatherproof for protection of motor and drive when located on the roof, and a drain connection should be provided in the fan housing.

CHAPTER 30

VENTILATION AIR CLEANING

30.1 INTRODUCTION

It is usual to treat the outside air supplied to all laboratories to remove some or all of the particulate matter naturally present. In some laboratories, such as clean rooms and microelectronic processing laboratories, it is essential to remove all particles, large and small, from the air supply to preserve the essential functions of the facilities that call for a dust-free environment. Other laboratories that call for a contamination-free air supply are analytical chemistry and biology laboratories engaged in measuring trace quantities of chemicals that occur in air, water, and tissues.

Ventilation supply air is characterized by low concentrations of particulate ($50–200$ $\mu g/m^3$) and gaseous contaminants, such as SO_2 ($0.2–0.8$ ppm), NO_x ($0.1–0.5$ ppm), and volatile organic compounds of all species ($0.5–1.0$ ppm). Particles in outside air are characterized by small average size (median aerodynamic diameter approximately 0.5 μm) but a very wide size range. Pollens and spores range from 10 to 40 μm in diameter, soot and wind-blown soil range from 1 to 20 μm, and sulfate and nitrate salts formed by reactions in the atmosphere between SO_x, NO_x, and ammonia are less than 0.5 μm. The importance of airborne dust sizes is related to the fraction(s) that must be removed prior to introducing ventilation air to the laboratory. As a general rule, the smaller the particle sizes that must be removed from the air, the

Guidelines for Laboratory Design: Health and Safety Considerations, Second Edition
By Louis DiBerardinis, Janet S. Baum, Gari T. Gatwood, Anand K. Seth, Melvin W. First, and Edward F. Groden.
ISBN 0-471-55463-4 Copyright © 1993 by John Wiley & Sons, Inc.

more difficult and more costly it becomes. Therefore, there is a wide selection of air-cleaning devices to respond to a wide range of needs for air purity from removal of pollens and large soiling particles to removal of all particles down to molecular size. Removal of gas-phase contaminants is made difficult because of the low concentrations that are normally present in the atmosphere. This is because gas-phase cleaning is accomplished by absorption or adsorption devices that remove the undesirable components from air, and both methods depend for effectiveness on the differential concentration of the compounds to be removed in the sorbent and in the air being treated. Because the concentration of undesirable gases and vapors is very low in ambient air, the available concentration difference driving force for absorption or adsorption is minuscule, making the task difficult and expensive.

30.2 AIR-CLEANING EQUIPMENT FOR LABORATORY SUPPLY VENTILATION

Filters, which are porous beds of glass or plastic fibers that separate particulate matter, solid or liquid, from the air passing through them, are widely used for cleaning ventilation air. Particles may also be removed from air supplies by electrostatic precipitation. These are two-stage devices that first charge airborne particles in a zone containing high-voltage discharge wires and then attract the charged particles to collecting plates of opposite sign. For ventilation air, particles are given a positive charge and attracted to negatively charged collecting plates because this arrangement produces less ozone than negative particle charging. Electrostatic precipitators are subject to electrical supply interruptions. When this occurs, the collecting plates lose their charge and the accumulated dirt and dust have a tendency to blow off. Therefore, it is usual to install filters downstream of electrostatic precipitators to catch the released dust and prevent it from traveling further downstream into the ventilated laboratories.

Neither filters nor electrostatic precipitators are effective for gas and vapor contaminants. For the class of contaminants that are commonly found in ambient air, the use of adsorbents is the air-cleaning method of choice because they are effective for low gas concentrations and are easy to service. Activated carbon (activated charcoal) is widely used to remove organic vapors from air. Other adsorbents are available for removing inorganic gases and vapors such as ammonia, hydrogen sulfide, mercury, and a large number of acid gases. Granular adsorbent beds are poor particle collectors, but they do collect some particles and they progressively degrade adsorbent effectiveness. Therefore, activated carbon and other adsorbent beds must be preceded by moderately-high-efficiency particulate filters to protect them from rapid inactivation by dirt accumulation.

30.2.1 Air Filters

Air filters are rated by the amount of particulate matter they are designed to remove from air, expressed as percent efficient. Filter ratings have a wide range: 5–10% for warm air residential heating systems; 35–45% for ventilation of schools, stores, and restaurants; 85–95% for fully air-conditioned modern hotels, hospitals, and office towers; 99.99+% for air supplied to clean rooms, microelectronic laboratories, and hospital operating rooms. Filter efficiency of 80% is recommended for hospital laboratories (ASHRAE, 1991).

The most widely used test methods for ventilation air filters are published by ASHRAE as Standard 52-76 (ASHRAE, 1976). It contains two different protocols. One uses a prepared test dust consisting of road dust, carbon black, and cotton fibers. In this procedure, the test dust is aerosolized by compressed air and blown into the filter at a concentration many times that normally found in ambient air. The filter is rated by the weight percent of dust retained. This is an obsolete test method that harks back to the days when coal was the only fuel. It has little relevance to today's air filter requirements. The second test method uses unaltered atmospheric air as the test medium and rates filter efficiency on the basis of the percent reduction in discoloration of simultaneous samples taken on white filter papers up- and downstream of the filter under test. Reductions in discoloration can be related to weight percent efficiency. In addition to dust collecting efficiency, the test measures filter resistance increase with dust deposition and dust holding capacity. Ventilation filters in the 35–95% efficiency range are evaluated by the atmospheric dust discoloration test.

High-efficiency filters are rated by a test method developed by the military for testing the service gas mask and adopted by the nuclear industry. It uses an artifical test aerosol of 0.3 μm diameter generated from high-boiling-point liquids that are heated to the vapor stage and then cooled under carefully regulated time and temperature to produce condensed droplets of uniform diameter. Filter efficiency is measured by a very sensitive light-scattering method that can measure as few as 10 particles per cubic centimeter of air. It is usually referred to as the *DOP test* because the original liquid used was dioctylphthalate. The particle size selected for this test (0.3 μm) is known as the *minimum filterable size* because particles both smaller and larger than 0.3 μm will be filtered at higher efficiency. The reason for there being a minimum filterable particle size is explainable from filtration theory. Not all high-efficiency filters have their minimum filterable size at 0.3 μm; the theory and empirical confirmation were developed for the military services' high-efficiency particulate air (HEPA) filter that has a minimum efficiency of 99.97% for 0.3-μm particles. Special HEPA filters for microelectronics laboratories are commercially available that give 99.9995% efficiency with the DOP test. They are all single-use disposable filters. As efficiency increases, so does the airflow resistance of the filters (energy consumption) and the filter cost. A sampling of the many types of ventilation filters that are commercially available is illustrated. The most widely used grades are (1) 20–35% dust spot efficiency (shown in Figure 30-1), used for shops, storage rooms, mechanical equipment rooms, and animal laboratories, (2) 55–70% dust spot efficiency (shown in Figure 30-2), used for teaching laboratories and laboratories that do not have special clean air requirements, (3) 80–95% dust spot efficiency (shown in Figure 30-3), used for hospitals, trace analytical laboratories, and photography laboratories, and (4) 99.97 + % DOP efficiency filters (shown in Figure 30-4), used for clean room laboratories, microelectronic laboratories, and wherever superclean air is a requirement for carrying out effective laboratory procedures.

30.2.2 Electrostatic Precipitators

Electrostatic precipitators are used in office buildings and a variety of commercial properties for cleaning ventilation air. They are less frequently used in laboratory air systems. They are rated for efficiency by the same dust spot methods used for ventilation filters. Commercial units are usually rated 85% or 95% dust arrestance efficiency. The 95%-efficient unit differs from the 85%-efficient unit in the retention time of the air within the unit, with longer retention time in the charging and precipitating stages resulting in higher efficiency. Figure 30-5 is a diagram of a two-stage electrostatic precipitator for ventilation air cleaning. The ionizing stage usually operates at 12,000 volts, whereas the collecting stage operates at 6000 volts. The ionizing stage contains a series of fine-diameter, highly charged wires that continuously ionize the air close to them and propel clouds of positive air ions across the air gap to ground. During the pasage, they collide with dust particles and give

FIGURE 30-1. 20–35% Dust spot efficiency filters. *Source:* Farr Co., El Segundo, California.

FIGURE 30-2. 55–70% Dust spot efficiency filters. *Source:* Farr Co., El Segundo, California.

FIGURE 30-3. 80–95% dust spot efficiency filters. *Source:* Farr Co., El Segundo, California.

FIGURE 30-4. HEPA filter. *Source:* Farr Co., El Segundo, California.

them a unipolar positive charge. In the collecting stage the positively charged dust particles deposit on grounded plates by electrostatic attraction. From time to time, the collecting plates must be cleaned by washing with a detergent to prevent collected dirt from building up to the point where it tends to become reentrained in the flowing airstream.

30.2.3 Adsorbents for Ventilation Air

Activated carbon is the usual adsorbent used in ventilation air systems to remove organic vapors, sulfur dioxide, nitrogen oxides, and ozone. It is purchased for service in the form of $\frac{1}{4}$ to $\frac{1}{2}$-in.-deep panels containing a packing of 80 to 120-mesh activated coconut-shell carbon. The panel sizes usually match the cross section of the particulate filters installed ahead of them. Filters impregnated with carbon granules are on sale as combination particulate filters and gas adsorbers. Anything less than a $\frac{1}{4}$-in. thickness of well-packed granules is useless as an adsorbent stage, and $\frac{1}{2}$-in. thickness is likely to give much more satisfactory service. Activated coconut-shell gas carbon can give 95–99% removal of most organic compounds.

Activated carbon, as well as other adsorbents, ultimately becomes saturated with the vapors it is installed to remove, and it must be replaced by fresh

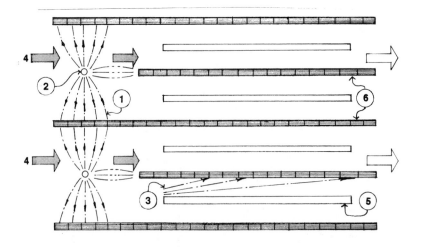

KEY
1   Path Of Ions
2   Wires At High Positive
    Potential
3   Theoretical Paths Of Charged
    Dust Particles
4   Air Flow
5   Intermediate Plates Charged
    To High Positive Potential
6   Alternate Plates Grounded
☐   Clean Air
▨   Contaminated Air

FIGURE 30-5. Electrostatic precipitator: Diagram.

carbon. For large installations, the spent carbon may be returned to the supplier for reactivation and reuse. For small operations, disposal may be more cost-effective.

Activated carbon gives no discernible signal when it reaches saturation. The air coming out of the carbon bed can be monitored for unacceptable gas penetration, or the advice of the supplier can be depended upon for selecting periodic intervals for routine replacement.

CHAPTER 31

LABORATORY HOODS AND OTHER EXHAUST AIR CONTAMINANT-CAPTURE FACILITIES AND EQUIPMENT

31.1 GENERAL COMMENTS

Frequent references have been made throughout the text to laboratory hoods of various designs and to alternative exhaust systems that provide control of toxic, corrosive, combustible, and reactive gases, vapors, and aerosols originating from the materials being used and the processes employed. Assembling information on each of the options into a single section should help make those having design responsibility for these items better informed and thereby improve selection of this class of equipment. Because most of this equipment is built-in during construction or renovation, selection errors tend to become chronic sources of complaint after the facility is put into service and unusually costly to remedy. Therefore, thorough familiarity with laboratory hoods and all manner of laboratory contamination-control exhaust systems is essential for everyone involved in the laboratory design process. It cannot be overemphasized that laboratory workers are trained from their earliest introduction to experimental science to regard their laboratory hood as their principal, and all-purpose, safety device. Therefore, they tend, unquestioningly, to accept the efficacy of all such facilities. When the safety devices do not perform their intended function adequately, serious harm is likely to result. The laboratory designer's responsibility for providing facilities adequate to safeguard laboratory workers' health and safety is a heavy

Guidelines for Laboratory Design: Health and Safety Considerations, Second Edition
By Louis DiBerardinis, Janet S. Baum, Gari T. Gatwood, Anand K. Seth, Melvin W. First, and Edward F. Groden.
ISBN 0-471-55463-4   Copyright © 1993 by John Wiley & Sons, Inc.

burden that cannot be delegated or evaded. The sections that follow describe each of the exhaust ventilation contamination-control devices that are commonly found in laboratories. In spite of significant differences in design, they perform an equivalent function. Therefore, selection is generally guided by tradition, cost, energy consumption, and similar factors, each of which may receive a more or less important rating by different laboratory users, designers, and owners. Regardless of the selection criteria used, it should be kept in mind that correct selection of blowers and duct construction materials and methods is essential for a full measure of safety and user satisfaction. The latter topics are treated in other sections: Chapter 29 for exhaust fans and Chapter 32 for exhaust ducts and accessories.

31.2   CONVENTIONAL BYPASS CHEMICAL FUME HOODS

31.2.1   Introduction

Bypass chemical fume hoods constitute the most widely used type of laboratory hood and are familiar to all laboratory workers. It is an improved version of an even older type that used to be referred to as a *fume cupboard* (a term still used in Britain), and, true to its name, the fume cupboard was an open-faced wood and glass, boxlike enclosure connected to an exhaust blower, usually located at the top rear. Many important modifications have been made to the basic design to improve airflow characteristics for the purpose of achieving greater fume-capturing efficiency and reducing energy requirements.

The most important modification to the old-fashioned fume cupboard was the addition of a vertical sliding sash covering the open front. The sliding sash made it possible to expose the entire open front for full access but also made it possible to lower the sash to provide a safety shield between experiment and experimenter. Lowering the sash during dangerous experiments also decreased the open-face area; this served to increase the inflow velocity through the remaining opening, thereby providing greater protection for the worker against an outflow of contaminants. However, when the sliding sash was lowered beyond a certain point, the inflow velocity (for the same air volume required to exhaust ventilate the hood when the sash was in the full open position) became so great that air turbulence effects at the edges of the opening caused backflow inside the hood, decreasing safety rather than enhancing it. In addition, excessively high velocities sweeping across the floor of the hood disrupted experiments by extinguishing burners, scattering powders, and cooling constant-temperature flasks.

31.2.2   Important Features of Bypass Chemical Fume Hoods

To correct the several deficiencies of the old fume cupboard, the following features were developed and are found in modern, efficient bypass chemical fume hoods:

- Bottom and side airfoils around the open face to produce turbulence free airflow into the hood.
- A mechanism to minimize excessive velocities (300 fpm) when the total opening is 6 in. or less. This may be accomplished by an air bypass or by switching the fan to a lower speed. A typical bypass hood is illustrated in Figure 31-1.

Other important characteristics of acceptable bypass chemical fume hoods are the following:

- The sash will be constructed of shatterproof material.
- The material of construction of the hood will be resistant to damage by the materials to be used in the hood.
- Non-asbestos-containing materials will be used.
- The hood will provide adequate containment. This can be evaluated by setting off a 30-s smoke candle or other heavy smoke generator inside the hood with the sashes in the fully open and then in the fully closed position. The smoke will be totally and quickly removed and there will be no reverse flow out of the hood. Refer to Section 31.10 for other details on evaluating hood performance.
- There will be an inward airflow across the entire opening. Reverse flow can be detected by passing a smoke stick or equivalent smoke generator across the entire parameter of the face opening and looking for flow out of the hood.
- No velocity measurement across the hood face with the sashes positioned to provide the maximum opening will be less than 80 fpm or greater than 120 fpm.
- The airflow through the hood will provide an average face velocity of 100 fpm at the maximum face opening.
- The airflow through the system will be monitored by an in-line measurement device (Pitot tube, orifice meter) or by static pressure measurements in the duct just downstream of the hood and calibrated to the specific hood system. The dial of the measurement instrument will be visible to the hood user.
- During renovations, when existing fume hoods are to remain, an attempt to modify them should be made, if necessary. This would include providing a bottom airfoil (if missing) and providing some means of controlling face velocity at all sash positions by means of a bypass or air volume control and providing an airflow monitoring system.

31.2.3 Model Hood Specification

To further illustrate the important characteristics of an acceptable bypass chemical fume hood, the *Laboratory Fume Hood Specifications and Performance Testing Requirements,* published by the Environmental Medical Ser-

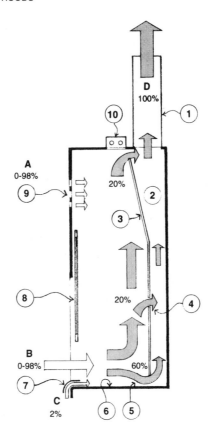

KEY

1 Exhaust Duct
2 Exhaust Plenum
3 Rear Baffle
4 Fixed Center Slot
5 Adjustable Bottom Slot
6 Work Surface
7 Airfoil Sill
8 Vertical Sliding Sash
9 Room Air Bypass, Does Not Open Until Sash Is 75% Closed.
10 Lights
☐ Clean Air
▨ Contaminated Air

AIR FORMULAS

A + B + C = D
B + C = D (Sash Fully Open)
A + C = D (Sash Fully Closed)

FIGURE 31-1A. Bypass hood: Section. (Approximate percentages shown for descriptive purposes only.)

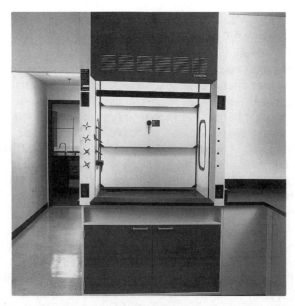

FIGURE 31-1B. Bypass hood. *Source:* Hamilton Industries Inc., Two Rivers, Wisconsin.

vice at the Massachusetts Institute of Technology (MIT, 1991), have been adapted as a model specification and are provided for use in developing a specification for fume hood purchase by the reader.

31.2.3.1 Standard Bypass Fume Hood

1. All fume hoods shall be of the airfoil design, with foils at the bottom and along both vertical sides of the face opening. The bottom airfoil will be raised approximately 1 to 1.5 in. to allow air to pass under the foil and across the work surface and to serve as the terminus for vertical sliding sash(es). The vertical foils are to be flush with the hood interior surface to minimize turbulence as air enters the hood.

2. The superstructure construction for all hoods shall be counter-mounted and not more than 65 in. high, 36 in. deep, outside dimensions, and of length indicated. Interior clear working height should not be less than 47 in. above work surface for full depth of hood from interior of lintel panel to face of baffle plenum. Double side-wall construction is to enclose all structural reinforcements, sash balance mechanisms, and mechanical connections for service outlets and controls (as indicated) and shall be of airfoil type not more than 4 in. wide. Access should be provided for the inspection–maintenance of the sliding sash operating mechanism and also for the installation of mechanical services.

3. The materials for the hood superstructure (unless otherwise specified) shall be $\frac{1}{4}$-in.-thick, nonflammable, acid-resistant material for the hood interi-

ors and laboratory steel, factory painted (baked enamel) for the exterior. The color is to be selected by the owner or architect from the manufacturer's standard colors (unless otherwise specified).

4. All hoods shall be of the "bypass" type. The bypass shall be located above the hood face opening and just forward of the sash when raised. All air exhausted must pass through the work chamber. The bypass must provide an effective line-of-sight barrier between the area outside the hood and the hood interior and must also provide an effective barrier capable of controlling transfer of flying debris during an explosion within the hood. It must assure essentially a constant volume of air at all sash positions. The bypass shall control the increase in face velocity as the sash is lowered to at least twice the design velocity but not more than three times the design velocity.

5. All hoods shall have a vertical sliding glass sash of 7/32-in.-thick combination safety sheet with metal frame to operate on stainless steel cables over the ballbearing pulleys with metal counterbalancing weights. (*Note:* Spring sash balances are not acceptable.) When in the closed position, the sash shall rest on top of the bottom airfoil; and when fully raised the height of the open face must be at least 32 in. from the top of the work surface. The sash must operate freely. Install a recessed finger grip(s) or drawer pulls (2 per sash) for raising and lowering sash.

6. The bottom airfoil shall be fabricated from 316 stainless steel.

7. All hoods shall have a removable baffle with two slots: one upper and one lower, adjustable, with hand-operated plastic adjusting knobs for the adjustment of the airflow. A protective stainless steel screen with the equivalent free area of the total bottom slot opening shall be provided for the bottom slot. This protective screen shall not impede the use of the cup sink.

8. All hoods shall have an exhaust plenum chamber and be equipped with a hood outlet opening sized for approximately 1500 fpm at design flow. This opening shall be located to the rear of the back baffle and centered in top of the exhaust plenum unless otherwise specified. The opening should be equipped with a 316 stainless steel duct stub extending at least 1 in. above the top of the hood.

9. Hood work surface shall be type 316 stainless steel and shall be of the recessed (dished) type with a 3/8-in. raised lip along all four edges (to retain spilled liquids) and a uniform edge thickness of 1 1/4 in. The work surface shall contain an integrally welded type 316 stainless steel 3- × 6-in. cup sink located near one of the rear corners unless otherwise specified, so that it is not obstructed by the protective screen at the bottom slot and will also receive the unobstructed discharge stream from the cold water outlet.

If the storage cabinet is to be vented, the raised surface of the work top, above the recessed area, shall contain one or two holes to receive the $1\frac{1}{2}$-in. I.D. vent pipe(s) from the acid storage cabinet(s) supporting the hood. The vent pipe can be constructed of lead, PVC, or another appropriate material compatible with the materials used in the hood. Vent pipe holes shall be located between the rear baffle and back panel of the hood, in the raised portion of the work top. The raised surface shall be provided all around the recessed pan area, and it shall be 2 to 4 in. wide across the front edge. The vent pipe should extend up behind the rear baffle.

10. At the top and front of each hood, install a pocket enclosure to receive the vertical sliding sash when in the up position. Each pocket enclosure may contain two removable access panels, one each on the front and rear faces, for access to the electrical junction box and lighting fixture (relamping and cleaning). Provide a removable end panel on each side of the hood, extending from the sash enclosure to the rear of the hood, for access to service fittings and service lines. The panels must be removable from the exterior of the hood, and re-installable from the exterior, without interference with or removal of any drop ceiling, furring from the top of the hood to the ceiling, or any adjoining end curbs of work tops.

11. All fume hoods shall be equipped with $\frac{1}{4}$-in.-thick, nonflammable, acid-resistant flush, removable access panels(s) on the side(s) of hood interior of sufficient size required for the installation of the various mechanical services.

12. The hoods shall be exhausted so as to maintain an average velocity of 100 fpm through the full open face (sash fully raised). The velocity must also be uniform and not vary more than 10 fpm from the average.

13. When two-speed exhaust fan systems are used, a microswitch to control the exhaust air fan motor shall be actuated by hood sash only. The microswitch shall be so located that when the exhaust is turned to low speed the design volume of air is still exhausted from the room while maintaining a average face velocity of 100 fpm through the reduced face area. When variable-air-volume hoods are used, an average face velocity of 100 fpm must be maintained as the face area is varied. A minimum total exhaust volume must be established for each hood. From this one can obtain the sash height below which the total exhaust volume will not change. A bypass will then need to be activated at this point (see Section 31.6).

14. Unless otherwise noted on drawing or job specification, piping from the hood service outlets, except for the drain and vent lines, shall be installed between the double-wall side walls and extend up and be connected to the respective services above the hood.

15. All hoods shall have a fluorescent lighting fixture operated by an exterior switch mounted on exterior face of one of the vertical foils. The fixture shall have two 40-W, fluorescent, rapid-start lamps. The fixture shall be hinged on one side and be accessible for relamping, cleaning, or other maintenance work from the exterior of the work chamber. Mount the fixture at top of hood, setting it on a fixed and gasketed $\frac{1}{4}$-in.-thick safety glass shield.

16. All hoods shall have one or more 15-A duplex grounded receptacle(s)

mounted on the exterior face(s) of the vertical foil(s). The opening and/or housing for the receptacle should have a built-in through-flow ground fault interruptor (GFI).

Note. The wiring within the hood shall be so installed that when a through-flow GFI is installed it also gives GFI protection to the lighting switch and lighting fixture. When a hood has more than one duplex receptacle, only one through-flow GFI is required, which is to give protection to any other receptacle in the hood.

17. The lighting fixture with its operating switch and the electrical receptacle(s) for each hood shall be on the same circuit. The wiring for these electrical items shall connect into a single 4- ×4-in. junction box so located above the hood structure that it is always easily accessible no matter in what position the hood is placed.

31.2.3.2 Hood Support Cabinets

1. Base cabinet for hood support shall be constructed and finished in a manner similar to hood.

2. Provide acid storage cabinets as hood support cabinet. Interior of cabinet shall be completely lined with $\frac{1}{4}$-in.-thick nonflammable, acid-resistant material including the interior sides of cabinet doors.

3. Each base cabinet shall be vented if used for storage of toxic materials. Each indivdual door shall have an air intake louver located at center and bottom of door. Each base unit shall be exhausted by means of a 1 1/2-in.-I.D. pipe vent that extends from the center and top of the rear wall (unless otherwise noted) of the storage compartment itself and up through the 2-in.-I.D. hole in the raised portion of the work surface (refer to Section 31.2.3.1, item 9) and into the hood (behind the rear baffle plate) and terminating a minimum of 2 in. above the work surface.

4. Provide a 12-in.-deep, full-width, noncombustible, and adjustable shelf in each base unit. The shelf must be of sufficient strength or depth so that excessive deflection does not occur when it is fully loaded. Provide a 2-in.-deep, liquid-tight pan which covers the entire bottom cabinet.

31.2.3.3 Shelves Inside Hoods

There has been a recent trend by laboratory hood users to install storage shelves inside the hood. While this is not encouraged, if it is allowed it should be done carefully. The shelves should be positioned so as not to effect airflow within the hood. The best location is at the rear of the hood interior, above the bottom slot and below the middle slot. The slots should not be blocked by any storage. Shelves on the side of the hood may adversely effect airflow and should be avoided. Ideally a performance test (Section 31.10) should be conducted to determine the effect of the shelf on hood containment.

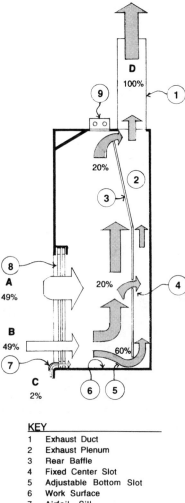

KEY

1	Exhaust Duct
2	Exhaust Plenum
3	Rear Baffle
4	Fixed Center Slot
5	Adjustable Bottom Slot
6	Work Surface
7	Airfoil Sill
8	Horizontal Sliding Sash
9	Lights
☐	Clean Air
▨	Contaminated Air

AIR FORMULAS

A + B + C = D

FIGURE 31-2A. Bypass hood with horizontal sliding sash: Section. (Approximate percentages shown for descriptive purposes only.)

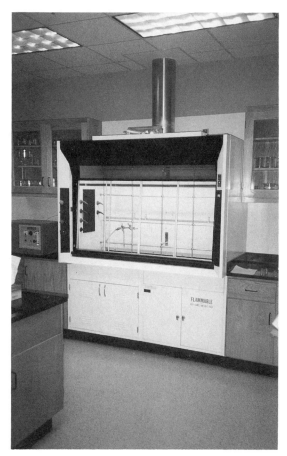

FIGURE 31-2B. Bypass hood with horizontal sliding sash. *Source:* Kewaunee Scientific Corporation, Brick, New Jersey.

31.2.4   Horizontal Sliding Sash Option

Although most chemical fume hoods are provided with vertical sliding sash, it is possible to utilize horizontal sliding sash, instead. The principal advantage of a horizontal sliding sash is that only half of the hood face can be left open at any time while giving full access to all parts of the hood interior. With a vertical sliding sash, it is necessary to open the sash to its full height to gain access to the upper parts of the hood for equipment setups and monitoring tall apparatus. A hood with horizontal sliding sash is illustrated in Figure 31-2. With a horizontal sliding sash, the full height of the opening is always available whenever the sash is open, facilitating ready hood access. Additional details of the horizontal sliding sash modifications for the chemical fume hood are contained in Chapter 26, concerned with energy conservation, because use of horizontal sliding sash is an important method for reducing the need for laboratory exhaust air volume.

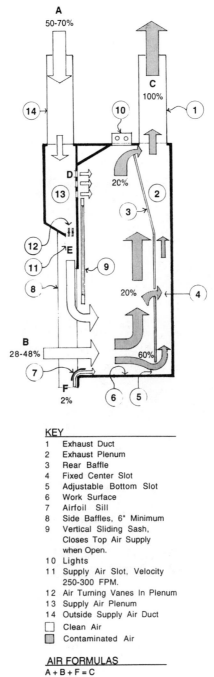

KEY

1	Exhaust Duct
2	Exhaust Plenum
3	Rear Baffle
4	Fixed Center Slot
5	Adjustable Bottom Slot
6	Work Surface
7	Airfoil Sill
8	Side Baffles, 6" Minimum
9	Vertical Sliding Sash, Closes Top Air Supply when Open.
10	Lights
11	Supply Air Slot, Velocity 250-300 FPM.
12	Air Turning Vanes In Plenum
13	Supply Air Plenum
14	Outside Supply Air Duct
☐	Clean Air
▨	Contaminated Air

AIR FORMULAS

$A + B + F = C$
$D + E = A$
$D + F = C$ (Sash Closed)
$B + E + F = C$ (Sash Open)
$A = E$ (Sash Open)

FIGURE 31-3A. Auxiliary air chemical fume hood: Section.

FIGURE 31-3B. Auxiliary air chemical fume hood. *Source:* Kewaunee Scientific Corporation, Brick, New Jersey.

31.3 AUXILIARY AIR CHEMICAL FUME HOODS

31.3.1 Introduction

Auxiliary air chemical fume hoods differ from bypass chemical fume hoods in only one important respect: A major portion of the air exhausted from the hood is provided from a supply air diffuser canopy attached to the hood just above the hood face. An auxiliary air chemical fume hood is shown in Figure

31-3. The purpose of this modification of a conventional bypass chemical fume hood (Section 31.2) is to reduce the demand for fully conditioned makeup air for hood service (see Chapter 26 on energy conservation). This is accomplished during cold weather by heating the auxiliary air to a lower temperature than the room HVAC supply air, and it is accomplished during hot weather by not cooling or dehumidifying the auxiliary supply air. The rationale is that the hood user ordinarily spends little time working in front of the hood and, in any event, will not be seriously discomforted by unconditioned air flowing from above into the open hood face. In hot climates, especially, the savings that can be realized by not cooling and dehumidifying all the hood exhaust air can be substantial. However, this type of chemical fume hood requires an auxiliary air supply system for each hood that is generally separate from the room HVAC supply air system, thereby increasing the number and complexity of the HVAC services to the laboratory.

31.3.2 Model Hood Specification

The requirements for auxiliary air chemical fume hoods that differ from, or are supplementary to, those outlined in Section 31.2 are the following by way of illustration. They are adapted from a hood specification schedule used by the Environmental Medical Services of the Massachusetts Institute of Technology for hood procurement (MIT, 1991):

1. All of the performance and construction requirements for bypass chemical fume hoods must be met. (Refer to Section 31.10 for performance tests.)

2. The supply plenum shall be so located that all of the auxiliary air shall be supplied exterior to the hood face and no leakage or passage of auxiliary air behind the sash will be allowed until the sash is lowered to the point of bypass opening. This system shall be capable of supplying up to 70% of the volume of the hood exhaust.

3. The supply air shall be tempered to 60–90°F during the ''winter'' season.

4. The capture efficiency of the supply air by the hood exhaust system shall be at least 95% within the supply air temperature range of 60–90°F.

5. The fume hood leakage within the above operating range shall not exceed 0.003% as determined by performance tests.

31.4 PERCHLORIC ACID FUME HOODS

Perchloric acid hoods are special because perchloric acid volatilizes and can condense on and react with organic materials used in the hood or duct construction. Perchlorates are thus formed which may be explosive. Perchloric acid is a powerful oxidizer and a high-boiling-point chemical that undergoes spontaneous and explosive decomposition. The same is true of many perchlorate salts. In conventional hood exhaust ducts, volatilized perchloric acid cools, condenses, and deposits in horizontal runs, creating, in time, a severe explosion hazard. To avoid deposition and accumulation of perchloric acid in hood exhaust ducts, perchloric acid hood systems have been developed. They feature (a) dedicated duct systems that have few, if any, horizontal runs and (b) no dead areas that can capture and accumulate condensed perchloric acid; furthermore, they are constructed of materials such as stainless steel with welded seams or neoprene gaskets at joints that do not degrade under perchloric acid attack. In addition, they should be located as close to the outside of the building as possible and have stainless steel fans.

The additional special characteristics of perchloric acid hood systems are that the interior of the hood is also constructed of materials that are not degraded by perchloric acid and that all of the exhaust ducts are continually water-washed into a sewer or sump to remove condensed perchloric acid and prevent accumulation of the acid or its salts. The water wash should be adjusted for pH before disposal to the sewer. Perchloric acid hoods must meet the same performance requirements as those outlined in Sections 31.2 and 31.3 for bypass and auxiliary air chemical fume hoods.

31.5 HOODS FOR WORK WITH RADIOACTIVE MATERIALS

Some radioactive materials, such as the long-lived emitters (for example, plutonium), have exceptionally low permissible exposure limits; consequently, hoods designated as radioactive hoods are usually operated at somewhat higher face velocities than are chemical fume hoods. Many users of radioisotopes recommend 150-fpm average face velocity for hoods, but most industrial hygiene engineers believe that this face velocity is too high because it promotes excessive air turbulence at the edges of the working opening. Instead, they recommend a maximum average face velocity of 125 fpm. If velocities of 125 to 150 fpm are used, the work procedures followed should be designed to avoid the hood leakage discussed in Section 31.2.3. These include working further into the hood and keeping the back hood slots unblocked. Whether 125- or 150-fpm face velocity is specified, the hoods that are purchased for service with radioactive materials should include all the features outlined in Sections 31.2 and 31.3. In addition, radioactive hoods should be equipped with 316, 18-gauge stainless steel ducts. All the interior hood surfaces should be constructed of a smooth, cleanable, nonporous material such as stainless steel.

Provisions should be made for HEPA filters and/or activated charcoal adsorbers to be installed at the hood air outlet when required by NRC regulations (NRC, 1991), and the fan should be selected to handle the increased static pressure produced by the air filtration system. Finally, provisions should be made to install a continuous effluent air radioactivity monitoring system when this is required by NRC regulations.

FIGURE 31-4. Minihood for radioactive material. *Source:* Atlantic Nuclear, Canton, Massachusetts.

In some cases it may be possible to install a "minihood" or enclosure inside of a fume hood that has an air cleaner and blower integral to it that discharges air into the fume hood. This hood can be used for small amounts of liquids or solids that are radiolabeled (Figure 31-4). A health physicist should be consulted before such a system is employed.

31.6 VARIABLE-AIR-VOLUME HOODS

This type of hood and its associated systems are a new and unique category of contaminant control system and are discussed in detail in Chapter 33. The general concept involves varying the quantity of air exhausted and supplied to a space based on the activity and needs of the space at any given time.

The selection of this type of hood must be made carefully, weighing its advantages and disadvantages. The significant advantage is that it can reduce the total quantity of supply and exhaust air to a space when not needed, thereby reducing total operating costs. The major disadvantages are that they require more sophisticated controls that have not been available for a long enough time period to be fully evaluated for their reliability and they require a different type of maintenance that traditional maintenance operators are not used to providing. Conditions for their selection and safe use are discussed in Chapter 33.

31.7 GLOVE BOXES

When the toxicity, radioactivity level, or oxygen reactivity of the substances under study is too great to permit safe operation in a chemical fume hood, resort must be made to a totally enclosed, controlled-atmosphere glove box. The special feature of a glove box, as the name suggests, is the total isolation of the interior of the box from the surrounding environment and the consequent need to manipulate items inside the box by means of full-length gloves sealed into a sidewall of the box. To prevent loss of materials from the inside of the glove box to the laboratory, the box is maintained under substantial negative pressure (0.25 to 0.50 in. w.g.) relative to the laboratory by means of an exhaust blower connected to the box interior. The atmosphere inside the box may be maintained sterile and dust-free by use of a constant air in-leakage through a HEPA filter. As shown in Figure 31-5, a diagram and photograph of a typical glove box, the interior may be further isolated from the laboratory environment by the use of air locks for passing items into and out of the box.

Stainless steel and safety glass are the preferred materials of construction for the interior of the box, to facilitate cleaning and decontamination. The interior should be finished smooth and free of sharp edges that could damage the gloves. All controls should be located outside the box for safety and ease of manipulation, as shown in Figure 31-5. When highly toxic or infectious

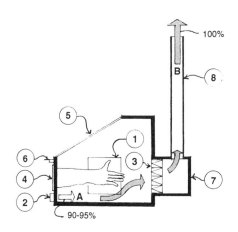

KEY

1	Air Lock Pass-Through
2	Constant Air In Leakage HEPA Filter
3	Roughing Filter
4	Glove Ports
5	Glass Window (Sealed)
6	Controls
7	Exhaust Blower
8	Exhaust Duct To Final Air Cleaning Filter.
☐	Clean Air
▨	Contaminated Air

AIR FORMULAS
A + Leakage = B

FIGURE 31-5A. Glove box: Section. (Approximate percentages shown for descriptive purposes only.)

FIGURE 31-5B. Glove box. *Source:* The Baker Company, Sanford, Maine.

materials are being used in the glove box, the exhaust air should be cleaned in two or more stages. A prefilter should be used to remove the major load of coarse particulate matter, followed by a HEPA filter with a minimum efficiency of 99.97% for 0.3-μm test aerosol particles. When highly toxic volatile chemicals are used inside the box, an activated charcoal adsorber stage may be added to the air cleaning train. If no toxic aerosols are present, the HEPA filter may be omitted but it is advisable to retain the prefilter to protect the activated charcoal from dirt. The specific size of the air-cleaning components will depend on the glove box size and the chemicals used. The blower must be selected to overcome the flow resistance of these added elements.

31.8 BIOLOGICAL SAFETY CABINETS

Three classes of biological safety cabinets are recognized (NSF, 1992):

Class I. A ventilated cabinet for personnel and environmental protection, with an unrecirculated inward airflow away from the operator. The cabinet exhaust air is treated to protect the environment before it is discharged to the outside atmosphere. This cabinet resembles a chemical fume hood with a filtered exhaust and is suitable for work with low- and moderate-risk biological agents where no product protection is required.

Class II. A ventilated cabinet for personnel, product, and environmental protection having (a) an open front with inward airflow for personnel protection, (b) downward HEPA-filtered, laminar airflow for product protection, and (c) a HEPA-filtered exhaust air for environmental protection. Class II cabinets are suitable for low- and moderate-risk biological agents.

Class III. A totally enclosed, ventilated cabinet of gas-tight construction. Operations in the cabinet are conducted through attached rubber gloves. The cabinet is maintained under negative air pressure of at least 0.5 in. w.g. Supply air is drawn into the cabinet through HEPA filters. The exhaust air is treated by double HEPA filtration, or by HEPA filtration and incineration. Class III cabinets are suitable for high-risk biological agents and are accompanied by much auxiliary safety equipment. There are only a handful of such facilities worldwide. They must be designed, installed, and certified by experienced biological safety professionals.

Class II biological safety cabinets are widely used and are available in four recognized types:

Type A. Cabinets (1) maintain a minimum calculated average inflow velocity of 75 fpm through the work area access opening, (2) have HEPA-filtered downflow air from a common plenum (i.e., a plenum from which a portion of the air is exhausted from the cabinet and the remainder supplied to the work area), (3) may exhaust HEPA-filtered air back into the laboratory, and (4) may have positive-pressure contaminated ducts and plenums. Type A cabinets are suitable for work with low to moderate risk biological agents in the absence of volatile toxic chemicals and volatile radionuclides.

Type B1. Cabinets (1) maintain a minimum (calculated or measured) average inflow velocity of 100 fpm through the work area access opening, (2) have HEPA-filtered downflow air composed largely of uncontaminated recirculated inflow air (3) exhaust most of the contaminated downflow air through a dedicated duct exhausted to the outdoor atmosphere after passing through a HEPA filter, and (4) have all biologically contaminated ducts and plenums under negative pressure, or surrounded by negative-pressure ducts and plenums. Type B1 cabinets are suitable for work with low- to moderate-risk biological agents. They may also be used with biological agents treated with minute quantities of toxic chemicals and trace amounts of radionuclides required as an adjunct to microbiological studies if work is done in the direct-exhausted portion of the cabinet or if the chemicals or radionuclides will not interfere with the work when recirculated in the downflow air.

Type B2. Cabinets (sometimes referred to as "total exhaust") (1) maintain a minimum (calculated or measured) average inflow velocity of 100 fpm through the work-area access opening, (2) have HEPA-filtered downflow air drawn from the laboratory or the outside air (i.e., downflow air is not recirculated from the cabinet exhaust air), (3) exhaust all inflow and downflow air to the outdoor atmosphere after filtration through a

HEPA filter without recirculation in the cabinet or return to the laboratory room air, and (4) have all contaminated ducts and plenums under negative pressure or surrounded by directly exhausted (nonrecirculated through the work area) negative pressure ducts and plenums. Type B2 cabinets are suitable for work with low to moderate risk biological agents. They may also be used with biological agents treated with toxic chemicals and radionuclides required as an adjunct to microbiological studies.

Type B3. Cabinets (1) maintain a minimum (calculated or measured) average inflow velocity of 100 fpm through the work-area access opening, (2) have HEPA-filtered downflow air that is a portion of the mixed downflow and inflow air from a common exhaust plenum, (3) discharge all exhaust air to the outdoor atmosphere after HEPA filtration, and (4) have all biologically contaminated ducts and plenums under negative pressure or surrounded by negative-pressure ducts and plenums. Type B3 cabinets are suitable for work with low to moderate risk biological agents treated with minute quantities of toxic chemicals and trace quantities of radionuclides that will not interfere with the work if recirculated in the downflow air.

The above descriptions were taken from National Sanitation Foundation Standard No. 49 (NSF, 1992). Standard No. 49 includes basic requirements for construction and certification testing of all Class II biological safety cabinets. The appearance of the words "laminar flow" in the title of the standard makes it necessary to caution against confusing Class II biological safety cabinets with laminar flow workbenches because, although the latter devices provide work protection, they fail to provide personnel or environmental protection, and therefore should never be used with toxic, infectious, or otherwise hazardous materials. Figures 31-6 and 31-7 identify the important parts and show the airflow patterns for Type A and Type B1 cabinets, the two most widely used types.

The heart of every cabinet is the HEPA filter, a pleated paper filter contained inside a rectangular wooden frame. The filter is clamped tightly to a specially prepared flange built into the cabinet frame, and a compressible gasket on the face of the wooden filter frame makes the required airtight seal. Threaded clamps are usually used, but some manufacturers use spring-loaded clamps.

Most cabinets use direct-drive blowers with forward curved impellers driven by permanent split-capacitor motors and controlled by a Triac speed controller. The Triac is a solid-state device that can be adjusted to vary the ac voltage at the output from essentially zero to full line voltage. The permanent split-capacitor motor is a brushless electric motor of fractional horsepower size. It requires an external capacitor that remains in the circuit during starting and while running. It is the largest ac motor that will run at variable speed, reducing rpm and horsepower as input voltage is reduced.

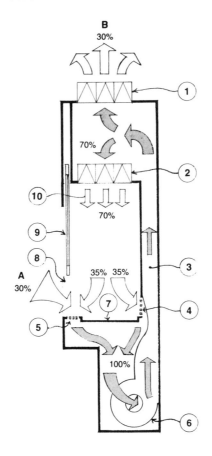

KEY
1 Exhaust Air HEPA Filter
2 Supply Air HEPA Filter
3 Exhaust Plenum
4 Rear Exhaust Grille
5 Front Exhaust Grille
6 Fan (1 Per Unit)
7 Work Surface
8 Front Opening
9 Vertical Sliding Sash
10 Filtered Recycled Air
11 Lights
☐ Clean Air
▨ Contaminated Air

AIR FORMULAS
A = B

FIGURE 31-6A. Class II Type A biological safety cabinet: Section. (Approximate percentages shown for descriptive purposes only.)

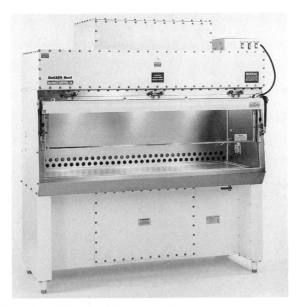

FIGURE 31-6B. Class II Type A biological safety cabinet. *Source:* The Baker Company, Sanford, Maine.

The forward curved blower impeller has a performance characteristic that prevents it from overloading the motor as the filters increase in resistance because the torque required to turn this blower increases rapidly with increasing airflow but only very little with increasing pressure. Therefore, as filter resistance rises, airflow tends to drop and the load on the motor to decrease, but this causes the motor to speed up and airflow remains close to its original value. Starting with new filters in a cabinet, airflow will not decline more than 10% even after filter pressure drop has increased 50%. But as filters increase above 50% of new resistance, the amount of pressure increase needed to produce a 10% drop in airflow decreases and more frequent manual adjustments are needed.

Use of UV lamps in biological safety cabinets was discontinued in the 1983 revision of NSF Standard No. 49 because of their limited effectiveness for decontaminating cabinets. Correct procedure, when using a biological safety cabinet, calls for washing down the work area with a suitable disinfectant upon completion of work. Without a good washdown, remaining soil is likely to shield organisms from the UV light, whereas with a good washdown, further disinfection of the cabinet surfaces is generally not needed, especially if the blower is left running continuously.

All cabinets contain electrical utility outlets. Some are useful only for powering small appliances, whereas others have full 15-A capacity. All units should have ground fault interrupters in the electrical utility outlet line. They measure leakage to ground and automatically cut off power when leakage exceeds 5 mA. The proper functioning of the ground fault interrupter must be verified for all units whenever cabinet certification tests are performed.

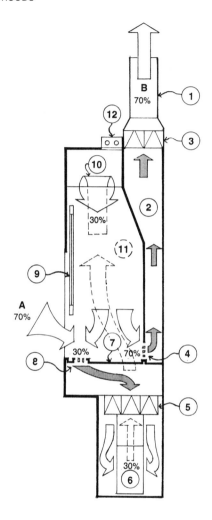

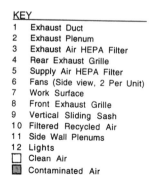

KEY

1	Exhaust Duct
2	Exhaust Plenum
3	Exhaust Air HEPA Filter
4	Rear Exhaust Grille
5	Supply Air HEPA Filter
6	Fans (Side view, 2 Per Unit)
7	Work Surface
8	Front Exhaust Grille
9	Vertical Sliding Sash
10	Filtered Recycled Air
11	Side Wall Plenums
12	Lights
☐	Clean Air
▨	Contaminated Air

FIGURE 31-7A. NCI-1 Type B biological safety cabinet. (Approximate percentages shown for descriptive purposes only.)

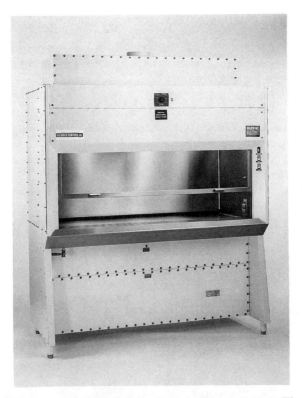

FIGURE 31-7B. NCI-1 Type B biological safety cabinet. *Source:* The Baker Company, Sanford, Maine.

Prior to sale, cabinets are submitted for certification to the National Sanitation Foundation by the manufacturers. Those that have passed the personnel, environmental, and product safety tests can be identified by a distinctive NSF medallion placed on the exterior of the cabinet. Field recertification by a competent technician is needed when a cabinet is first installed, at annual intervals thereafter, and whenever a cabinet is moved to a new location or is serviced internally. It is standard safety practice to sterilize the entire internal structure of a working cabinet with formaldehyde gas each time it is necessary to service interior parts. NSF Standard No. 49 contains detailed instructions for performing the field certification tests and for formaldehyde sterilization.

31.9 CAPTURE HOODS

The most efficient and cost-effective form of contaminant control is local exhaust ventilation (LEV). This involves capture of the chemical contaminant at its source of generation. The laboratory chemical fume hood (Section

31.2) is a specialized form of capture hood that totally encloses the emission source. Often, total enclosure of the source is not possible, or is not necessary. A capture hood controls the release of toxic materials into the laboratory by capturing or entraining them at or close to the source of generation, usually a work station or laboratory operation. Considerably less air volume is required than for the standard chemical fume hood.

To work effectively, the air inlet of a capture hood must be placed near the point of chemical or biological release. The distance away will depend on the size and shape of the hood and the velocity of air at the intake slot or face, but should usually be not more than 12 in. from the generation source. Design face or slot velocities are typically in the range of 500 to 2000 fpm. Many design guidelines exist for this class of exhaust hoods in Chapter 3 of the *Industrial Ventilation Manual* (ACGIH, 1992). Two types of capture hoods that find frequent application in the laboratory are "canopy" hoods and "slot" hoods.

FIGURE 31-8. Canopy hood.

Canopy hoods are used primarily for capture of gases, vapors, and aerosols released from permanent laboratory equipment. Examples include ovens, gas chromatographs, autoclaves, and atomic absorption spectrophotometers. They are usually found in analytical and biological laboratories and some pilot plant operations. This type of hood is generally used when the process to be controlled is at elevated temperatures or the emissions are directed upward. Many equipment manufacturers recommend specific capture hood configurations that are suitable for their units. An example can be seen in Figure 31-8.

Slot hoods are used for control of laboratory bench operations that cannot be performed inside a containment hood (chemical fume hood) or under a canopy hood. Laboratories that may find use for slot ventilation include clinical (histology pathology), anatomy, teaching, and general chemistry. Typical operations include slide preparation, microscopy, biological specimen preparations, mixing, and weighing operations. An example is shown in Figure 31-9.

FIGURE 31-9. Slot hood.

Another laboratory operation that needs special consideration is the weighing of highly toxic materials. Because of the nature of the materials used, it is advantageous to do this in a ventilated enclosure. A chemical fume hood is not the best location because its relatively high airflow may disrupt the weighing process and it is not an efficient use of ventilated space. A ventilated weighing station has been developed as part of the National Toxicology Program (Hoyle, 1987) and is shown in Figure 31-10.

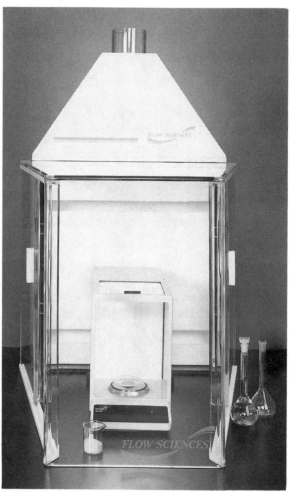

FIGURE 31-10. Weighing station. *Source:* Flow Sciences Inc. Wilmington, North Carolina.

31.10 PERFORMANCE TESTS

31.10.1   Introduction

From the initial selection of a hood to its continuing use in the laboratory, the owner will need some method of evaluating its performance. The method involves selecting the conditions under which the hood will be tested, choosing an appropriate challenge test, and developing acceptance criteria that will adequately determine if the hood meets the user's protection needs. There are a variety of performance tests currently used. We will discuss several, indicating their appropriateness for specific applications and their advantages and disadvantages. Each user must determine which test will accurately determine if a specific hood will provide the needed protection.

Factors to consider when choosing a performance test include (1) reason for testing, (2) type and quantity of chemicals or biological agents to be used in hood, (3) types of operations and equipment to be used in hood, (4) number and type of users, (5) diversification of hood use, both in the short term (months) and long term (years), (6) location of hood within the facility, (7) type of hood (conventional or auxiliary air), and (8) ease of performance of test.

The conditions under which one would test a hood's performance characteristics normally fall into two categories: (1) selection of the type of hood to be purchased and (2) evaluation of hoods in use within the facility. The first category involves testing in a controlled, laboratory-type setting, whereas the second involves an evaluation in the laboratory under practical use conditions. Both will be discussed here.

Performance tests involve measurement of the hood flow characteristics (face velocity and air quantity) and the efficiency of the hood in containing an artificial challenge gas or aerosol generated within the hood.

31.10.2   Tests for Selection of a Hood

Prior to purchasing one or more hoods, the user must determine under what conditions the hood may be used, what type of hood to select (bypass vs. auxiliary air), and what protection factor is needed, for example, what is the maximum allowable loss of containment (leakage) that is acceptable. (Refer to Section 2.3.4.4 for assistance in defining hood use and types.)

Considerable research on performance tests has been conducted since 1970, and several test protocols have been proposed for adoption as a standard. They involve a variety of challenge chemicals, methods of challenge, and criteria for acceptance. Some of the proposed challenge chemicals are uranine dye (Chamberlin, 1978), Freon (ASHRAE, 1985), and sulfur hexafluoride (Chamberlin, 1982). These tests have been employed under various conditions by several organizations. The reader is encouraged to review the above references in detail before making a selection. In addition, the current literature should be searched since this area is receiving considerable research attention.

31.10.3 Field Performance Tests for Fume Hoods after Installation

After hoods have been installed but before their use, they should be tested to verify adequate performance. Generally, this involves measurement of total volume flow and face velocity across the hood opening and comparison to design guidelines (see Section 2.3). Face velocity measurements should be made at 9 to 12 points equally distributed across the opening of the hood (Chamberlin, 1982; BSI, 1979; SAMA, 1975; ACGIH, 1992). In addition, observation of airflow patterns should be made by generating a source of smoke across the face opening. It has also been common practice to conduct these tests at regular intervals throughout the year on operational hoods.

For many years, health and safety professionals have recognized that this procedure may not be the most accurate field assessment of hood performance. More recently, other field techniques have been attempted using some variation of the ASHRAE test referred to in Section 31.10.2 (Fuller, 1979; Mikell, 1981). While these can provide much information after installation and before use, they are difficult to use on a regular basis since they are time-consuming and can intrude on operations performed in the hood. A new method has been developed to provide an easy, quick, and unobtrusive assessment of fume hood performance (Ivany, 1986; DiBerardinis, 1991). This test involves the release of a tracer gas through a diffuser inside the hood under normal operating conditions. Measurements of the tracer gas are made outside the hood to determine hood leakage. Leakage can then be related to chemical exposure standards. This test can be used as an after-installation/before-use acceptance criterion and as a regular interval performance test.

Whichever performance test is selected, it is important to include the requirements in the design documents with criteria for acceptance.

CHAPTER 32

EXHAUST AIR DUCTS
AND ACCESSORIES

Fume hood and capture hood ducts differ from HVAC ducts in that the materials that pass through them are often highly corrosive and toxic. Consideration should be given to the fact that such ducts will have to be serviced or replaced during the life of the average laboratory building. Therefore, safety to personnel making repairs or replacement should not be overlooked. ASHRAE HVAC Applications, Chapter 14, (ASHRAE, 1991) also provides helpful guidelines.

There is no consensus on the best material to use for exhaust duct construction. Table 32-1 provides guidelines regarding chemical compatibility of duct material. The following is a review of a number of code compliance requirements.

- BOCA (1992) specifies steel ducts for the removal of dust and vapors.
- NFPA-90A specifies that ducts shall be constructed of steel, aluminum, or other inert noncombustible materials. In addition, duct materials meeting the requirements of UL181, Classes 0 and 1, are acceptable (UL, 1984). Class 0 materials have a surface-burning characteristic of zero. Class 1 materials have a flame-spread rating not over 25 and a smoke-developed rating not over 50.
- SMACNA (1974) recommends that NFPA-91 and local building codes be consulted for fire resistance, flame-spread ratings, and smoke-generating

Guidelines for Laboratory Design: Health and Safety Considerations, Second Edition
By Louis DiBerardinis, Janet S. Baum, Gari T. Gatwood, Anand K. Seth, Melvin W. First, and Edward F. Groden.
ISBN 0-471-55463-4   Copyright © 1993 by John Wiley & Sons, Inc.

TABLE 32-1. Commonly Used Duct Materials

Material	Limitations of Use
Glazed ceramic pipe	Rarely used today because of installation and sealing difficulties.
Epoxy-coated stainless steel	Extensive experience not yet available but appears promising as a versatile material.
Stainless steel	May be attacked by some chemicals, especially hydrochloric acid. Care should be used in selecting type. Stainless No. 316 is one of the more resistant alloys.
Monel metal	May be attacked by some chemicals, such as halide salts and acids.
Synthetic or cementitious "stones"	Absorb moisture, attacked by strong alkaline chemicals.
Reinforced plastic, principally glass fiber-reinforced polyester (FRP)	Various resins have different chemical and fire resistance. Care should be taken to select chemically resistant resins for final interior layer.
Aluminum	Limited resistance to many chemicals. Care should be used to install only in systems that do not experience corrosive chemical exposure.
Galvanized steel	Limited resistance to corrosion by a wide variety of materials used in research and teaching laboratories. Not recommended.
Black steel	Useful only with dry and noncorrosive dusts.

characteristics. NFPA-91 applies only to transmission of nonflammable fumes and vapors.

Many chemicals encountered in laboratories are flammable. For this reason, many fire departments oppose the use of rigid PVC ductwork due to the possibility of forming toxic fume degradation products in a fire. Transite has come into disfavor as a duct material because of its asbestos content, but it is not clear that transite sheds respirable fibers during normal duct usage. Table 32-1 is a tabulation of commonly used duct materials. We recommend that exhaust ducts serving hoods which exhaust radioactive materials, volatile solvents, and strong oxidizing agents (perchloric acid) be fabricated of stainless steel for a minimum distance of 10 ft from the hood outlet.

In general, high-alloy stainless steels (e.g., 316) and glass-fiber-reinforced polyester (FRP) have proven to be the most satisfactory duct materials for corrosion, impact, and vibration resistance, as well as for ease of fabrication and installation. Rigid polyvinylchloride (PVC) has excellent corrosion resistance but is brittle and, therefore, has inferior impact and vibration-cracking resistance. Because plastic ducts have thicker walls than stainless steel of the same inside diameter, they occupy somewhat more space in duct and utility chases. Rigid PVC is nonflammable by reason of its high chlorine content, but may produce hydrogen chloride gas as a fire degradation product. Polyester resins are free of chloride but can be formulated with phosphate and other fire-retardant additives that make them self-extinguishing, and this type of polyester should be insisted upon for all FRP ducts.

Considerable care must be exercised to evaluate the types of materials that are likely to be used in laboratories prior to making a selection of duct materials. Similar considerations will be a guide to the features that will be provided in the design of the building for ease of servicing and replacing ducts. It should be kept in mind, as well, that laboratory usage may change over a period of time and that considerable conservatism should, therefore, be exercised when selecting chemical fume hood duct materials to avoid inadvertent future failures. High-velocity air movement in ducts is desirable to ensure that solids do not deposit in the joints, cracks, or corners of the duct system. A minimum suggested design velocity is 2000 fpm. Higher conveying velocities (3500–4500 fpm) are usual. A minimum of turns, bends, and other obstructions to airflow is also desirable. Where perchloric acid is to be used, the duct configuration should be free of bends and horizontal runs and permit thorough washdown of all interior duct surfaces.

A typical duct specification is likely to include some or all of the language that follows. "All fume hood and local exhaust ducts shall be constructed of round piping with the interior of all ducts smooth and free of obstructions. All joints shall be welded or epoxy sealed airtight. The use of flexible piping for spot exhaust points shall be kept to minimum lengths and shall be equipped with tightly fitting, easily removable end caps for use when the exhaust point is not in use. Flexible tubing shall be noncollapsible and constructed entirely of metal or of a wire coil covered with multiple plies of flame-proofed, impervious fabric."

32.1 DUCT ACCESSORIES

32.1.1 Dampers and Splitters

When dampers are used, they should be equipped with indicating and locking quadrants and the damper blade should be riveted to the supporting rods. Dampers should be made of the same material as the ducts in which they are installed but should be two gauges heavier. Cast or malleable brackets, riveted to the sides of the duct, should be sufficiently long to extend full width of the branch ducts to which they are attached. Opposed blade dampers should have each blade sealed with foam rubber or felt in order to form an effective seal between blades with the damper in the fully closed position.

32.1.2   Fire Dampers and Fire Stops

Ten-gauge galvanized steel sleeve-type horizontal and vertical fire dampers should be installed in all ducts that penetrate fire walls or floors. The assembly should consist of 18-gauge galvanized, formed-steel blades with interlocking joints to form a continuous steel curtain when closed. The assembly should have a maximum depth of 4 in. and be suitable for horizontal or verticale airflow as required. Fire and smoke control dampers should meet or exceed NFPA 91 (NFPA, 1992). Fire dampers installed in ducts with either dimension under 12 in. should be constructed and designed so that the blade stock in the open position will be completely outside the airstream.

Unless required by code we do not recommend the use of the fire dampers in hood exhaust ducts.

32.1.3   Sheet Metal Access Doors

Hinged-type sheet metal access doors are needed in the ductwork at each automatic damper and control device and at each fire damper to give access to the fusible link. Access doors should be of the same material as the ductwork on which they are installed and should be constructed to be sealed airtight.

32.1.4   Flexible Connections

The inlets and outlets of supply units and exhaust fans should be connected to ductwork with flexible, airtight connectors for noise and vibration suppression. Those in exposed locations should be constructed of materials suitable for outdoor installation.

32.1.5   Sealing Duct Penetrations

Wherever ducts pass through walls, floors, or partitions, the spaces around them should be sealed with metal, mineral wool, or other noncombustible material and be of equal fire rating as the construction it penetrates.

CHAPTER 33

VARIABLE-AIR-VOLUME SYSTEMS

33.1 INTRODUCTION

As discussed in Section 31.6, variable-air-volume (VAV) ventilation control concepts are receiving serious consideration for installation in all types of laboratory HVAC systems. The more complex the system becomes and the greater the variety of functional demands made upon it, the more desirable do the special features of VAV appear to be. In addition to almost unlimited flexibility of function, there is the promise of worthwhile energy savings over the useful life of the system that more than fully compensate for a higher installation cost. The less desirable aspect of VAV controls is the much greater complexity of the HVAC system, especially the sensors and controllers, and the difficulties associated with establishing and maintaining correct functionality over a period of many years; unlike people, mechanical systems do not get better as they grow older. No advanced VAV control systems have been in laboratory service for a long enough period or in a wide enough range of applications to fully evaluate their operating characteristics over a complete life cycle. Earlier systems had problems with reliability of operation and repeatability of volume control. Recent systems have been improved. Nevertheless, the inherent advantages of VAV control systems can often overcome the disadvantages of higher control cost, greater complexity, and lack of

Guidelines for Laboratory Design: Health and Safety Considerations, Second Edition
By Louis DiBerardinis, Janet S. Baum, Gari T. Gatwood, Anand K. Seth, Melvin W. First, and Edward F. Groden.
ISBN 0-471-55463-4   Copyright © 1993 by John Wiley & Sons, Inc.

long-term experience, and more and more such systems are going into new and remodeled laboratory buildings.

For example, VAV controls may operate by modulating supply and exhaust fan speeds to match demand requirements, keeping in mind that fans must always operate within the stable region of their fan curves, or by modulating local volume control dampers with constant-speed or variable speed system fans. Small, simple systems, such as laboratory hoods, served by a single exhaust fan, are easily adaptable to modulating fan-speed controls. But, when systems become large, each modulating step becomes a small part of the total, the required number of interacting sensors and controllers multiplies, and control precision tends to become lost as large fans attempt to modulate in small steps. Therefore, large, complex systems tend to operate best with constant speed fans. Modulating controllers require fan modulation with static pressure control. Inasmuch as the energy savings associated with VAV control systems are primarily associated with reducing the discharge of large volumes of tempered and humidity-controlled air, the extra energy expended by operating main system fans at constant speed when the air volume demand is reduced will be trivial.

A very significant increase in system complexity occurs when laboratories and laboratory service areas must be maintained constantly at preselected pressure gradients relative to surrounding indoor spaces as well as to the out-of-doors. To maintain preselected differential room pressure gradients at all times when employing variable-air-volume systems means that the controllers for each space must regulate both supply and exhaust air simultaneously over the entire demand range. This means, for example, that when one 6-ft chemical fume hood exhausting 1000 cfm is reduced by a VAV system in a high-toxicity laboratory, which must be maintained at a negative pressure relative to surrounding laboratories, offices, and public spaces to prevent spread of contamination, the laboratory air supply must simultaneously and automatically decrease by 1000 cfm. Or, if the hood is kept on but the sash is lowered to a work opening of 1 ft, the supply and exhaust air volumes must both decrease simultaneously by 600 cfm.

Yet additional levels of complexity are introduced into VAV control systems when (1) building code requirements set a minimum number of hourly air changes that must be met regardless of whether any of the local exhaust systems are in service and (2) preset heating, cooling, and humidity needs must be satisfied. To function well, some systems of this nature demand high-tech computer programs that incorporate continuous feedback and rapid modular adjustment capacity.

VAV systems can be described as either general comfort ventilation or contaminant control ventilation. VAV systems for general ventilation have been in use for several decades and, therefore, there are more data available on their performance and operation. The main focus of this chapter is to the use of VAV systems for contaminant control. They may consist of either (1) a single-exhaust fan with hood sash control or variable-speed motor to provide

100 fpm over the full range of sash opening, (2) two or more hoods in a laboratory with individual exhaust fans with controls, (3) multiple hoods on a single-exhaust variable-speed fan with many variable-air-volume boxes delivering variable supply air to the laboratory as hoods go on and off and providing the required pressure relationship between other laboratories, corridors, and offices.

33.2 VARIABLE-AIR-VOLUME HOODS

Variable-air-volume hoods can be any chemical fume hood described in Sections 31.2 through 31.6, except that they will not generally have a bypass. The main concept is to provide a constant face velocity (usually 100 fpm) across the hood face opening by varying the volume as the hood opening is changed. When the flow is decreased the operating costs are reduced.

In order to determine when a VAV system is applicable to a given situation, one must fully understand the advantages and disadvantages of these systems.

33.2.1 VAV System Advantages

Some of the advantages of the variable-air-volume manifolded chemical fume hood exhaust systems are as follows:

1. It reduces energy utilization by varying the air volume with respect to demand.

2. It reduces the number of exhaust fans. By combining hoods in a central system, one can reduce the need to one or two large central fans.

3. One or more central fans can be installed as a backup to provide continuous ventilation if the other fans fail. In this manner, some ventilation is guaranteed to all fume hoods at all times. If one installs a single fan to a fume hood, there is no ventilation in that one fume hood if the fan fails.

4. As the number of fume hood stacks are minimized, those remaining can be inexpensively increased in height, thus reducing the possibility of re-entry of exhaust contaminants into the building. Architectural treatment can be provided to make them less conspicuous and integrate them with the building design.

5. If a bypass damper is provided at the inlet side of exhaust fans on the roof, the air exhausted can be diluted, thereby minimizing the chemical concentration of air exhausted.

6. As more fume hoods are connected, the concentration of a particular chemical will be diluted.

7. Because the total hoods in use at any time is estimated to be no more than 30–35% (Lentz, 1989; and Moyer, 1983-1987), the central mechanical system can be reduced in size for cost savings.

8. Reducing the number of individual stacks, increases usable floor space, and reduces the need for shafts.

9. In a small laboratory with a large number of fume hoods, all of which are not used at all times, a VAV system may allow reduction of make-up air supplied in the laboratory.

10. Because the exhaust fan is usually located some distance from the laboratory there is less noise.

11. System can be more flexible for future hood installations.

33.2.2 VAV System Disadvantages

With all these advantages, when should this type of system be used? To answer that question, one must look at the disadvantages:

1. A major disadvantage is fact that, to balance the system, there will be either manual or automatic dampers in the exhaust stream. These dampers can fail. And, one or more fume hoods can be without dedicated exhaust capacity. This can be a more serious condition than a exhaust fan failing.

2. Mixing incompatible chemicals in the exhaust stream may result in an unsafe situation.

3. When the control equipment used is a hot-wire element in the exhaust stream, it may be hazardous with certain chemicals being exhausted. Additionally, the element may become corroded and nonfunctional.

4. The control system may not always deliver the amount of air it is scheduled to deliver.

5. These systems are complex and require more sophisticated maintenance.

It is difficult to condemn or recommend any one approach. The present generation of automated controls can make a manifold system of multiple exhaust hoods to one fan system work. Such was not the case some time ago when these controls were not available. VAV systems are finding applications in the use of manifold systems. Their use provides advantages.

33.3 REQUIREMENTS FOR VAV SYSTEMS TO BE ACCEPTED

In order for variable-air-volume hoods to be accepted, the following conditions must be met:

1. The hood meets current fume hood specifications, with the exception that they are not the bypass type. However, a partial by-pass may be needed

to meet minimum exhaust requirements for the lab to maintain face velocity below 300 fpm at 6" sash height.

2. The hood has no air-cleaning (HEPA or charcoal) or stack-sampling devices in the exhaust system (stack sampling is done on high-level-radiation hoods, biological cabinets have HEPA filtration. Varying the volume may interfere with the operation of this equipment or the current monitoring techniques used.)

3. The hood has no other auxiliary equipment (high-velocity–low volume spot exhaust systems) on the same exhaust fan. This type of equipment is generally designed for a fixed volume and may not function properly if the volume of flow is decreased.

4. The laboratory will remain under negative pressure with respect to the corridor or adjoining rooms even at the minimum exhaust rate, if negative pressure was part of the original laboratory design. When the exhaust quantity is reduced, the supply quantity must be reduced by the same volume.

5. The laboratory will maintain (8) eight air changes per hour with hood(s) in the minimum exhaust rate position.

6. There are no extenuating circumstances based on hood and/or laboratory use which preclude the use of a VAV system. Examples of these circumstances might be (a) odorous compounds used on the lab bench and (b) excessive heat generation in the laboratory from process equipment or high-release compounds which require full exhaust rate dilution. All of these conditions might produce hazardous conditions if the exhaust volume is reduced.

7. An airflow monitoring/alarm device must be installed at the hood to provide operating information to the hood user.

8. An override capability must be provided to allow the user to have maximum exhaust regardless of sash position.

9. It is important for the hood user to know if the hood is a variable-air-volume hood and to understand how it operates.

10. Control response time and stability should be reviewed to provide consistent repeatable performance.

33.4 VARIABLE-VOLUME EXHAUST SYSTEMS OPERATIONAL CONCEPT

The concept here is to maintain a constant face velocity at the fume hood sash at varying sash openings while varying the total volume of exhaust flow. It is done by modulating a damper or controller in the exhaust duct connected to the fume hood. The control systems used are classified into three types: (1) velocity measurement in the ductwork, (2) velocity measurement of the hood in the annular space between the outside and inside casing and (3) direct sash position measurement.

Measurements of velocity pressures is difficult in the exhaust ductwork. The air velocity typically is very low, with correspondingly low velocity pressure. In addition, the devices may become corroded if the exhaust stream contains material that will attack the sensor. Care is advised in the selection.

Air-velocity measurements in the annular space have been improved during the last several years, and several commercial devices are now available to maintain the correlation with the fume-hood face velocity. They are sensitive to changes in air-flow patterns in the room as well as in the hood.

Varying the sash position works as follows: It could be connected by a cable to a potentiometer or other control system that transmits a signal to the volume control which opens or closes the damper as required, thus increasing or decreasing the total volume flow in proportion to the face-opening. There are some devices that are excellent and some that are not. Care should be exercised in selecting the method of controls. In all cases, the signal controls a damper in the exhaust ductwork to ensure that the correct air quantity is being discharged from that one fume hood. The control systems can be either pneumatic or electronic. Electronic control systems have improved and are continually being upgraded to provide better control capabilities. An electronic control system is our recommendation.

The damper in the exhaust air or terminal box must be selected carefully. All moving parts in the damper will come into contact with the contaminated exhaust air. The construction of the damper should be either a stainless steel damper with Teflon bearings or a coated damper. The damper can be either a simple barreled plate damper or a more elaborate terminal box. Refer to Section 33.5.1, which discusses terminal boxes. A terminal box is recommended because it has better control devices and has been in use long enough to demonstrate that it provides consistent results. It, however, takes more space, so in some cases where space is an issue, a damper must be used.

33.5 VAV SYSTEM CONTROLS AND COMPONENTS

Control systems, in their simplest manifestation, consist of two parts, a sensor and a control device. A sensor is a device that responds to changes in physical condition (such as temperature or pressure) and transmits a signal to a control device, an item of hardware that responds to changes in the sensor signal by restoring the measured environmental conditions (in this case quantity of airflow) to correspond with the sensor's set point.

33.5.1 Terminal Box

There are many different methods and controls available for VAV systems, and their selection will depend on the complexity of the systems. Terminal boxes are a necessary part of a VAV system. HVAC Systems and Equipment (ASHRAE (1992) describes the principle and discusses types of terminal boxes) which can be summerized as follows:

The term *terminal box* commonly refers to a factory-made assembly for air distribution purposes. This terminal box, without altering the composition of the treated air from the distribution system, manually or automatically fulfills one or more of the following functions: (1) controls the velocity, pressure, or temperature of the air; (2) controls the rate of airflow; (3) mixes airstreams of different temperatures or humdities; and (4) mixes, within the assembly, air at high velocity and/or high pressure with air from the treated space.

The six types of terminal boxes can be classified in the following manner:

Volume Response

- *Constant Flow.* A constant flow of air is always provided.
- *Variable Flow.* The volume of air is varied as needed.

System Pressure Response

- *Pressure-Dependent.* Airflow through terminal box varies in response to system pressure.
- *Pressure-Independent.* Airflow does not vary.

Method of Control

- *Self or Internal or Mechanical Volume Control.* No outside source of power is needed. Terminal box volume control is activated by the static pressure in the primary duct system.
- *External Volume Control.* An electric or pneumatic actuator is the power source. It consists of (i) velocity pressure or differential pressure sensors, (ii) pneumatic or electric motors, and (iii) linkages.

Four of the commonly used boxes in laboratories are as follows:

a. *Actuator Spring and Cone Type.* This is usually a pressure-independent variable air volume terminal box with electric or pneumatic actuator control. See Figure 33-1.
b. *Spring and Cone Type.* Same as above, without an actuator, but with integral springs.
c. *Butterfly Damper.* This is a pressure-dependent, external power box and requires an additional sensing system to function.
d. *Box Type.* The box is usually insulated. The inlet is usually round and the outlet is rectangular. Care must be taken in the installation of box-type terminal boxes. When installed in return-air ductwork, it is a common mistake to pipe in the rectangular side. See Figure 33.2.

When selecting terminal boxes, the following must be evaluated (Krajnovich, 1986):

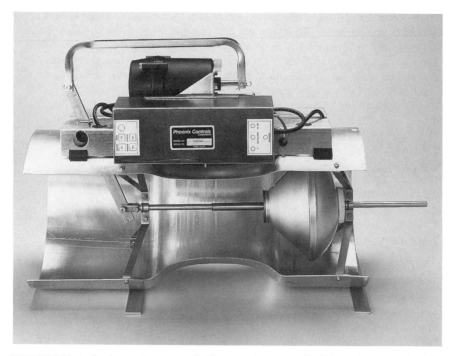

FIGURE 33-1. Spring and cone variable air volume terminal box. *Source:* Phoenix Controls, Newton, Massachusetts.

a. *Linearity.* "Linearity" refers to the relationship between the acuator pressure (controlled by the sensor) and the flow rate. Without linearity, control becomes difficult. Figure 33.3 shows the ideal versus typical linearity achieved in field use.

FIGURE 33-2. Box type terminal box. *Source:* Tutte and Bailey Division: Hart & Cooley, Holland, Maine.

b. *Hysteresis*. Hysteresis is a measure of the difference in terminal box performance while increasing the input versus decreasing the input. Figure 33-4 illustrates an example of hysteresis. A small value of hysteresis is imperative for consistent operation. A large hysteresis value yields two vastly different output values for identical input values.

Variable-air-volume terminal boxes must provide the necessary range of CFM at maximum airflow, and health and safety standards must be attainable at minimum airflow requirements of the exhaust system. In addition, temperature, humidity, and pressure balance must be maintained. It is important in selecting a terminal box that design linearity, hysteresis, and pressure independence characteristics are met for consistent trouble free operation. Boxes should be corrosion resistant, factory calibrated and should be designed to be insensitive to inlet and exit conditions.

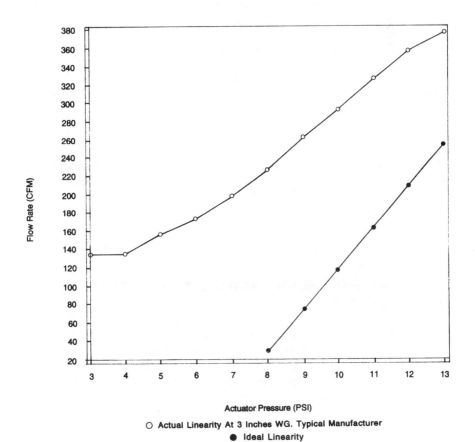

O Actual Linearity At 3 Inches WG. Typical Manufacturer
● Ideal Linearity

FIGURE 33-3. Linearity graph showing ideal versus actual for typical terminal box. From Krajnovich (1986).

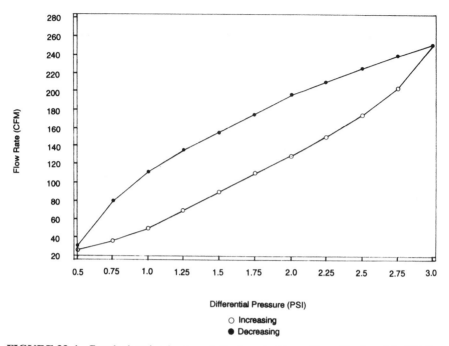

FIGURE 33-4. Graph showing hysteresis for terminal box. From Krajnovich (1986).

In addition, adequate consideration must be given to response time and stability for a VAV hood control system. Ahmad (1990) provides an excellent discussion of the response time and Avery (1992) describes the reasons for the instability of VAV systems.

33.6 VARIABLE-AIR-VOLUME (VAV) SYSTEM FAN CONTROLS AND COMPONENTS

In a variable-air-volume system, the volume modulation can only occur when one or a combination of methods control the supply or exhaust fan. It is important that both supply and exhaust fans are controlled to ensure that the fans continue to operate at the stable region of each fan curve. The control system consists of two parts:

a. Sensing system
b. Fan control hardware

33.6.1 Sensing System

The sensing system usually measures duct static pressure at a location. The location (based on duct layout) is typically between 75% and 100% of the distance between the first and last air terminal (Figure 33.5). This control point static pressure is also called ''reference static.'' Care must be taken in selecting the ''reference static'' sensor location. Multiple static sensors (Figure 33.6) are required when more than one duct runs from the supply fan. The sensor with the lowest static requirement will control the fan to meet the set-point. Selection of either a single-point sensor or a multiple-point sensor depends upon the duct distribution layout. This sensing element (either pneumatic, electric, or electronic) transmits a signal to the controller, which actuates the fan control hardware.

33.6.2 Fan Control Hardware

There are many devices available for controlling the fan for volume modulation. Some are more complex, more expensive, and more efficient than others. The selection of the particular control system is based upon the end result required. The following options are available in ASHRAE HVAC Applications Chapter 41.14 (ASHRAE 1991).

- Fan inlet or discharge dampers
- Fan scroll or inlet box dampers
- Fan inlet vanes
- Fan runaround or scroll bypass
- Controllable pitch blades
- Variable speed

33.6.3 Fan Inlet or Discharge Dampers

Either parallel or opposed blade dampers are provided in either fan inlet or discharge duct. This option is fairly inexpensive but not very energy-efficient, and noise levels could be objectionable. The basic principle is to introduce an additional pressure drop in the airstream. The fan is therefore forced to operate against a larger pressure. The design engineer must be aware of the fan curve so that when the volume of air generated by the fan is varied, it still operates within the stable region. In general, fan inlet or discharge dampers are easy to install and are low-first-cost items.

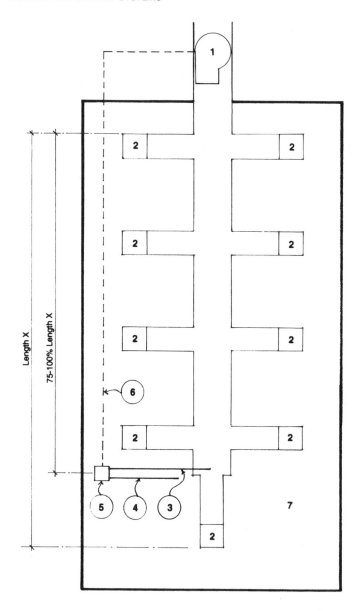

KEY

1 Supply Fan
2 Air Terminals
3 Duct Static Monitor
4 Reference Static Monitor
5 Control
6 Control/Fan Link
7 Laboratory

FIGURE 33-5. Duct static control diagram with single sensor.

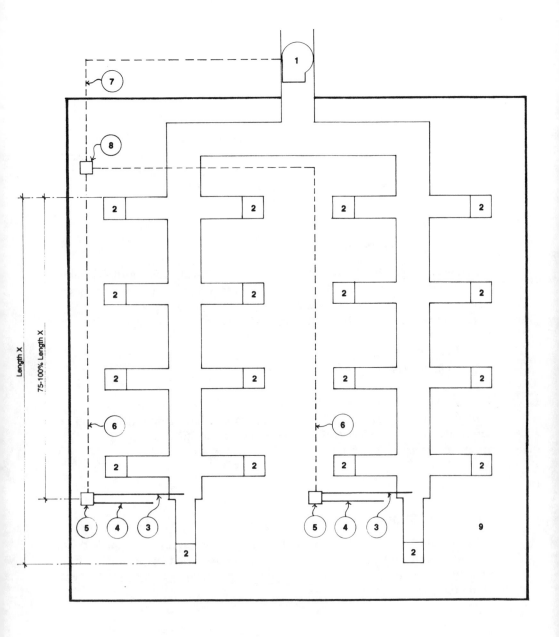

KEY

1	Supply Fan
2	Air Terminals
3	Duct Static Monitor
4	Reference Static Monitor
5	Control
6	Control/Selector Link
7	Selector/Fan Link
8	Selector
9	Laboratory

FIGURE 33-6. Duct static control diagram with multiple sensors.

Fan Scroll or Inlet Box Control (AMCA, 1990)

Inlet box dampers may be used to control the airflow volume through the system. Either parallel or opposed blade types may be used. The parallel blade type is installed with the blades parallel to the fan shaft so that, in a partially closed position, a forced inlet vortex will be generated. The effect on the fan characteristics will be similar to that of inlet fan vane control, although it is less noisy. The opposed blade type is used to control airflow volume by changing the system pressure by manipulating the damper.

33.6.5   Inlet Vane Control

Variable vanes mounted in the fan inlet can be used to modulate airflow. They are arranged to generate a forced inlet vortex which rotates in the same direction as the fan impeller (pre-rotation).

Inlet vanes may be of two different basic types:

1. Integral (built in)
2. Cylindrical (add on)

This is a very efficient method of fan control. The application, however, is limited to centrifugal type fans. The system can be retrofitted to existing centrifugal fans. Noise control and energy operating efficiency are within the acceptable range.

33.6.7   Fan Runaround or Scroll Bypass

A bypass duct arrangement with dampers is provided across the fan. The fan handles a constant flow, but the system flow is varied. This approach is usually not recommended. Even though the system flow can be varied, there is limited energy savings.

33.6.7   Controllable Pitch

This is usually applied in axial fans for makeup air systems. The pitch of the blades is varied by an actuator connected to a gear assembly which is capable of rotating the blades within prescribed limit. This is a fairly efficient system. It is, however, limited to axial fan systems. Care should be taken to ensure that the fan only operates in its stable region; otherwise, destruction of equipment is very possible. This option usually cannot be retrofitted.

33.6.8   Variable Speed

By varying the motor rpm, the speed of the fan is changed. This approach is gaining in popularity because of the many options available for achieving true variable speed. Branda (1984) discusses the options which are:

1. Adjustable voltage inverter (AVI)
2. Pulse width modulator (PWM)
3. Adjustable current sensor inverter (ACZ)
4. Wound rotor slip recovery system
5. Eddy current drive
6. DC motors

There are various advantages and disadvantages to each option. A variable-speed drive is the most efficient manner in which to control fan motor speed and thereby increase or decrease fan capacity. However, it requires (a) availability of manufacturer support for the spare parts and (b) trained technicians for startup and maintenance. Before a specific application, the engineer should thoroughly review the various systems available and then choose. The variable-speed drive is electrically a nonlinear load. Electric harmonic currents may result. Section 1.5.5 should be consulted.

33.6.9   Selecting the Fan Control Method

The selection of an appropriate fan control method is very complex. The following points should be considered: (1) first cost, (2) is it a retrofit or a new

TABLE 33-1. Matrix for Assisting the User in Selecting the Appropriate Type of Fan Control Device[a]

Fan Control	First Cost	New	Retrofit	Efficiency	Centrifugal Fan	Axial Fan	Service and Spare Parts	Noise
Fan inlet or discharge damper	1	5	1	4	1	NA	1	5
Fan inlet vanes	3	1	2	3	NA	NA	2	2
Fan runaround or scroll bypass	2	2	NA	5	1	NA	3	3
Controllable pitch	3	4	5	2	NA	1	4	2
Variable speed	1	1	1	1	2	2	5	1

[a] Scale: 1, most recommended; 5, least recommended. NA, not applicable.

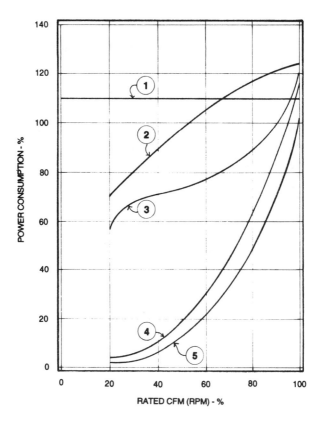

KEY
1   Constant Volume
2   Outlet Damper
3   Variable Inlet Vane
4   Variable Speed Drive
5   Centrifugal Fan Curve

FIGURE 33-7. Power consumption efficiency for various fan control methods.

application, (3) efficiency of operation, (4) type of fan, (5) availability of spare parts and trained technicians, and (6) noise levels. The matrix shown in Table 33.1 has been provided to aid the user in selecting the fan controller. The electrical power consumption and efficiency of several fan control options are shown in Figure 33.7.

PART VI

APPENDIXES

APPENDIX I

EMERGENCY SHOWERS*

1. Pull ring should not exceed 77 in. from the floor except where disabled persons are involved, and then the maximum should be determined functionally (Figure I-1).
2. The shower head should be at least 84 in. from the floor (Figure I-1).
3. The shower head should be an "Emergency Deluge Shower" as manufactured by the Speakman Company (Speakman, 1992), or its equivalent.
4. The horizontal distance from the center of the shower head to the pull bar should not be greater than 23 in.
5. The shower should provide at least 30 gal/min flow with the operating valve in the open position.
6. Tempered water showers should be equipped with a mixing valve with an "antiscald" feature such as manufactured by Powers, Series 420 Hydroguard (Powers, 1992).
7. Tempered water showers should be preset at a temperature between 70°F and 90°F.
8. The valve for the shower should be quick acting, such as a ball valve, and should remain open after the initial pull until manually closed.

* For additional information see the American National Standards Institute standard on emergency showers (ANSI.Z358.1, 1990).

Guidelines for Laboratory Design: Health and Safety Considerations, Second Edition
By Louis DiBerardinis, Janet S. Baum, Gari T. Gatwood, Anand K. Seth, Melvin W. First, and Edward F. Groden.
ISBN 0-471-55463-4   Copyright © 1993 by John Wiley & Sons, Inc.

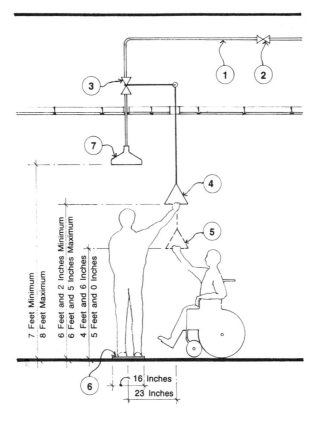

KEY

1 1 Inch I.P.S. or Larger
2 Tagged Shut-Off Valve
3 Opperating Valve
4 Hand Pull
5 Wheelchair Hand Pull Extension
6 Area of Floor Marked Yellow
7 Deluge Shower

FIGURE I-1. Deluge shower: diagram.

APPENDIX II

EMERGENCY EYEWASH UNITS

1. Eyewash units shall be approved as such by the manufacturer. Haws Drinking Fountain Company's "EMERGENCY EYEWASH," or its equivalent, should be used.
2. The water supply must be potable and capable of providing 3 to 7 gal/min depending on the type of eyewash used.
3. For each floor of the building or for each group of labs there should be at least one tempered eyewash unit.
4. Tempered water eyewash units should be equipped with a mixing valve with an "antiscald" feature such as manufactured by Powers, Series 420 Hydroguard. (Powers, 1992)
5. Tempered water eyewash units should be preset to a temperature of 70°F.
6. For disabled persons' use, the hand-held eyewash spray on a hose is the recommended unit.

Guidelines for Laboratory Design: Health and Safety Considerations, Second Edition
By Louis DiBerardinis, Janet S. Baum, Gari T. Gatwood, Anand K. Seth, Melvin W. First, and Edward F. Groden.
ISBN 0-471-55463-4   Copyright © 1993 by John Wiley & Sons, Inc.

APPENDIX III

EXCESS FLOW CHECK VALVES

Excess flow check valves, also known as "flow limit valves," are simple devices which can provide emergency control of gas flow for many laboratory operations. The valves are designed to shut off gas flow if a preset flow rate is exceeded. This could prevent the flow of toxic or flammable gases into an area when other conditions have resulted in failure of point-of-use control systems (i.e., a needle valve). Such control is particularly important where compressed gases are piped through a building or from one room to another.

These valves are limited to certain gases. Malfunction can occur if they are used with incompatible gases. Gas equipment manufacturers must be involved in the proper selection of this equipment. An alternative exists, namely an excess flow switch which is sometimes used in conjunction with the valve. The switch is not dealt with in this appendix but is available through the same source as the valve.

Figure III-1 is a functional schematic of an excess flow check valve shown in the open position to demonstrate how it functions.

Figure III-2 shows a flow limit valve which can operate at pressures between 130 and 3000 psig and which delivers flow rates from 100 to 10,000 standard cubic feet per minute. This particular unit is normally installed in the delivery line between the cylinder and process.

Guidelines for Laboratory Design: Health and Safety Considerations, Second Edition
By Louis DiBerardinis, Janet S. Baum, Gari T. Gatwood, Anand K. Seth, Melvin W. First, and Edward F. Groden.
ISBN 0-471-55463-4   Copyright © 1993 by John Wiley & Sons, Inc.

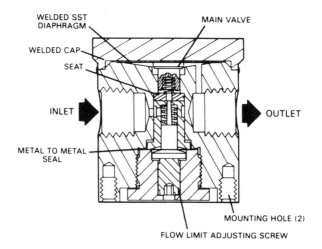

WELDED SST
DIAPHRAGM

MAIN VALVE

WELDED CAP

SEAT

INLET

OUTLET

METAL TO METAL
SEAL

MOUNTING HOLE (2)

FLOW LIMIT ADJUSTING SCREW

FUNCTIONAL SCHEMATIC
(valve shown open)

FIGURE III-1. Functional schematic of excess flow check valve.

This appendix has been included because many laboratory designers and users are unaware of the valves' existence and use. Their use must be considered in the design stages since it may affect decisions regarding tank farms and gas piping.

RESET
OPEN BYPASS
NORMAL OPERATION
CLOSE BYPASS

FIGURE III-2. Excess flow check valve: commercial.

APPENDIX IV

SIGNS

Signs for laboratories are used for many purposes. Among them are:

1. Identifying exits and safety equipment and procedures.
2. Identifying electrical, piping, plumbing, and other facility-type equipment.
3. Identifying hazardous materials, equipment, and special conditions.

The first two purposes have been well defined, and standard methods exist for indicating exits, pipe content, electrical runs, and panels in ANSI/EIA 359A, (ANSI, 1984), and ANSI Z53.1, 1979 (ANSI, 1979). Hazardous materials, equipment, and special conditions are covered by agency requirements for radioactive materials, lasers, biological hazards, and the like. The problem that led to the development of the sign system shown below (Section IV–1) is related to three concerns: (1) people's real use of the facilities (for example, laboratory personnel sometimes nap in laboratory lounges); (2) emergency responders' need to know about facility layout and presence of hazardous materials; and (3) the availability of a logical and manageable information system to facilitate dealing with the first two.

Two surveys were made of laboratories across the country, including university, government, industrial, commercial, and nonprofit establish-

Guidelines for Laboratory Design: Health and Safety Considerations, Second Edition
By Louis DiBerardinis, Janet S. Baum, Gari T. Gatwood, Anand K. Seth, Melvin W. First, and Edward F. Groden.
ISBN 0-471-55463-4   Copyright © 1993 by John Wiley & Sons, Inc.

ments, to determine what sign information systems were in use. The results showed that no uniform system seemed to have worked well, including NFPA 704 (NFPA, 1992). However, many useful ideas were gleaned from the study (Gatwood, 1985), resulting in the "General Policy" discussed under Section IV.1.

The study showed the following:

1. Respondents using NFPA 704 were experiencing difficulties with training, maintenance, quantity decision, improper information, and lack of information.

2. Fire department and emergency response personnel can be expected to know that laboratories generally contain some flammable liquids, toxic chemicals, compressed gases, and other hazards.

3. Fire department and emergency response personnel may already have some familiarity with laboratories through in-service inspections, site visits to grant annual permits/licenses, plan review, and special reviews by owner/occupant.

4. Signs on doors alone may be insufficient to help a firefighter, particularly when fire and/or smoke render them unreadable.

5. The sign system needs to be as simple and yet as effective as possible.

6. Fire department and emergency response personnel should not have to learn, understand, or commit to memory unique systems. The information should be self-evident.

7. The need for frequent changes to signs should be eliminated, given the administrative difficulty and poor history of maintaining signs.

8. Experiments and their associated materials often move from room to room within a research group.

9. Other jurisdictional agencies require signs that may already be present and should be acceptable to the sign plan. Examples are: biohazards, radiation, strong magnetic fields, lasers, UV, explosives, and high-voltage signs.

10. Emergency response and fire department personnel can be expected to enter each building at a prescribed location, giving them access to an annunciator panel and building information.

IV.1 GENERAL POLICY

At the primary fire department and emergency response personnel entrance to each building containing one or more laboratories there should be a Type I sign (Figure IV-1) indicating the type of laboratories, the common hazards expected to be encountered, and a notation about special hazards and their location within the building. At this building location, if appropriate, there

```
BUILDING   NAME_____

BUILDING   USE_____

ZONE   (IF   APPLICABLE)_____


THIS BUILDING/ZONE CONTAINS WORKING QUANTITIES OF THE FOLLOWING:
```

MATERIALS	BEING USED	"SPECIAL CONDITION IN ROOM NO."
Flammable Liquids		
Water Reactives		
Hazardous Biological Agents		
Highly Toxic Chemicals		
Compressed Gas		
Explosives		
High Voltage		
Lasers		
Strong Magnetic Fields		
Radioactives		
Radiation		
Microwave Radiation		
Other		

```
DATE POSTED_____    FIRE DEPARTMENT___(Name)_____
```

FIGURE IV-1. Type 1 sign at building entrance.

may be supplemental information such as building diagrams, zone diagrams, voice alarms, and protection system(s) information.

A copy of this Type I sign, where appropriate, should be posted at each zone entrance.

A second sign (Type II, Figure IV-2), should be posted on the room door of any room containing a special condition or hazard, as noted on the sign at the entrance to the building or zone.

What constitutes a special hazard or condition can be defined in general terms, such as an "appreciable quantity of water-reactive metals in which a fire should not be fought with water" or "an area of a building where persons might be expected to be napping." But the specific situations would be determined on an individual basis by management and the fire department. Questions of quantity should, of course, be a part of that procedure, and signs

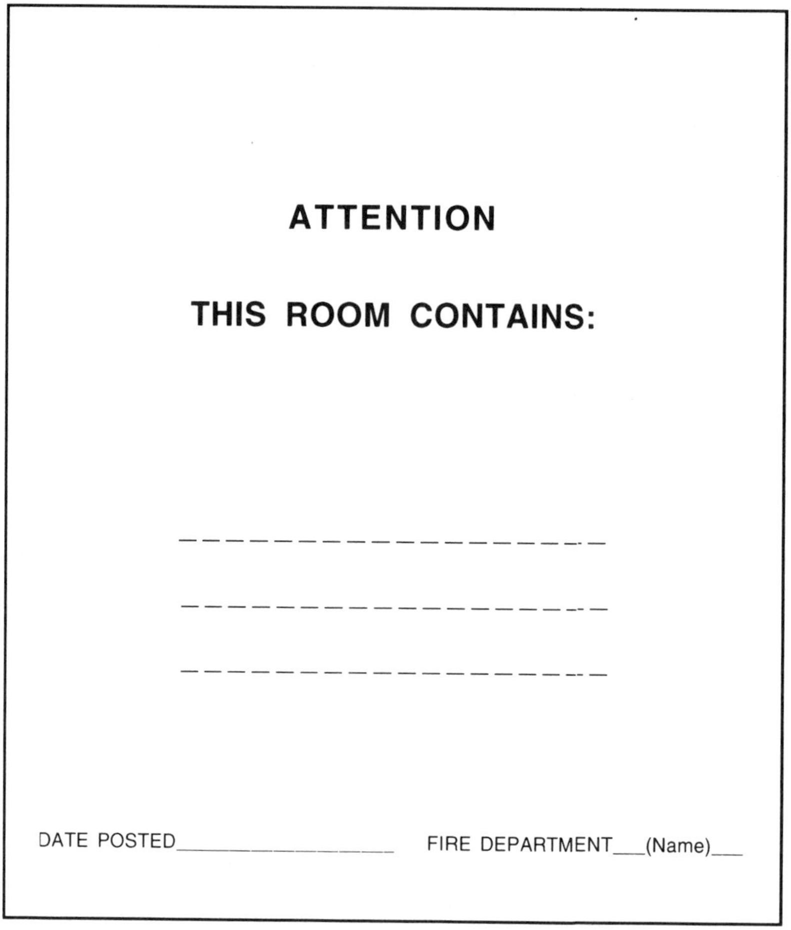

ATTENTION

THIS ROOM CONTAINS:

‗ ‗ ‗ ‗ ‗ ‗ ‗ ‗ ‗ ‗ ‗ ‗ ‗ ‗ ‗ ‗ ‗

‗ ‗ ‗ ‗ ‗ ‗ ‗ ‗ ‗ ‗ ‗ ‗ ‗ ‗ ‗ ‗

‗ ‗ ‗ ‗ ‗ ‗ ‗ ‗ ‗ ‗ ‗ ‗ ‗ ‗ ‗ ‗

DATE POSTED_____   FIRE DEPARTMENT___(Name)___

FIGURE IV-2. Type 2 sign for doors of rooms containing special hazards.

indicating those special cases would, as mentioned, be located on the room door, at the zone entrance, and at the building's primary fire department and emergency personnel entrance.

Before adoption of this system, communication with the local authority having jurisdiction, such as the local fire department, should be initiated. The sign formats shown may be modified to adapt to a set of specific needs.

APPENDIX V

MATRIX

This appendix has been prepared as a summary reference for key safety and health issues that should be addressed in the common-type laboratories. It can be used as a checklist for laboratory design or to refer the reader to the particular section in the book for a further discussion of the item. For each item we have provided guidance as to its applicability in each type of laboratory.

- Those where HP is indicated are the items we believe are applicable to the design and in many cases are probably required by regulation. They require careful consideration, and rejection of their application to a particular design should be made at the highest level of authority on the project.
- Some items may not be required by regulation but we believe represent good safety practice and are indicated by RP. For example, fire suppression systems are preferred in a pilot plant but may not be feasible.
- The applicability of some items will require a special evaluation of the particular design and needs of the laboratory and are indicated by SE. For example, the use of emergency showers in a clean room will depend on the types of chemicals used in the room.

Guidelines for Laboratory Design: Health and Safety Considerations, Second Edition
By Louis DiBerardinis, Janet S. Baum, Gari T. Gatwood, Anand K. Seth, Melvin W. First, and Edward F. Groden.
ISBN 0-471-55463-4   Copyright © 1993 by John Wiley & Sons, Inc.

MATRIX OF BUILDING CONSIDERATIONS FOR SPECIFIC LABORATORY TYPES

Chapters by laboratory type: 3 General Chemistry Lab, 4 Analytical Chemistry Lab, 5 High Toxicity Laboratory, 6 Pilot Plant, 7 Physics Laboratory, 8 Clean Room Laboratory, 9 Controlled Environment Rm, 10 High Pressure Laboratory, 11 Radiation Laboratory, 12 Biological Laboratory, 13 Clinical Laboratory, 14 Teaching Laboratory, 15 Gross Anatomy Laboratory, 16 Pathology Laboratory, 17 Team Research Laboratory, 18 Animal Laboratory, 19 Microelectronics Laboratory, 20 Print Making Studio

BUILDING CONSIDERATIONS	PARAGRAPH NUMBER	3	4	5	6	7	8	9	10	11	12	13	14	15	16	17	18	19	20
Evaluate Distribution of Mechanical Equipment and Services	1.2.2.4	HR	HR	HR	HR	HR	HR	HR	HR	HR	HR	HR	HR	HR	HR	HR	--	HR	HR
Directional Airflow in Laboratories	1.3.1 2.3.2	HR	HR	HR	HR	HR	HR	HR	HR	HR	HR	HR	HR	HR	HR	HR	--	HR	HR
Supply Air Systems	1.3.4	HR	HR	HR	HR	HR	HR	HR	HR	HR	HR	HR	HR	HR	HR	HR	--	HR	HR
Supply Air Velocity and Quality	1.3.4.2	HR	HR	HR	HR	HR	HR	HR	HR	HR	HR	HR	HR	HR	HR	HR	--	HR	HR
Evaluate Location of Supply Air Intake To Building	1.3.4.3	HR	HR	HR	HR	HR	HR	HR	HR	HR	HR	HR	HR	HR	HR	HR	--	HR	HR
Location of Supply Air to Room	1.3.4.2 2.3.3.1.2	HR	HR	HR	HR	HR	HR	HR	HR	HR	HR	HR	HR	HR	HR	HR	--	HR	HR
HVAC Controls/Alarms	1.3.8	RP	RP	HR	RP	RP	RP	RP	RP	HR	HR	SE	SE	HR	HR	RP	--	HR	RP
Temperature Control	1.3.8.1 2.3.2	HR	HR	HR	HR	HR	HR	HR	HR	HR	HR	HR	HR	HR	HR	HR	--	HR	HR
Humidity	1.3.8.1	SE	SE	SE	SE	SE	SE	SE	SE	SE	HR	SE	SE	HR	HR	SE	--	HR	SE
Emergency Power Supply	1.4.1	HR	HR	HR	HR	HR	HR	HR	HR	HR	HR	HR	HR	HR	HR	HR	--	HR	HR
Fire Detection System	1.4.4.1	HR	HR	HR	HR	HR	HR	HR	HR	HR	HR	HR	HR	HR	HR	HR	--	HR	HR
Fire Suppression System	1.4.4.2	RP	SE	RP	RP	SE	SE	RP	RP	RP	RP	HR	HR	HR	HR	SE	--	HR	RP
Fire Alarm System	1.4.5	HR	HR	HR	HR	HR	HR	HR	HR	HR	HR	HR	HR	HR	HR	HR	--	HR	HR
Lighting	1.5.1	HR	HR	HR	HR	HR	SE	SE	HR	HR	HR	HR	RP	RP	RP	RP	--	SE	SE
Plumbing	1.5.3	HR	HR	HR	HR	HR	HR	HR	HR	HR	HR	HR	HR	HR	HR	HR	--	HR	HR
Laboratory Egress	2.2.1	HR	HR	HR	HR	HR	SE	HR	HR	HR	HR	HR	HR	HR	HR	HR	--	SE	SE
Furniture Location	2.2.2	RP	RP	RP	RP	RP	RP	RP	RP	RP	RP	RP	RP	RP	RP	RP	--	RP	RP
Handicapped Access	2.2.5	HR	HR	HR	HR	HR	HR	HR	HR	HR	HR	HR	HR	HR	HR	HR	--	HR	HR
Laboratory Chemical Fume Hood	2.3.4.4	HR	HR	HR	SE	HR	SE	SE	SE	HR	SE	HR	HR	HR	HR	HR	--	SE	SE
Perchloric Acid Hood	2.3.4.4.3	SE	SE	SE	SE	SE	NA	NA	NA	SE	SE	SE	NA	NA	SE	SE	--	SE	NA
Biological Safety Cabinet	2.3.4.4.4	NA	NA	SE	NA	NA	NA	NA	NA	NA	HR	HR	NA	NA	HR	NA	--	SE	NA
Local Exhaust Ventilation	2.3.4.4.5	SE	RP	HR	HR	SE	SE	SE	SE	RP	SE	HR	SE	SE	HR	RP	--	RP	RP
Emergency Gas Shut-Off	2.4.1.1	HR	HR	HR	HR	HR	H R	HR	HR	HR	HR	HR	HR	HR	HR	HR	--	HR	HR
Ground Fault Circuit Interruptors	2.4.1.2	HR	HR	HR	HR	HR	HR	HR	HR	HR	HR	HR	HR	HR	HR	HR	--	HR	HR
Master Electrical Disconnect Switch	2.4.1.3	HR	HR	HR	HR	HR	HR	HR	HR	HR	HR	HR	HR	HR	HR	HR	--	HR	HR
Emergency Showers	2.4.1.4	HR	HR	HR	HR	SE	SE	SE	HR	HR	HR	HR	HR	HR	HR	HR	--	SE	HR

- There are also some items that are definitely not recommended nor applicable to a particular laboratory, and they are indicated by NA. In some cases NA is used to discourage the use of certain materials in a type of laboratory. For example, flammable liquid storage is not recommended in a physics lab in order to discourage the use of flammable liquids in this type of lab.

BUILDING CONSIDERATIONS	PARAGRAPH NUMBER	General Chemistry Lab 3	Analytical Chemistry Lab 4	High Toxicity Laboratory 5	Pilot Plant 6	Physics Laboratory 7	Clean Room Laboratory 8	Controlled Environment Rm 9	High Pressure Laboratory 10	Radiation Laboratory 11	Biological Laboratory 12	Clinical Laboratory 13	Teaching Laboratory 14	Gross Anatomy Laboratory 15	Pathology Laboratory 16	Team Research Laboratory 17	Animal Laboratory 18	Microelectronics Laboratory 19	Print Making Studio 20
Emergency Eye Wash	2.4.1.5	HR	HR	HR	HR	SE	SE	SE	HR	HR	HR	HR	HR	HR	HR	HR	--	HR	HR
Chemical Spill Control	2.4.1.6	RP	RP	HR	RP	RP	RP	RP	RP	RP	RP	HR	HR	HR	HR	RP	--	HR	RP
Construction Methods and Materials	2.4.2	HR	HR	SE	HR	HR	HR	SE	SE	HR	HR	HR	HR	HR	HR	HR	--	SE	HR
Experiment Alarm Systems	2.4.4	RP	RP	RP	RP	RP	RP	RP	RP	RP	RP	NA	SE	SE	NA	RP	--	RP	NA
Hazardous Chemical Disposal	2.4.5	HR	HR	HR	HR	HR	HR	HR	HR	HR	HR	HR	HR	HR	HR	HR	--	HR	HR
Chemical Waste Treatment	2.4.5.2	HR	HR	HR	HR	HR	HR	HR	HR	HR	HR	HR	HR	HR	HR	HR	--	HR	HR
Flammable Liquid Storage	2.4.6.3	HR	HR	HR	RP	NA	NA	NA	RP	RP	RP	HR	HR	HR	HR	RP	--	HR	HR
Special Hazard Chemicals	2.4.6.4	RP	RP	HR	RP	HR	RP	RP	RP	RP	RP	NA	NA	NA	NA	RP	--	HR	RP
Compressed Gas Cylinders	1.4.8 2.4.7	RP	RP	RP	RP	RP	RP	RP	RP	RP	RP	RP	RP	RP	RP	RP	--	RP	RP
Emergency Cabinet	2.4.8	RP	RP	RP	RP	RP	RP	RP	RP	RP	RP	HR	HR	HR	HR	RP	--	HR	RP
Change Room	5.2.2	NA	NA	HR	NA	NA	HR	NA	NA	SE	NA	NA	NA	SE	SE	NA	--	HR	NA
Work Surfaces	5.2.3	RP	RP	HR	RP	RP	RP	RP	RP	HR	HR	RP	RP	RP	RP	RP	--	SE	RP
Floors And Walls	5.2.4	RP	RP	HR	RP	RP	RP	RP	RP	HR	HR	RP	RP	RP	RP	RP	--	HR	RP
Handwashing Facilities	5.2.5	RP	RP	HR	RP	RP	RP	RP	RP	HR	HR	HR	HR	HR	HR	RP	--	HR	RP
Access Restrictions	5.2.6	RP	RP	HR	RP	RP	HR	RP	HR	HR	HR	RP	RP	HR	HR	RP	--	HR	RP
Glove Box	5.3.3	NA	NA	HR	SE	NA	SE	NA	NA	HR	NA	NA	NA	NA	NA	NA	--	SE	NA
Filtration of Exhaust Air	5.3.6	SE	SE	HR	SE	SE	SE	SE	SE	SE	SE	SE	SE	SE	SE	SE	--	SE	SE
Protection of Laboratory Vacuum System	5.4.2	SE	SE	HR	SE	SE	SE	SE	SE	SE	SE	SE	SE	SE	SE	SE	--	SE	SE
Radioisotope Hood	11.3.2	NA	SE	SE	SE	SE	NA	NA	SE	HR	SE	SE	NA	NA	SE	SE	--	SE	NA

KEY

HR Highly Recommended And Often Required By Regulation.
RP Recommended Practice.
SE Special Evaluation Needed.
NA Not Applicable

If the reader has any doubts about the particular applicability of an item, they should refer to the section in the text for the rationale of the selection.

Since the animal laboratory is a unique type of laboratory, it is not used in this matrix.

REFERENCES

ACGIH (1989)   *Hazard Assessment and Control Technology in Semiconductor Manufacturing,* Lewis Publishers, Inc., 121 South Main St., P.O. Drawer 519, Chelsea, MI, 1989.

ACGIH (1992)   American Conference of Governmental Industrial Hygienists, 1992, *Industrial Ventilation, A Manual of Recommended Practices,* 22nd ed., Committee on Industrial Ventilation, Cincinnati, OH, 1992.

ADA (1990)   Americans with Disabilities Act, 28 *Code of Federal Regulations, Title III,* Public Accommodations and Commercial Facilities, Part 36, 1990.

Ahmed (1990)   Ahmed, O., and Bradley, S., "An Approach to Determining the Required Response Time for a VAV Fume Hood Control System," *ASHRAE Transactions,* 1990, Vol. 96, Pt. 2.

AIA (1987)   *Architects Handbook of Professional Practice,* Vols 1–4, American Institute of Architects, Washington, D.C., 1987.

AIA (1992)   *Minimum Construction Guidelines for Health Care Construction,* American Institute of Architects, 1992.

Alereza (1984)   Alereza, T., and Breen, J., III, "Estimates of Recommended Heat Gains Due to Commercial Appliances and Equipment," *ASHRAE Transactions,* **90** (2A), 25–58, 1984.

Guidelines for Laboratory Design: Health and Safety Considerations, Second Edition
By Louis DiBerardinis, Janet S. Baum, Gari T. Gatwood, Anand K. Seth, Melvin W. First, and Edward F. Groden.
ISBN 0-471-55463-4   Copyright © 1993 by John Wiley & Sons, Inc.

AMCA (1974)   *Laboratory Methods of Testing Fans for Rating,* AMCA Standard 210-74, Arlington Heights, IL, 1974.

AMCA (1986)   *Classifications for Spark Resistant Construction,* Standard 99-0401-86, Air Movement and Control Association, Inc., Arlington Heights, IL, 1986.

AMCA (1990)   *Std #201-90 Fans and Systems*
Std #202-99 Trouble Shooting
Std #203-90 Field Performance Management
Std #200-87 Air Systems

ANSI (1979)   *Safety Color Code for Marking Physical Hazards,* Standard Z53.1, American Standards Institute, New York, 1979.

ANSI (1980)   *Standard for Testing Nuclear Air-Cleaning Systems,* ANSI/ASME N510, American Society of Mechanical Engineers, New York, 1980.

ANSI (1982)   Photography (film and slides). *Practice for Storage of Black-and-White Photographic Paper Prints,* Standard PH 1.48-82, American National Standards Institute, New York, 1982.

ANSI (1984)   *Standard Colors for Identification and Coding,* Standard ANSI/EIA 359A, 1984.

ANSI (1984)   Photography (film). *Safety Photographic Film,* Standard PH 1.25-84, 1984.

ANSI (1985)   *Processed Safety Film-Storage,* Standard PH 1.43-85, 1985.

ANSI (1986)   *Providing Accessibility and Usability for Physically Handicapped People,* ANSI Standard 117.1-1986.

ANSI (1988)   *Specifications for Stability of Ammonia-Processed Diazo Films,* Standard IT 9.5-88.

ANSI (1990)   *Emergency Eyewash and Shower Equipment,* ANSI Standard Z358.1-1990.

ANSI (1990)   *Method of Measuring Floor Area in Office Buildings,* ANSI Standard Z65.1-1990.

ANSI (1992)   *Laboratory Ventilation,* ANSI/AIHA Standard Z9.5, American Industrial Hygiene Association, Merrifield, VA, 1992.

ANSI/ASHRAE (1989)   *Ventilation for Acceptable Indoor Air Quality,* ANSI/ASHRAE Standard 62-1989, American Society of Heating, Refrigerating and Air-Conditioning Engineers, Atlanta, GA.

ASHRAE (1925)   *Guide,* American Society of Heating, Refrigerating and Air Conditioning Engineers, Atlanta, GA, 1925.

ASHRAE (1976)   *Methods of Testing Air Cleaning Devices Used in General Ventilation for Removing Particulate Matter,* Standard 52-76.

ASHRAE (1977)   Fundamentals, ASHRAE, 1977

ASHRAE (1985)   *Method of Testing Performance of Laboratory Fume Hoods,* ASHRAE Standard 110-1985 RA.

ASHRAE (1985)   *Energy Conservation in Existing Buildings—Commercial,* ASHRAE 100.3-1985.

ASHRAE (1989)   *Fundamentals*, 1989.

ASHRAE (1989)   *Energy Efficient Design of New Buildings*, ASHRAE 90.1-
                   1989.

ASHRAE (1989)   *Energy Conservation in Existing Facilities—Industrial*, ASH-
                   RAE 100.4-1989.

ASHRAE (1990)   *Refrigeration*, 1990.

ASHRAE (1991)   *Energy Conservation in Existing Buildings—Institutional*,
                   ASHRAE 100.5-1991.

ASHRAE (1991)   *HVAC Applications*, 1991 Chapter 12, Industrial Air Condi-
                   tioning, Chapter 14, Laboratories, Chapter 16, Clean
                   Spaces, Chapter 20, Photographic Material, Chapter 34,
                   Testing, Adjusting & Building, Chapter 41, Automatic
                   Control

ASHRAE (1992)   GPC-4P, *Draft for Preparation of Operating and Maintenance
                   Documentation for Building Systems*, 1992.

ASHRAE (1992)   *HVAC Systems and Equipment*, 1992 Chapter 18, Fans, Chap-
                   ter 20, Humidification, Chapter 25, Air Cleaners for Particu-
                   late Contaminants

ASME (1989)   American Society of Mechanical Engineers, *ASME Unfired
                   Pressure Vessel Code*, Section VIII, ASME, Fairfield, NJ,
                   1989.

ASSE (1977)   Weaver, A., and Britt, K., "Criteria for Effective Eyewashes
                   and Safety Showers," *Professional Safety Magazine*,
                   pp. 38–54, June 1977, Society of Safety Engineers, Des
                   Plaines, IL.

ASTM (1992)   ASTM E06.25, *Standard for Building Area Measurement* (sub-
                   mitted for committee ballot, 1992) American Society for
                   Testing Materials, Washington, DC, 1992.

Atkinson (1991)   Atkinson, Koven, Feinberg Engineers, *Systems Report*, **1**
                   (3) 1500 Broadway, New York, N.Y. 10036, 1991.

Avery (1992)   Avery, G., "The Instability of VAV Systems," *Heating Piping
                   and Air Conditioners*, Penton Publishing, Cleveland, OH,
                   Feb., 1992.

Billings (1989)   Billings, C.E. and Greenley, P.L., "Tracer Study of Toxic Gas
                   Release and Containment in Microelectronics Research
                   Laboratories" presented at American Industrial Hygiene
                   Association Meeting, May 1989.

BOCA (1991)   The Building Officials and Code Administrators International,
                   Inc., *BOCA Code*, Chicago, IL, 1991.

Branda (1984)   Branda M.R., "A Primer on Adjustable Frequency Inven-
                   ters," *Heating/Piping/Air Conditioning*, August, 1984.

BSI (1979)   British Standards Institution, *Draft Standard for Safety Re-
                   quirements for Fume Cupboards, Performance Testing,
                   Recommendation on Installation and Use*, BSI Docu-
                   ment 79/52625 DC, London, United Kingdom, August
                   1979.

Buffalo Forge (1983)	Jorgensen, R., Ed., *Fan Engineering, An Engineer's Handbook on Fans and Their Applications,* 8th ed., Buffalo Forge, Buffalo, NY, 1983.
Burgess (1985)	Burgess, W. A., and Forster, F., "The Evaluation of Gas Cabinets Used in the Semiconductor Industry," in *Ventilation '85,* H. D. Goodfellow, Ed., Elsevier, Amsterdam, 1986.
California (1988)	*California Uniform Fire Code,* Article 80.
Caplan (1978)	Caplan, K. J., and Knutson, G. W., "Laboratory Fume Hoods: A Performance Test," *ASHRAE Transactions,* **84**(1), 1978. pp 511-537
Carnes (1984)	Carnes, L., "Air-to-Air Heat Recovery Systems for Research Laboratories", *ASHRAE Transactions,* **90**(2), 1984. pp 327–340
CDC (1988)	Centers for Disease Control and National Institutes of Health, *Biosafety in Microbiological and Biomedical Laboratories,* HHS Publication No. (CDC) 88-8395, U.S. Government Printing Office, Washington, DC, 1988.
CFR (1984)	*Code of Federal Regulations, Title 9,* Subchapter A, Animal Welfare, Office of Federal Register, Washington, DC, 1984.
Chamberlin (1978)	Chamberlin, R., and Leahy, J., *Laboratory Fume Hood Standards,* EPA Report 68-01-4661, 1978.
Chamberlin (1982)	Chamberlin, R., and Leahy, J., *Development of Quantitative Containment Performance Tests for Laboratory Fume Hoods,* EPA Contract No. 68-01-6197, June 1982.
CMR (1979)	Codes of Massachusetts Regulations, *Cross Connections—Drinking Water,* 310 CMR 22.22, Massachusetts Department of Environmental Quality Engineering, Boston, MA, Sept. 20, 1979.
CMR (1990)	Massachusetts Department of Environmental Protection, 310 CMR 22.22, 1990.
Crawley (1984)	Crawley & Crawley, Chapter 26, Scherberger, L. and Harter, E. *Industrial Hygiene Aspects of Plant Operations,* Vol. 2, Macmillan, NY, 1984.
Davis (1987)	Davis, S. J., and Benjamin, R., *Heating Piping/Air Conditioning,* Renton Publishers, Cleveland, OH, 1987.
Degenhardt (1983)	Degenhardt, R., and Pfost, J., "Fume Hood System Design and Application for Medical Facilities," *ASHRAE Transactions,* **89**(2B), 558–570, 1983.
DeRoss (1979)	DeRoss, R., et al., *Hospital Ventilation Standards and Energy Conservation: A Summary of Literature with Conclusions and Recommendations,* FY79 Final Report, University of Minnesota, 1979.
DiBerardinis (1983)	DiBerardinis, L. J., et al., "Storage Cabinet for Volatile Toxic Chemicals," *American Industrial Hygiene Association Journal,* **44** (8), 583–588, 1983.

DiBerardinis (1991)   DiBerardinis, L. J., First, M. W., Ivany, R. "Field Results of an In-Place, Quantitative Performance Test for Laboratory Fume Hoods," *Applied and Environmental Hygiene,* **6** (3), 227–231, 1991.

DOT (1992)   *Hazardous Materials Transporation,* U.S. Department of Transportation, 40 CFR100, 1992.

Edgerton (1989)   Edgerton S. A., Kenny D. V., Joseph D. W., "Determination of amines in indoor air from steam humidification", *Environmental Science & Technology,* **23**(4), 1989.

EPA (1992)   *General Regulations for Hazardous Waste Management,* U.S. Environmental Protection Agency, 40 CFR 260, 1992.

EPA (1992)   *Standards for the Tracking and Management of Medical Waste,* U.S. Environmental Protection Agency, 40 CFR 259, 1992.

Exxon (1980)   DiBerardinis Associates, *Guidelines for Laboratory Design: Health and Safety Considerations,* Manual prepared for Exxon Corporation, Research and Environmental Health Division, E. Millstone, NJ 1980.

Fawcett (1984)   *Hazardous and Toxic Materials,* Wiley, New York, 1984.

First (1977)   First, M. W., "Control of Systems, Process and Operations" 3rd ed., Vol. 4, *Air Pollution,* A. C. Stern, Ed., Chapter 1, Academic Press, New York, 1977.

FM (1992)   Factory Mutual, *Factory Mutual Approval Guide 1992,* Norwood, MA, 1992.

Fuller (1979)   Fuller, F. H., and Etchells, A. W., "The Rating of Laboratory Hood Performance," *ASHRAE Journal,* 49–53, October 1979.

Gatwood (1985)   Gatwood, G. T., and Fresina, J., Questionnaire survey on laboratory hazards warning systems conducted in 1983 and reported to the Cambridge, MA Fire Department in October 1985.

GSA (1979)   *Public Buildings Service Guide Specifications, Airflow Control Systems,* General Service Administration PBS (PCD): 15980, Washington, DC, 1979.

GSA (1988)   Federal Standard 209D, *Standardization of Definitions and Air Cleanliness Classes,* Naval Publications Center, 5801 Tabor Avenue, Philadelphia, PA, 1988.

HHS (1988)   *Biosafety in Microbiological and Biomedical Laboratories,* Center for Disease Control, U.S. Public Health Service, 1988.

Hills (1990)   Hills B., Lushniak B., Sinks T., "Work place exposure to the corrosion—Inhibiting chemicals from a steam humidification System," *Applied Occp Environmental Hyg.* (S10), October, 1990.

Hoyle (1987)   Hoyle, R. E., Murray, T. H., Snyder, C. A., Prokopetz, A. T., and Walters, D. B., *Design and Evaluation of an Exhausted*

Enclosure for an Analytical Balance, NIEHS/NTP, Research Triangle Park, NC, 1987.

IES (1987)   *IES Lighting Handbook,* Application Volume, Illuminating Engineering Society of North America, 345 East 47th Street, New York, 1987.

Inglis (1980)   Inlis, J. K., Pergamon Press, Oxford, Maxwell House, Fairview Park, Elmsford, NY, 1980.

Ivany (1989)   Ivany, R., First, M. W., and DiBerardinis, L. J., "A New Method for Quantitative, In-Use Testing of Laboratory Fume Hoods," *American Industrial Hygiene* J. **50**(5), 275–280, 1989.

JCAHO (1991)   Joint Commission on Accreditation of Health Care Organizations, One Renaissance Blvd., Oakbrook Terrace, IL, 1991.

Jones (1966)   Jones, A. R., and Hopkins, D. J., "Controlling Health Hazards in Pilot Plant Operations," *Chemical Engineering Progress, Applied Industrial Hygiene J.* **62** (12), 59–67, 1966.

Kodak (1989)   *Disposal and Treatment of Photographic Effluent,* Publication No. J-55, Eastman Kodak Company, Rochester, N.Y. 14650, 1989.

Kodak (1990)   *Photolab Designs for Professionals,* Publication No. K-13, Eastman Kodak Company, Rochester, N.Y. 14650, 1990.

Krajnovich (1986)   Krajnovich L., Hittle D. C., "Measured Performance of Variable Air Volume Boxes". *ASHRAE Transactions,***92,** 203–214, 1986.

LANL (1987)   Lisa Woodrow, *An Evaluation of Four Quantitative Laboratory Fume Hood Performance Test Methods,* LA-11143-T, November 1987.

LANL (1991)   Fairchild, C. I., et al., *Health-Related Effects of Different Ventilation Rates in Plutonium Laboratories,* LA-11948-MS, Los Alamos, NM, January 1991.

Lentz (1989)   Lentz, M., and Seth, A. "A Procedure for Modeling Diversity in Laboratory VAV Systems," *ASHRAE Transactions,* **95** (1A), 1989. pp. 114–120.

Macomber (1989)   Macomber, J. D., "You Can Manage Construction Risks," *Harvard Business Review,* Harvard Business School, Boston, MA, 1989.

Mass (1990)   Massachusetts State Building Code, 780 CMR, 1990.

Mass (1992)   Massachusetts Register, Architectural Barriers Boards, *1982 Rules and Regulations.*

Mayo (1989)   Mayo, D. W., Pike, R. M., and Butcher, S. S., *Microscale Organic Laboratory,* 2nd ed., Wiley, New York, 1989.

McCann (1979)   McCann, M., *Artist Beware,* Watson-Guptill Publication, New York, 1979.

McQuiston (1977)   McQuiston, F. C., and Parker, J. D., *Heating, Ventilating and Air Conditioning Analysis and Design,* Wiley, New York, 1977.

MDA (1992)	MDA Scientific, Inc., 405 Barclay Blvd., Lincolnshire, IL
Mikell (1981)	Mikell, W. G., and Hobbs, L. R., "Laboratory Hood Studies," *Journal of Chemical Education,* **58** (5), A165–A170, 1981.
MIT (1991)	*Laboratory Fume Hood Specifications and Performance Testing Requirements,* Environmental Medical Service, Massachusetts Institute of Technology, Cambridge, MA, 1991.
Moyer (1983)	Moyer, R., "Fume Hood Diversity for Reduced Energy Conservation," *ASHRAE Transactions,* **89** (2B), 1983, pp. 552–537.
Moyer (1987)	Moyer, R., and J. Dungan. "Turning Fume Hood Diversity into Energy Savings," *ASHRAE Transactions,* **93** (2B): 1822–1834, 1987.
NAS (1976)	National Academy of Sciences, *ILAR News,* Vol. XIX, No. 4, Summer 1976, NAS, "Long-Term Holding of Laboratory Rodents", Institute of Laboratory Resources, Washington, D.C.
NEBB (1988)	*Procedural Standards for Certified Testing of Cleanrooms, 1988.* National Environmental Balancing Bureau, 8224 Old Courthouse Road, Vienna, VA.
NFPA (1992)	National Fire Protection Association, Batterymarch Park, Quincy, MA. Individual items are:
	NFPA 10 Portable Fire Extinguishers, 1990
	NFPA 12 Carbon Dioxide Extinguishing Systems, 1989
	NFPA 12A Halon 1301 Fire Extinguishing Systems, 1989
	NFPA 13 Installation of Sprinkler Systems, 1989
	NFPA 30 Flammable and Combustible Liquids Code, 1990
	NFPA 45 Fire Protection for Laboratories Using Chemicals, 1991
	NFPA 55 Proposed Standard "Standard for the Storage, Use and Handling of Compressed and Liquified Gases in Portable Cylinders," 1993 Edition
	NFPA 68 Venting of Deflagrations, 1988
	NFPA 70 National Electric Code, 1990
	NFPA 72 Installation, Maintenance and Use of Protective Signaling Systems, 1990
	NFPA 72E Automatic Fire Detectors, 1990
	NFPA 90A Installation of Air Conditioning and Ventilation Systems, 1989
	NFPA 91 Installation of Blower and Exhaust Systems for Dust, Stock and Vapor Removing or Conveying, 1990
	NFPA 99 Health Care Facility Code, 1990
	NFPA 101 Safety to Life from Fire in Buildings and Structures, 1991
	NFPA 318 Standard for Protection of Cleanrooms, 1991
	NFPA 704 Identification of the Fire Hazards of Materials, 1990
NIH (1981)	*NIH Guidelines for the Laboratory Use of Chemical Carcinogens,* U.S. Department of Health and Human Services, May 1981.

NIH (1985)   *Guide for the Care and Use of Laboratory Animals,* Publication No. 85-23, National Institutes of Health, Bethesda, MD, 1985.

NIOSH (1981)   McManus K., Baker D., *Health Hazard Evaluation,* Report No. HETA 81-247-958, NIOSH, 1981.

NIOSH (1983)   Fannick N., Lipscomb J., McManus K., *Health Hazard Evaluation—Johnson Museum,* Cornell University, Ithaca, N.Y., Report No. HETA 83-020-1351, NIOSH, U.S. Department of Health and Human Services, Public Health Service, Cincinnati, 1983.

NIOSH (1985)   *Hazard Assessment of the Electronic Component Manufacturing Industry,* NIOSH Technical Report, Contract 210-80-0058, NTIS PB 86-104049, 1985.

NRC (1991)   U.S. Nuclear Regulatory Commission, 10 CFR20, "Standards for Protection against Radiation," Final Rule, May 1991.

NSC (1985)   "Controlling Perchloric Acid Fumes," Research and Development Section Fact Sheet, National Safety Council, Chicago, IL, May 1985.

NSF (1992)   National Sanitation Foundation Standard No. 49 for Class II (Laminar Flow) Biohazard Cabinetry, NSF, Ann Arbor, MI, 1992.

Nuffield (1961)   Nuffield Foundation, *The Design of Laboratory Buildings,* Oxford University Press, Oxford, 1961.

OSHA (1987)   *Design and Construction of Inside Storage Rooms,* General Industry Standard 29 CFR 1910.106, page 144, OSHA 2206, Nov. 7, 1987

OSHA (1992)   *General Industry,* OSHA Safety and Health Standards (29 CFR 1910). 134, pg. 274. OSHA 2206, Revised, 1992.

29 CFR 1910.1450 "Occupational Exposures to Hazardous Chemical in Laboratories," January 31, 1990.

29 CFR 1910.37 "Means of Egress, General," 1992.

Peterson (1983)   Peterson, R., Schoter, A., and Martin, D., "Laboratory Air System Further Testing," pp. 57-598. *ASHRAE Transactions,* **89**(2B), 1983.

Pipitone (1984)   Pipitone, D. A., Ed., *Safe Storage of Laboratory Chemicals,* Wiley, New York, 1984.

Powers (1992)   Powers Process Controls, 3400 Oakton St., Skokie, IL.

Ruys (1990)   Ruys, T., *Handbook of Facilities Planning,* Van Nostrand Reinhold, New York, 1990.

SAMA (1975)   Scientific Apparatus Makers Association, *Standard for Laboratory Fume Hoods, Laboratory Equipment II,* Washington, DC, 1975.

Schuyler Waechter (1987)   Schuyler, G., and Waechter, W., *Performance of Fume Hoods in Simulated Laboratory Conditions,* Report No. 487-1605 by Rowan, William, Davies and Irwin, Inc., (under contract to Health and Welfare Department, Canada), 1987.

Shaw (1985)　　　　Shaw, S., *Overexposure: Health Hazards in Photography*, Friends of Photographs Inc., Carmel, CA, 1985.

SMACNA (1984)　　Sheet Metal and Air Conditioning Contractors National Association, Inc., *Thermoplastic Duct (PVC) Construction Manual*, P.O. Box 221230, Chantilly, Tyson, VA, 1974.

SMACNA (1985)　　Sheet Metal and Air Conditioning Contractors National Association, Inc., *HVAC Duct Construction Standards: Metal and Flexible*, P.O. Box 221230, Chantilly, Tyson, VA, 1985.

Speakman (1986)　　Speakman Company, Safety Equipment Division, Wilmington, DE.

Sterling (1985)　　Sterling E.M., Arundel A., and Sterling T. D., "Criteria for human exposure to humidity in occupied buildings," *ASHRAE Transactions* **91** (1), 1985.

Stoeker (1958)　　Stoeker, W. F., *Refrigeration and Air Conditioning*, McGraw–Hill, New York, 1958.

Tescom (1987)　　Tescom Corporation, Elk River, MN.

Thuman (1977)　　Thuman, A., *Plant Engineers and Managers Guide to Energy Conservation*, Van Nostrand, New York, 1977.

UBC (1988)　　*Article 80, "Hazardous Materials"*, Uniform Building Code, International Conference of Building Officials, 5360 South Workman Mill Road, Whittier, CA, 1988.

UL (1984)　　Underwriters Laboratory, *Factory Made Air Ducts and Connectors*, Standard No. 181, Underwriters Laboratories, Inc., Northbrook, IL, 1984.

USP (1990)　　*United States Pharmacopeia*, Vol. XXI, 1990.

Walters (1980)　　Walters, Douglas, B., Ed., *Safe Handling of Chemical Carcinogens, Mutagens, Teratogens and Highly Toxic Substances*, Vol. 1, Ann Arbor Science, Ann Arbor, MI, 1980.

Yaglou (1925a)　　Yaglou, C. P., "Comfort Zones for Men at Rest and Stripped to the Waist," *Transactions of the American Society of Health and Ventilating Engineers*, **33**, 165, 1925.

Yaglou (1925b)　　Yaglou, C. P., and Miller, W. E., Effective Temperature with Clothing, *Transactions of the American Society of Health and Ventilating Engineers*, **31**, 89–99, 1925.

Young (1991)　　Young, "Storage of Laboratory Chemicals," *Improving Safety in the Chemical Laboratory*, 1991. pg 209, table 12.1.

INDEX

This index is organized mainly by chapter. It includes headings that may appear more than once in other chapters. For best results the reader should look up topics by type of laboratory (such as Chemistry, Analytical etc.) or under generic terms (Laboratory or Building). The index also has a number of key words that are cross-referenced from the entire book.